Innovation in Com

Innovation is the creation of new, technologically feasible, commercially realizable products and processes and, if things go right, it emerges from the ongoing interaction of innovative organizations such as universities, research institutes, firms, government agencies and venture capitalists.

Innovation in Complex Social Systems uses a 'hard science' approach to examine innovation in a new way. Its contributors come from a wide variety of backgrounds, including social and natural sciences, computer science, and mathematics. Using cutting-edge methodology, they deal with the complex aspects of socio-economic innovation processes. Its approach opens up a new paradigm for innovation research, making innovation understandable and tractable using tools such as computational network analysis and agent-based simulation.

This book of new work combines empirical analysis with a discussion of the tools and methods used to successfully investigate innovation from a range of international experts, and will be of interest to postgraduate students and scholars in economics, social science, innovation research and complexity science.

Petra Ahrweiler is Professor of Technology and Innovation Management at University College Dublin (UCD), Director of UCD's newly established Innovation Research Unit IRU, and Research Affiliate of the Engineering Systems Division at MIT, USA.

Routledge studies in global competition

Edited by John Cantwell
University of Reading, UK
and
David Mowery
University of California, Berkeley, US

Innovation in Complex Social Systems

Edited by Petra Ahrweiler

Routledge
Taylor & Francis Group

LONDON AND NEW YORK

First published 2010
by Routledge
2 Park Square, Milton Park, Abingdon, Oxon OX14 4RN

Simultaneously published in the USA and Canada
by Routledge
711 Third Avenue, New York, NY 10017

Routledge is an imprint of the Taylor & Francis Group, an informa business

Typeset in Times by Wearset Ltd, Boldon, Tyne and Wear
First issued in paperback in 2013

British Library Cataloguing in Publication Data
A catalogue record for this book is available from the British Library

Library of Congress Cataloging in Publication Data
Innovation in complex social systems/edited by Petra Ahrweiler.
p. cm. – (Routledge studies in global competition; v. 49)
Includes bibliographical references and index.
1. Social systems. 2. Technological innovations–Economic aspects. I.
Ahrweiler, Petra, 1963–
HM701.I56 2010
306.4'6–dc22

2009038186

ISBN13: 978-0-415-63236-2 (pbk)
ISBN13: 978-0-415-55870-9 (hbk)
ISBN13: 978-0-203-85532-4 (ebk)

Contents

PART II

The actors and networks of innovation (empirical research)

PART III
The systemic aspects of innovation (modeling)   233

Innovation is computable

Figures

Tables

Contributors

Petra Ahrweiler, Professor of Technology and Innovation Management at University College Dublin, Ireland; Director of the UCD Innovation Research Unit (IRU), Complex and Adaptive Systems Laboratory UCD CASL; Research Affiliate at the Engineering Systems Division, Massachusetts Institute of Technology, USA.

Thomas J. Allen, Co-Director, Leaders for Global Operations (LGO) and System Design and Management (SDM) Programs, Massachusetts Institute of Technology, Sloan School of Management.

William Allen, Business Development Manager, Fujitsu.

Luca Ansaloni, PhD Student, Department of Social, Cognitive and Quantitative Sciences, University of Modena and Reggio Emilia.

Bjørn Asheim, Professor of Economic Geography and Deputy Director, CIRCLE (Centre for Innovation, Research and Competence in the Learning Economy), Department of Social and Economic Geography and CIRCLE, Lund University.

Michael J. Barber, Researcher, Austrian Institute of Technology, Department of Foresight and Policy Development.

Christoph Barmeyer, Professor of Intercultural Communication, University of Passau.

Florian M. Becke, Senior Start-up Consultant, CAST Center for Academic Spin Offs Tyrol.

Riccardo Boero, Professor of Macroeconomics, Department of Economic and Financial Sciences 'G. Prato', University of Torino.

Hannes Brauckmann, Former Researcher at the UCD Innovation Research Unit (IRU), UCD Innovation Research Unit (IRU), Complex Adaptive Systems Laboratory CASL, University College Dublin.

John L. Casti, Senior Research Scholar (IIASA); cofounder and director (Kenos Circle), International Institute for Applied Systems Analysis (IIASA) and Kenos Circle.

Federico Cecconi, Researcher at the Laboratory of Agent-based Social Simulation (LABSS), Institute of Cognitive Sciences and Technologies (ISTC); Associate professor at Libera Università Maria Ss. Assunta (LUMSA), Institute of Cognitive Sciences and Technologies (ISTC), National Research Council.

Tommaso Ciarli, Research Scientist, Evolutionary Economics Group, Max Planck Institute of Economics.

Bernd Ebersberger, Professor for Innovation Management and Economics, Management Center Innsbruck.

Henry Etzkowitz, Research Fellow, Stanford University, Michelle R. Clayman Institute for Gender Research.

Francesca Giardini, Research Associate at the Laboratory of Agent-based Social Simulation (LABSS), Institute of Cognitive Sciences and Technologies (ISTC), Institute of Cognitive Sciences and Technologies (ISTC), National Research Council.

Nigel Gilbert, Professor of Sociology, Director of the Centre for Research in Social Simulation, University of Surrey.

Peter Gloor, Research Scientist, Center for Collective Intelligence (CCI), Massachusetts Institute of Technology, Sloan School of Management.

Horst Hanusch, Professor of Economics, Institute of Economics, University of Augsburg.

Martin Heidenreich, Professor of Sociology, Director of the Jean Monnet Centre for Europeanisation and Transnational Regulations (CETRO), Institute for Social Sciences, University of Oldenburg.

Harald F.O. von Kortzfleisch, Professor of Management of Information, Innovation, Entrepreneurship and Organization (Mi2EO) at the Management Institute of the Computer Science Faculty of the University of Koblenz-Landau.

Knut Koschatzky, Professor of Economic Geography at Leibniz University Hanover and Head of Competence Center 'Policy and Regions' at Fraunhofer Institute for Systems and Innovation Research (ISI).

Andreas Krueger, Postdoctoral Scientist, Innovation Research Unit (IRU).

T. Austin Lacy, Doctoral student, Institute of Higher Education at the University of Georgia.

Riccardo Leoncini, Professor of Economics, Department of Economics, Alma Mater University of Bologna.

Philipp Magin, PhD student at the Management Institute of the Computer Science, Faculty of the University of Koblenz-Landau and Strategy and

Marketing, Director of Purigion UG (haftungsbeschraenkt), University Koblenz-Landau and Purigion UG (haftungsbeschraenkt).

Julie Michel, Research Assistant, Chair of Economics and Social Policy at the Faculty of Economics and Social Sciences, University of Fribourg, Switzerland.

Sandro Montresor, Professor of Economics, Department of Economics, Alma Mater University of Bologna.

Rajneesh Narula, Director, John H. Dunning Centre for International Business.

Terhi Nokkala, Research Fellow, Centre for Research in Social Simulation, Department of Sociology, University of Surrey.

Uwe Obermeier, PhD Researcher, UCD Innovation Research Unit (IRU), Complex Adaptive Systems Laboratory CASL, University College Dublin.

Rory O'Shea, College Lecturer, University College Dublin.

Andreas Pyka, Professor of Innovation Economics, Institute of Economics, University of Hohenheim.

Marina Ranga, Research Fellow, Stanford University, Michelle R. Clayman Institute for Gender Research.

Ornit Raz, Research Affiliate, Engineering Systems Division (ESD), Massachusetts Institute of Technology, Department of Engineering.

Federica Rossi, Post-Doctoral Researcher, University of Torino.

Margherita Russo, Professore Associato of Economic Policy, University of Modena and Reggio Emilia.

Stefania Sardo, PhD Student, Ca' Foscari University (VE).

Pier Paolo Saviotti, Director of Research, Laboratoire d'Économie Appliquée de Grenoble (Grenoble Applied Economics Laboratory (GAEL)).

Ramon Scholz, Research Fellow, Economy Department of Jacobs University, Bremen.

Flaminio Squazzoni, Assistant Professor of Economic Sociology, Department of Social Sciences, University of Brescia.

Marco Valente, Assistant Professor of Economics, University of L'Aquila.

Marco Villani, Professor of Engineering and Computer Science, Responsible for the Modelling and Simulation Laboratory of the department, Department of Social, Cognitive and Quantitative Sciences, University of Modena and Reggio Emilia.

Josh Whitford, Assistant Professor of Sociology, Columbia University.

Acknowledgments

This book follows the international conference 'Innovation in Complex Social Systems' (10–12 December 2008) at University College Dublin (UCD), where the authors of this volume met to discuss the state of the art in complexity-oriented innovation research. The conference was hosted by UCD's newly found Innovation Research Unit (IRU) and was jointly sponsored by the sixth Framework Programme of the European Commission project NEMO (network models, governance, and R&D collaboration networks), ICT Ireland Skillnet, and UCD Research. The editor gratefully acknowledges this financial support of the conference, without which this book would not have been possible. Furthermore, I would like to thank all authors of this book for their enthusiasm concerning this project and their great contributions – among them most prominently my long-time cooperation partners Nigel Gilbert and Andreas Pyka for their large part in developing the research program which is emerging in this book.

Chapter 16 is reprinted with permission from Nigel Gilbert, Petra Ahrweiler and Andreas Pyka, 'Learning in Innovation Networks: Some Simulation Experiments', *Physica A: Statistical Mechanics and its Applications*, 378 (1), 'Social Network Analysis: Measuring Tools, Structure and Dynamics', *Social Network Analysis and Complexity*, 1 May 2007, pp. 100–9, ISSN 0378-4371, DOI: 10.1016/j.physa.2006.11.050.

1 Innovation in complex social systems

An introduction

Petra Ahrweiler

Innovation policymakers, business managers and the public often expect that the current investments in R&D, higher education institutions, science-industry networks etc. will immediately produce a flow of products and processes with high commercial returns. The disappointments and legitimatory problems arising from missing outputs are considerable and show the limits of steering, control and policy functions. Although there is not a fundamental apprehension against the importance of knowledge and innovation, the responsible innovation managers mention a frustration with the too messy and complicated features of the innovation process, which simply 'does not seem to compute'. Innovation, the creation of new, technologically feasible, commercially realizable products, processes and organizational structures (Schumpeter 1912, Fagerberg *et al.* 2006), is emerging from an ongoing interaction process of innovative organizations such as universities, research institutes, firms, government agencies, venture capitalists and others. These organizations generate and exchange knowledge, financial capital, and other resources in networks of relationships, which are embedded in institutional frameworks on the local, regional, national and international level (Pyka and Kueppers 2003). Innovation is an emergent property from these interactions on the micro level – if the combination of actors and organizations, their compatible capabilities, and their cooperative behaviors match. No equation will predict this match or warn of a mismatch beforehand.

This book now has something new to say about innovation. Its contributions make full use of cutting-edge methods coming from the natural and social sciences, from computer science, and mathematics to deal with the complex aspects of socio-economic innovation processes – and this without leaving out the messy features of empirical reality and the 'human element', but indeed taking full account of it. Its approach, as highlighted by its title, opens up a new paradigm for innovation research: the contributions analyze innovation in complex social systems while making innovation understandable and tractable using tools such as computational network analysis and agent-based simulation.

This introduction will address the question of how a coherent research program might arise from the title of the book, by discussing theoretical and methodological issues around researching innovation in complex social systems... It will also outline a perspective on innovation, which is more or less shared by the authors of this volume, and finally, the chapter will take on the usual task of an introduction and present the concept and structure of this book while outlining briefly the role of the individual contributions.

Innovation networks

Innovation in knowledge-intensive industries happens in networks. Though this is commonplace, in the meantime, some illustration might be useful to realize fully why this is the case and what this implies. The following fictive story, located in the biotechnology-based pharmaceutical industry, is based on an empirical case, but with invented labels.

In the interdisciplinary Life Sciences Research Cluster of the University of Unisa, two PhD students from biochemistry and pharmacy have discovered something while working for their interdisciplinary research project. They are sure that their scientific discovery will turn out to be useful for developing a new drug to cure AIDS. They now want to found a firm for developing this idea into a product. The University of Unisa offers help with this plan through its Technology Transfer Unit, which contributes some knowledge on entrepreneurial activities. Notably, it points the young entrepreneurs to the Science, Technology, Innovation (STI) Policy agency at the regional government, GovBIO. This agency has a government fund to support entrepreneurship in the life sciences. Using this governmental start-up money, the university spin-off company UnisaLAB is located in the Science Park of the university, which offers the young firm the advantage of using the university's lab further on in the product development process, and to use the university's hospital for the extensive clinical testing periods of the new drug.

However, UnisaLAB is immediately confronted with big problems: drug development is enormously expensive (about €600 million on average; this was definitely not the amount of money they got from GovBIO), it will last between ten and 12 years on average including all the clinical testing, and it is very risky (most projects fail). Furthermore, UnisaLAB needs to obtain an expensive license from another biotech firm to carry out a part of the development process, for which there is no competence at the university or inside the firm. Here, UnisaLAB realizes that they have no experience with IPR protection and fear that such multifold cooperations will enable the new knowledge to diffuse to competitors. Finally, UnisaLAB has no own expertise in commercializing and marketing a product, until recently its founders and their crew were simply scientists. UnisaLAB (a dedicated biotech firm, or DBF) is unable to cope with these issues alone. However, there is a multinational company (a large diversified firm, or LDF) in the vicinity, LDF-Tech, which is keen to exploit the new technology UnisaLAB hopes to develop. They offer to commercialize the technology and minimize the problems of the development process.

In the meantime, some good has come from UnisaLAB's university connection to cutting-edge research: the ever innovative founders have started on a second innovative product realizing that the market in biotech industries is quickly changing and product cycles are very small. Though UnisaLAB has managed, unlike many other DBFs, to stay independent in the partnership with LDF-Tech without being acquired, the management of UnisaLAB now decides against a renewal of the partnership for developing the second product. They have learnt through the networking with LDF-Tech how to commercialize and market a new product. All they want now is the money to finance the development phase for the second one; they can cope with the rest now. For this reason they contact the venture capital firm VCBio to provide the funds for their new project. However, there is some part of the development process, which involves the DBF from which they had obtained the license for the first product. Consequently, they decide to co-patent the second product idea together with the other DBF and form an R&D alliance for the project time.

There are many possible realizations and activations of empirical innovation networks (for the biotech industries of the United Kingdom and Germany, see e.g. Ahrweiler *et al.* 2006). They are the organizational 'hardware' of innovation, which arrange themselves, which can be combined, designed and composed in various settings.

Furthermore, there are the processes, the 'software' running on these organizational structures. They are all about the availability of knowledge, about finding the right partners, about getting the financial resources in time, and about smart coordination smoothing the micro dynamics between the involved network participants and others (cf. various frameworks about such cooperative arrangements around knowledge production in networks, e.g. 'Mode 2' from Gibbons *et al.* 1994, or the 'Triple Helix' concept from Etzkowitz and Leydesdorff 1997).

Which node of the network acts as the originator, which as the transmitter, which as the enabler or receiver, can change. The actual network shape depends on participants, sectors, locations, and many other factors. The roles and processes are necessary, the actors and structures can vary. For example, in our fictive story the inventive idea would not need to come from a university background – very often it comes from a large diversified firm (LDF), and this is where the DBF spins off from.

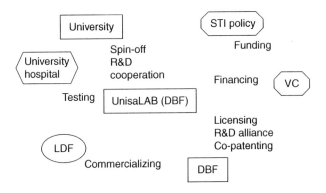

Figure 1.1 Innovation network from UnisaLAB perspective: organizations and links.

Policymakers and managers of firms, universities and other participating organizations try to find out as much as possible about the structures and processes responsible for innovation. The managers want to know how to position their organization optimally in these networks – the policymakers are concerned with the bird's eye perspective on the well-being and competitiveness of the overall network on the different policy levels. Those practitioners turn to science for insights into the mechanisms and processes producing these network structures and for guidance how to optimize their performance. What can scientists tell them?

To provide descriptions and explanations for why and how innovation happens, we need to analyze its structures and processes. The structural 'hardware' consists of inter-organizational innovation networks. Network analysis of innovation networks is one of the most vibrant interdisciplinary research activities we can observe in the moment. Work here is so abundant that it would need an entire chapter to summarize just the most important discussions and results. The studies mentioned here are just placeholders for their many equivalents. All parts of innovation networks have been of interest so far: there are studies concerning the binary combinations of involved actors (university–university, university–SME, university–MNE, SME–MNE, SME–SME, etc.); and about all possible links between these actors – R&D alliances (e.g. Siegel *et al.* 2003), spin-off activity (e.g. Smith and Ho 2006), licensing (e.g. Thursby and Kemp 2002) and all other possible link types. We find studies on university–industry links (cf. Ahrweiler *et al.* forthcoming) and all sorts of work on inter-firm networks (e.g. Schilling and Phelps 2005, Porter *et al.* 2005). For example, the relationship mentioned in the innovation narrative above, i.e. the DBF–LDF link of commercializing knowledge in the biotechnology-based pharmaceutical industry, provoked many comparative studies for different national contexts (Oliver and Liebeskind 1998, Hagedoorn and Roijakkers 2002, Pyka and Saviotti 2005).

Most of these studies have been carried out by economists or other social scientists. However, due to a rising interest in physics in the past ten years concerning complex networks, there has been much overlap and co-publication between physics and the social sciences from hybrid backgrounds such as econophysics or sociophysics.

> Research by physicists interested in networks has ranged widely from the cellular level, a network of chemicals connected by pathways of chemical reactions, to scientific collaboration networks, linked by co-authorships and co-citations, to the world-wide web, an immense virtual network of websites connected by hyperlinks.
>
> (Powell *et al.* 2005: 1132)

Networks consisting of nodes and edges (or actors and relations, or units and links, etc.) are a ubiquitous phenomenon, where general insights apply to their topologies, structural properties and measures (Albert and Barábasi 2002, Newman 2003). Network analysis methods (Wasserman and Faust 1994) have profited immensely from progress in physics concerning the field of graph theory and complex networks.

On a general level, innovation networks show features of both so-called scale-free networks (Barábasi and Albert 1999), and of small worlds (Watts and Strogatz 1998, Watts 1999). We have, ourselves, already studied some aspects concerning these two general features in more detail (Pyka *et al.*2009, Ahrweiler *et al.* forthcoming). This area connects to interesting debates, i.e. whether strong ties (such as friendship, contracts, face-to-face interaction) or weak ties (such as access to information through loose contacts) are good for innovation (Granovetter 1973, Uzzi 1997, Burt 1992, 2004, Ahuja 2000, Walker *et al.* 1997, Verspagen and Duysters 2004). What special network topologies do, or do not do, for knowledge flows has been widely discussed in this research area (Cowan *et al.* 2007, Gloor 2006, Sorensen *et al.* 2006) occupied both by physicists and social scientists, often in interdisciplinary co-authorship relations. Can we be satisfied with this contribution of science to describe and explain the innovation process? It seems as if we would just need to accumulate all these network analyses into different parts of the collaborative structure to provide the full picture.

However, suppose you were the manager of the Science Park at Unisa University from the narrative above. You would have a few successful research clusters, one of them in the life sciences consisting of academic institutes and companies as partners. You would have witnessed some success stories realizing that interdisciplinary and inter-organizational cooperation is beneficial for innovation and scientific output. In that case, you would be interested in questions like:

- How is interdisciplinary communication processed in and between research clusters, what are the effects, what are the barriers, what would be supportive?
- How is inter-organizational communication processed in and between research clusters, what are the effects, what are the barriers, what would be supportive?
- How is knowledge communicated and exchanged in the clusters?
- How do new research topics, ideas, themes emerge in the network?
- How are new research projects set up by the network?
- How do research groups coordinate heterogeneous disciplinary backgrounds?
- What is the added-value of cooperation between researchers of a cluster?
- How does scientific cooperation relate to innovation and scientific output?

Some of these questions can be answered by network analysis. Others – mostly those referring to *new knowledge*, and those referring to *agency* (i.e. the possibility of actors to move intentionally in the action space) – seem to need some different analytical tool. These network analyses deal with knowledge as 'flow substance' in a way, which does not discriminate knowledge very much from what flows in other types of networks, such as energy or information. It is structure that matters – not the particularities of the flow substance, i.e. knowledge. Indeed, there are many valid insights to gain from this structural perspective:

e.g. where there is no link, nothing can flow – be it knowledge or anything else; or: linkages indicate groups who share – be it knowledge or anything else. This makes network topology a significant issue.

Furthermore, the structural focus does not favor the agency dimension: this would imply more complex node properties or more heterogeneous link types for each node – be they people or organizations. A real-world actor moves in an action space, which consists of many dimensions (actors are constantly inventing, constructing, anticipating, changing, developing etc. their action space, not merely moving around in a given world). The notion 'actor' is telling in this respect: it is originally used for being on the stage in a theatre performing multiple roles. Actors in different roles would need rich node descriptions concerning properties, behaviors, and states, and/or a richer link structure, which manifests what the actor does in relation to others. In network analyses, instead, the dimensions of nodes are rather limited – if an organization is in an EU R&D network, it is doing funded EU research with other organizations – whatever roles it performs besides, do not matter, nor how these different roles feedback on the respective R&D network tie. At the moment, multi-level networks are a research challenge for people interested in complex networks. Indeed, many valid insights again can be got from reducing node properties and link types per network under investigation.

However, we need to add a class of special attributes concerning our topic here. Otherwise, we will never be able to describe innovation processes adequately and help practitioners to deal with their problems.

a   Though we can provide time series analyses to show network dynamics (cf. Powell *et al.* 2005 analyzing the field evolution for US biotech industry), our analyses so far focus on structures and states. We need to add a perspective focusing on what happens between the states we capture, i.e. the causal processes and mechanisms producing the structures and states we analyze.

b   So far, our analyses deal with knowledge in a manner akin to handling a bag of potatoes. Knowledge can be divided into units, which can be owned, transmitted, distributed etc. The most important feature of knowledge in innovation networks, however, is that it enables action (innovation, new product/process development). We need a representation of knowledge, which opens up this agency dimension.

c   Most analyses, so far, focus on innovation diffusion (Cowan and Jonard 2004, Abrahamson and Rosenkopf 1997, Valente 1996). This is, in a way, a consequence of (*b*). It is easier to say how the potatoes are rolling through the network, than to explain by which causal processes they are doing what they are doing, and where they have come from in the first instance. We need a focus on innovation generation, on the emergence of the new. This is what innovation is all about.

If we can add this procedural perspective to our analyses, we will be able to say more. Above, we have just told a narrative to point at the knowledge-related processes and the evolving action space of the network participants. However, this is definitely not the best we can do. Again, an inter- or this time a transdisciplinary initiative offers a better conceptual framework to help us. This is complexity science (Bar-Yam 1997, 2004, Braha *et al.* 2008, Casti 1995, Flake 1999, Stewart 1989, Waldrop 1992).

Complexity science perspectives locate innovation processes in turbulent environments with high uncertainty and ambiguity: they assign to innovation processes characteristics such as multi-scale dynamics with high contingency and non-linearity, emergence, pattern formation, path dependency, recursive closure, and self-organization (Frenken 2006, Lane *et al.* 2009). Such concepts (cf. Arthur 1989, 1998) are of rising importance to describe and explain innovation processes, building on mathematical concepts originating from physics and engineering science (Gell-Mann 1994, Kauffman 1993, 1995, Prigogine and Stengers 1984, Holland 1995). Business studies and management science have started with this: areas such as strategic organizational design (e.g. Anderson 1999, Brown and Eisenhardt 1998, Dooley and Van de Ven 1999, Eisenhardt and Bhatia 2002, McKelvey 1999), supply chain management (e.g. Choi *et al.* 2001) and innovation management (e.g. Buijs 2003, Chiva-Gomez 2004, Cunha and Comes 2003) have extensively applied complexity science key concepts.[1]

However, since we now speak about procedural aspects and qualitative properties of knowledge and agency, rather than merely about quantitative features of certain structures (Weidlich 1991), we need to make sure that this conceptual framework and the transdisciplinary cooperation between natural and social science holds.

Complex social systems

If we want to transfer concepts, instruments and tools of complexity science to the social sciences, we must ensure that this transfer is appropriate and not merely on the level of metaphors. We need to show that natural systems, for which complexity science provides adequate descriptions, and social systems have enough in common that the same perspectives apply for analysis. There have been some attempts to show this (e.g. Eve *et al.* 1997, Bridgeforth 2005), but the issues we want to discuss now have never been resolved by any of them sufficiently (cf. review of Byrne's 'Complexity Theory and the Social Sciences' (1998) in Ahrweiler 1999).

We find some assurance that the notion 'system' applies to the social realm, and especially to our topic – innovation. 'If anything, modern innovation theory demonstrates that a systemic perspective on innovation is necessary' (European Commission/DG Research 2002: 25). However, when it comes down to definitions, we are left with rather vague intuitions such as:

> The dominant mode of innovation is systemic. Systemic innovation is brought about through the fission and fusion of technologies; it triggers a series of chain reactions in a total system. ... The interactive process of

information creation and learning is crucial for systemic innovation. ... The characteristic trait of the new industrial society is that of continuous interactive innovation generated by the linkages across the borders of specific sectors and specific scientific disciplines.

(Imai and Baba 1991: 389)

To be more precise, the definition of 'innovation system' starts to approximate what we have called 'innovation network' so far. According to Beije (1998), an innovation system

Can be defined as a group of private firms, public research institutes, and several of the facilitators of innovation, who in interaction promote the creation of one or a number of technological innovations [within a framework of] institutions which promote or facilitate the diffusion or application of these technological innovations.

(Beije 1998: 256)

Are networks and systems the same thing? We will follow up on this question in the coming sections.

The notion of 'innovation system' is quite common since the 'national innovation systems' (NIS) approach was introduced to innovation research in the 1980s. This framework (Lundvall 1992, Nelson 1993) again focuses on actors and their interactions embedded in a national institutional infrastructure. It concentrates on 'the systemic aspects of innovation [and of] diffusion and the relationship to social, institutional and political factors' (Fagerberg 2003: 141). While relying on the same system concept but differentiating, elaborating and complementing the NIS approach, recent research targets sectoral systems of innovation (Malerba 2002), technological systems, and regional innovation systems (Fornahl and Brenner 2003).

However, is the system concept these approaches refer to valid for usage in a complexity science context? For general systems theory, the consensual starting point of a systems analysis is the difference between system and environment (Klir 1978). Systems are not only sometimes and not only adaptively oriented towards their environment – they simply do not exist without their environment and they are structurally oriented towards it. They constitute and reproduce themselves through the production and maintenance of a difference to the environment. Systems consist of inter-related elements, which are not further dissolvable for the system using them for its constitution and reproduction (cf. Luhmann 1987: 35). For the purposes here,[2] we will start the discussion based on a very crude assumption: the natural sciences may have more reason to speak about systems, where elements are bound together by laws and forces, which can be observed in a controlled way, and which can be described by mathematics. This leads to 'real' systems with elements, which cannot choose whether they are in or out. The systems, be it the solar system, the immune system, or a thermodynamic system, are really out there – they are constituted by their elements (often self-reproducing, self-organizing), they structure themselves internally, their operations are observable, and both – elements and operations – are clearly distinguished from the environment, i.e. everything which is not the system.

Coming to the social sciences, we cannot ignore the 'human element' at work in social systems, which decomposes itself into concepts such as freewill, creativity, subjectivity, irrationality, personality, etc. Here we can rightfully have some doubts about how to control the inventive minds of observers concerning system boundaries, and about the existence of any laws and forces to be described by mathematics. We need to sort this out for social systems to apply successfully the same tools and methods for description and explanation.

Dealing with the particularities of social systems, we seem to have three interrelated problems: a problem of how 'real' social systems are; a problem concerning the exact nature of their elements; and a problem of how they structure themselves internally. All these problems alienate social systems from natural systems so much that we cannot simply turn to evolutionary models, mathematics, statistical mechanics, and numerical simulations for adequate descriptions.

The first problem indicates that something like 'social systems' *might not be real*, i.e. might only exist as relatively arbitrary ascriptions of social scientists. Whether the former German chancellor still belongs to the 'political system' or does now belong to the 'economic system', is an issue of (self-) ascribing membership of an individual to an observer-specified context. People can move freely in and between numerous different groups and contexts, contexts can be constructed differently by different observers, and different observers can ascribe people differently to any given contexts. To call these constructions of context and ascription 'systems' and maintain that the people we ascribe membership to for our constructions act in a way, that we can apply sophisticated natural science methods to describe and explain their actions, seems to be a mere whim. This is a first important difference between networks and systems: networks are formed by membership – systems need to be 'just out there'.

So far, we take it silently for granted that *people are the supposed-to-be elements* of social systems (cf. Maturana 1991: 35, Hejl 1992: 271f.). However, this again leads us into perils if we want to apply methods borrowed from the natural sciences. A system element is a unit which is not further decomposable for a system, and which can be analyzed with respect to its states and its properties. Can people then be proper elements of social systems in that sense? The concept of an *individuum* with all these 'human element' features, with all the cognitive, intellectual, emotional, spiritual, etc. dimensions, seems to be intractable as a not further decomposable unit. Even the most fervent promoters of methodological individualism share the following consensus: the 'notion of "individuum" refers to … a highly artificial creation …. The action theory individual is an individuum's abstraction' (Prewo 1979: 308). For social systems theory, people would not feature as good candidates for the element role. It would need to be an abstraction from 'the whole' of a human being, and this is exactly what social systems theory tried to turn to. This is a second important difference between networks and systems: social networks consist of people or organizations – systems cannot.

What is interesting and observable for social science is the action of an individual, which is why this component was singled out as *individuum* abstraction in social systems theory. Instead of conceptualizing individual actors as elements of a system, *action* then served as the unit of analysis (cf. for the history of this reasoning Schwemmer 1987: 252–7). However, then the next problem, i.e. that of *system constitution and system structuration*, hit home. The tradition of 'open social systems' assumed that action systems constitute themselves in relation to their environment, and structure themselves internally through environmental transfers and through achievements of distinguishing themselves from their environment. Talcott Parsons, for example, attempted to understand each internal structuring of a social system as a forward-running inner-differentiation that increases the system's own complexity. The development of more 'internal environments' leads to an increasing need for integration between the differentiating social sub-systems, which fulfill necessary functions for the overall system (cf. Parsons 1961).

However, Parsons' structural functionalism could not indicate how action systems actually organize their achievements of internal structuring through the inclusion of external factors. This remained unclear, among other reasons, because a system structure for the selection and processing of external factors would already be required in existence. The structural functionalism of open systems could not explain how system constitution, system differentiation, and system complexity related to environmental transfers and internal operations (cf. Luhmann 1987, Schmid and Haferkamp 1987).

Looking for a solution, social systems theory turned to general systems theory, which conceived systems as 'operationally closed' and assigned internal achievements of structuration as achievements of the system alone (von Foerster and Zopf 1962). Von Foerster (1984) laid the foundation of the so-called second-order cybernetics, which endeavored:

> To make plausible viewing the behavior of a system … as a recursively running series of equally weighted states that may be interpreted as the prevailing solutions to the system equations that describe this same process. If the prevailing system behavior could actually be treated with help from such recursive equations, then the independent, self-relating development of structure would be undeniable, and self-organization should be expected in all cases.
>
> (Schmid and Haferkamp 1987: 10)

Following up on general systems theory, Niklas Luhmann (1987) presented a theoretical framework for social systems, which managed to capture solutions for all mentioned problems. He claimed the 'reality' of social systems in the same way natural systems are real, and he worked out for social systems the 'idea of a self-referential, recursively operating, repeated procurement of system-constituting elements' (Schmid and Haferkamp 1987: 11). In his critique of Parsons he started with showing that even action is not a proper candidate for being an element of social systems. Nearly the same sort of criticism applies as mentioned above. Actions do not belong to certain systems; again they are subject to ascriptions. We as (scientific) observers ascribe actions to people (actors), we ascribe actions to contexts in a way that we identify an action as belonging to a certain sub-system, e.g. by saying this is an economic action belonging to the economic system. This is no more 'real' and independent from all our different constructions as it was while we assigned people to different contexts. Again, we cannot claim that our inventiveness is rewarded by producing structures and dynamics which can be analyzed via 'hard' science methods.

As elements of social systems, we need something, which is really out there, and something that constitutes and structures internally the system it is part of. From general systems theory we have learnt that self-referential closed systems determine their boundaries through their mode of operation. Only similar operations are connectable in the system and contribute to the reproduction of the system; operations foreign to the system have no constituting or structuring effect.

For Luhmann, communication is the element and the modus operandi of social systems fulfilling these requirements. Communication produces communications and relates them to other communications. Elements and operations are the same. Communication systems, i.e. social systems, distinguish themselves from psychic and other systems by this modus operandi. With this, issues such as

> Subjectivity, the availability of consciousness, the foundation of consciousness will be understood as the environment of social systems, and not as their self-reference. Only with this distancing do we win the possibility of working out a truly ‚independent' theory of social systems.
>
> (Luhmann 1987: 234)

Communication, in Luhmann's theory, is a threefold selection process of information, signaling/messaging and understanding. This communication concept, where data processing and information sending is complemented by understanding, i.e. successfully connecting meaning to the transmitted, provides a kind of underlying knowledge society view: communication is also 'sharing knowledge', if we define knowledge as a meaningfully organized set of information(s).

Communications are 'real', but not observable. Action is a sub-case of communication in this conceptual framework – it is, when we assign certain parts of the communication process, those parts, which are observable, to certain actors and contexts. However, this already contributes our observer efforts (see above). The underlying 'real' element structure consists of communications without which nothing could be observed at all. 'The most important consequence of this analysis is: that communication cannot be directly observed, but can only be inferred. In order to be observed or to observe itself, a communication system must therefore be "flagged" as an action system' (Luhmann 1987: 226).

Society as the overall communication system, which produces its own elements, structures itself internally into various functional sub-systems (it creates new system/environment differences internally), such as the economic system or the science system. The capability of communication to connect or to reject connection following functional requirements creates these sub-systemic structures and 'enables self-structuration and self-differentiation' (Schwemmer 1987: 247). This is why communications can never work integratively; on the contrary 'communication works ... necessarily differentiatively' (Luhmann 1987: 200). Functionally differentiated systems have own selection mechanisms, which are semantically structured by communication media and so-called codes. Luhmann has worked out for all social sub-systems the selection standards, mechanisms, media and codes on how and under which conditions meaningful communications are selected as belonging to the same systemic context or are rejected as foreign. These selection mechanisms and semantics have been subject to evolutionary processes.

For example, the economic system (Luhmann 1994) in its beginning contained the sum of all communications with reference to 'ownership' concerning the distribution of scarce resources following the lead difference 'to have/to have not'. With the evolutionary introduction of money as a general medium to condition selection processes, the economic system was re-coded: it now contains all operations/communications, which are processed via the payment of money following the lead difference 'to pay/not to pay'. The communications, which exist within the system and its environment are perceived following this binary scheme. Using this code for the selection of communications enables the reduction of complexity, structuration, and the self-referential autopoiesis of the system: payments are made which have needed liquidity and which are generating liquidity – the system is closed and reproduces itself. For the science system (Luhmann 1992), the lead difference for selecting communications as belonging to this special sub-system is the binary scheme 'true/false'. This code operationalizes the communication medium 'truth'.

Social systems as communication systems are *per se* complex. Following Luhmann, we define as complex:

> A related set of elements where due to immanent restrictions of their connection capacity not every element can at any time be connected to any other.... Complexity in this sense implies selection requirement, selection requirement implies contingency, and contingency implies risk. Every complex phenomenon results from a selection of the relations between its elements, which are needed for constituting and maintaining. The selection places and qualifies the elements though – for them – other relations would have been possible.
>
> (Luhmann 1987: 46f.)

This is a third important difference between networks and systems: networks are open and can be characterized by their structures – systems are closed and can only be characterized by their processes.

With this framework of social systems theory, we have found the necessary iso-morphism between complex natural and complex social systems. As in the natural sciences, Luhmann's social systems theory provides us with real systems consist-ing of elements, which cannot choose whether they are in or out. Social systems, like the solar system, the immune system, or a thermo-dynamic system, are really out there – they are constituted by their elements, they structure themselves inter-nally, their operations are observable, and both – elements and operations – are clearly distinguished from the environment, i.e. everything which is not system.

Innovation networks in complex social systems

This explains why we do not like the term 'innovation system' very much but rather prefer the notion 'innovation networks' to refer to the structural com-ponents of innovation. The 'software' of inter-organizational innovation net-works seems to be a set of communications, which includes actions, belonging to the economic and/or the science system. Social self-organization theory of science (Krohn and Kueppers 1989) already postulated to distinguish meticu-lously between these two levels:

> The discovery of the 'organization fields', 'ecological communities' or 'institution networks' is not equivalent to the disclosure of the causes of innovation. ... At its core the point is not to unite causal innovation to net-works, but rather to see in networks the organizational conditions for the dynamics of innovation
>
> (Krohn 1995: 31)

On the procedural level, we will not attempt here to decide whether the set of communications leading to innovation establish an own social sub-system as a shared set of science and economic communications. Since internal differentia-tion is the evolutionary mode to increase complexity, which is tractable for the sub-systems, de-differentiation and fusion seem to be a strange strategy for evolving systems. This was already critically mentioned by Luhmannians against network research (Kaemper and, Schmidt 2000), though the biggest part of this critique was somehow misleading, because network research concerns struc-tures, and social systems theory concerns processes. Innovation in complex social systems 'is procedurally self-organizing and structurally network-forming' (Kowol and Krohn 1995: 78).

However, we can argue that in both sub-systems – the science system and the economic system with their respective media and codes – we can observe the same process of internal differentiation as a result of evolutionary requirements. Within the science system – under pressure to prove itself 'useful', to be financed, and to contribute to innovation – there are communications adopting parts of the eco-nomic code as a second-order binary scheme (i.e. with respect to applied research or technology development); within the economic system – under pressure to inno-vate – there are communications adopting true/false communications as a second-order selection mechanism to detect the *emergence of the new* for providing first mover advantages.

Both adoptions would enable the respective outer sub-system – science and the economy – to survive and continue its actual autopoiesis, would increase the internal complexity of each sub-system by introducing a new internal difference (the original code/the second-order code), and would fit the overall framework of differentiation and evolution. Innovation barriers in this framework could be explained, because innovation communications actually happen in both sub-systems (the science and the economic system), and do indeed *not* share one integrated systemic context. The task for analysis would be to observe and explain the co-evolution of these two internal differentiations of science and the economy together with their functions and mechanisms for the respective sub-systemic context. Innovation would be an emergent property on the level of the overall system, society.

What happens to firms, universities and government agencies in this conceptualization of social systems? The function systems need to provide decision mechanisms for the connectivity of communications. Following Luhmann again, organizations produce such decisions for social (sub-)systems. Organizations in his theory are 'just' communications of decisions providing the contexts for other decisions (cf. Luhmann 1997: 826ff.). With this, they are autopoietic social systems, too. They produce decision options, which would not exist otherwise. Their first decision is that about membership, which can be followed by a multitude of further decisions.

Luhmann's conceptualization of social systems seems to be compatible with a computer metaphor of social systems. The communication system leading to innovation as conceptualized above can be modeled using an artificial communication system, i.e. the computer. The innovation-related communications of the science and the economic system can be organized in algorithms for 'agents', which are organizations enabling decisions for connecting communications.

Using this framework, we can conceptualize complex social systems close to complex natural systems and start to apply complexity science tools and instruments such as evolutionary modeling, agent-based simulation etc. to analyze their dynamics.

The emerging research program

Finally, we have collected all the bricks to build a coherent framework for research. To put it in a nutshell, innovation research from a complex social systems perspective combines empirical research on innovation networks and on systemic features of innovation with modeling using computational methods such as network analysis, agent-based modeling and social simulation. This allows us to implement and test innovation policy or innovation management scenarios in a realistic and evidence-based, but controlled way. The computational laboratory *in silico* can inform policy and management on optimal network structures for innovation performance adapted to contextual conditions.

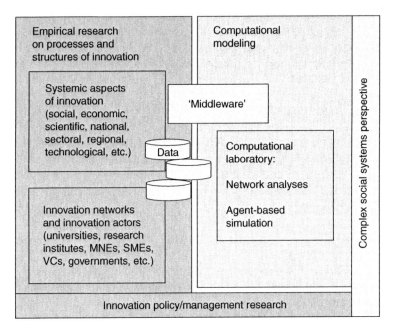

Figure 1.2 Conceptual framework for complexity-adapted innovation research.

The state of the art tells us that we are in a situation where everything helps to realize this conceptual framework. We need to employ a broad thematic perspective: studies are required on single actors as contributors to innovation as much as research on certain bilateral cooperations in innovation networks, or work on the overall innovation process covering each systemic angle. Research needs to cover R&D investment, knowledge production and invention, firm level innovation and entrepreneurship, demand-driven features of innovation, policy and governance issues, and much more.

The broad perspective includes methodological aspects of research as well. On the one hand, realistic simulation and computation will ask for handling large databases coming from quantitative research. We will need extensive empirical data collection and analysis for mapping the knowledge profiles of innovation networks. This will involve scientometric, econometric and sociometric studies mostly consisting of statistical research. Then, we need to develop and improve 'middleware tools' (e.g. Python) to bridge the gap between data and modeling by intelligent interfaces.

On the other hand, to inform agent-based simulations about decision contexts, strategies and behavior of innovative agents, we will need qualitative case studies on innovative organizations providing the required in-depth knowledge.

History-friendly models are formal models which aim to capture – in stylized form – qualitative theories about mechanisms and factors affecting industry evolution, technological advance and institutional change put forth by empirical research in industrial organization, in business organization and strategy, and in the histories of industries. They present empirical evidence and suggest powerful explanations. Usually these 'histories' … are so rich and complex that only a simulation model can capture (at least in part) the substance, above all when verbal explanations imply non-linear dynamics.

(Malerba *et al.* 1999: 3–4)

However, for this task we need not only empirical researchers familiar with collecting, analyzing and interpreting data from the socio-economic realm, we need modeling experts to set up the models and simulations. Specialists in computational network analysis will help to investigate the multi-level networks (knowledge networks, human networks, organization networks), we have to deal with. They will be looking, for example, at (co)patent networks, (co)publication and citation networks, contractual inter-organizational networks, and communication networks, which are established between innovative actors. The triangulation of different types of computational network analysis will allow the matching of 'codified' networks (documented by contracts or other formalized relationships) on 'living' network structures (based on the actual amount of communication between the network actors).

Agent-based modelers will build the simulation environment of the laboratory *in silico*. Agent-based simulations (Gilbert and Troitzsch 2005, Gilbert 2007) provide computational demonstrations of production algorithms: they show whether a specific communication/action pattern on the micro level is sufficient to produce a macro-level phenomenon such as innovation. Where the aim is to understand the processes and mechanisms in innovation networks (Tesfatsion and Judd 2006) and to identify access points for policy intervention – even suggest designs and scenarios – this is the approach of choice (Ahrweiler *et al.* 2004, Gilbert *et al.* 2007). The aim of simulation modeling is not primarily to predict specific system behavior or to reproduce statistical observations, but rather to gain a dynamic and detailed description of a complex system where we can observe the consequences of changing features and parameters. Innovation is an emergent property of a complex social system involving heterogeneous agents and evolving rule sets. Our simulations will serve as a laboratory to experiment with social life in a way that we cannot do empirically due to methodological reasons (cf. Ahrweiler and Gilbert 2005). Using this tool, we can understand innovation dynamics in complex social systems and find their potential for design, intervention and control.

Summarizing, we need empirical research to collect, analyze and interpret data from the socio-economic realm, and we need modeling to set up our simulations. At best, there is an emerging breed of researchers combining these methodological competences. We will not attempt to make this a disciplinary issue; it is just a combination of methodological competences that is required.

The overall approach of this broad thematic and methodological framework is the perspective that innovation happens in complex social systems, that networks provide the relevant morphology, and that we can use a combination of empirical research and modeling to understand innovation structures and processes, either on the micro and on the systems level.

This is how this book is intended to work as well. The rather programmatic introduction does not represent the umbrella under which all the authors have agreed to meet. The meeting place is condensed in the broad framework of the paragraph above. The introduction offers a potential and hopefully plausible theoretical background for instantiating this framework. Others are certainly possible and will be represented by the authors of this book.

The contributions of this book

This book comprises programmatic contributions of the leading international experts of innovation research and discusses issues of immediate concern to innovation policymakers and innovation business managers. On the theoretical side, it will provide systematic knowledge on the nature and characteristics of innovation processes to keep up with the complexity, with the non-linearities and the self-organizing features of innovation performance. With this, it will further demonstrate the embeddedness of socio-economic innovation research in complexity science and computational approaches. On the practical side, it will identify points of intervention and support for innovation and for collaborative networking with partners and stakeholders. The book attempts this:

*   by presenting results from the empirical analysis of innovative actors such as universities, SMEs, and MNEs while focusing on their respective contributions to the innovation process;
*   by illuminating the systemic context of innovation in an evolutionary framework; and
*   by discussing the tools and methods with which innovation networks in complex social systems can be successfully investigated to extend and apply our knowledge of them most effectively.

Part I of the book is dedicated to theoretical contributions regarding the systemic aspects of innovation. They complement what is said in the introduction coming from a complexity science and neo-Schumpeterian background.

John L. Casti[3] explores in 'When normal isn't so normal anymore' why life-changing technologies appear far more often than we think. From a broad complexity science perspective, he examines the outlook for innovation and technology in an era of declining economic growth. Horst Hanusch and Andreas Pyka introduce 'A neo-Schumpeterian approach towards public sector economics', which deals with dynamic processes causing qualitative transformation of economies basically driven by the introduction of novelties in their various and multifaceted forms. This approach is concerned with all facets of open and uncertain developments in socio-economic systems, e.g. the public and monetary side of an economic system. Pier Paolo Saviotti's chapter is concerned with the co-evolution of technologies and institutions as a general phenomenon of innovation accompanying the evolution of socio-economic systems. This co-evolution can be considered a simplified example of a systemic approach in which different variables interact and create feedback loops. The non-linear interactions shape in a fundamental way structural change and economic development.

Part II of the book is concerned with the actors of innovation embedded in mostly regional networks of relationships. Complementing again what was said about innovation networks in the introduction, Part II is started with a broader view on the role and importance of an actor-centered perspective and is organized around the main participants of innovation networks: multinational enterprises, small and medium businesses, and universities. Thomas J. Allen *et al.* present an empirical study of scientific communication among biotechnology companies, which supports the belief that geographic clustering does produce increased scientific exchange among companies. Critical to the formation of cluster-based scientific communication networks is the presence of both universities and large firms from the same industry.

Taking up this focus on regions and geographical proximity, the case study of Bernd Ebersberger and Florian M. Becke analyzes the distribution of competences concerning regional innovation systems in Austria. Turning to different types of firm actors in regional innovation systems, Bjørn Asheim tells us 'Innovation is small: SMEs as knowledge explorers and exploiters'. Traditionally, the challenge of SMEs has been to exploit knowledge for innovative purposes. In order to achieve this, a range of public policy instruments such as brokers, mobility schemes, technology centers, clusters and regional innovation systems (RIS) have been launched. The challenge of an RIS is to upgrade these SMEs to higher value-added production through innovation. SMEs as knowledge explorers are based on an analytical (science) knowledge base and apply an STI (science, technology, innovation) mode of innovation.

This chapter on the role of SMEs for different network policies is followed by an 'Innovation is big' view of two contributions dealing with multinational firms and their role for innovation. Since formal R&D activities tend to be expensive, scarce-resource thirsty, risky, and have a fairly large minimum efficiency scale, it is no wonder, that large MNEs tend to account for a considerably large share of the formal R&D activity undertaken in many industries. Rajneesh Narula and Julie Michel discuss 'Reverse knowledge transfer and its implications for European policy' highlighting some of the challenges large firms face in maintaining complex global networks at optimal level of efficiency. Martin Heidenreich *et al.* show that MNEs are indeed an important channel for the international transfer of technological knowledge across national and cultural and institutional boundaries. They point out that it is therefore necessary to analyze MNEs not only as knowledge-based networks, but also as an arena for cross-border power and exchange relations and as organizations of socio-institutionally embedded organizations.

Focusing on the role of policy actors, Federica Rossi *et al.* apply a complex system perspective to public innovation policy and private technology brokering.

Their contribution explores the economic and policy implications of two kinds of interventions directed at supporting innovation processes: an innovative public policy in support of networks of innovating firms, and a private initiative supporting innovation in the mechanical engineering industry thanks to the creation of a technology broker.

Last but not least, a number of chapters in this section deal with the role of universities for innovation. This is started off by Henry Etzkowitz and Marina Ranga discussing a strategy, centered on the entrepreneurial university, to address the current economic crisis. Universities and other higher education institutions that are major producers of knowledge are seen in a new light as potential direct contributors to economic advance in addition to their traditional supporting roles. Using data for a sample of US universities from the National Center for Education's (NCE's) Integrated Postsecondary Education Data System (IPEDS), (which contains extensive data for all US research institutions), the National Science Foundation's Integrated Science and Engineering Resources Data System (NSF WebCASPAR), and the US Licensing Activity Survey of the Association of University Technology Manager (AUTM), T. Austin Lacy tests the presence of scale economies in both traditional university outputs and the more recent, entrepreneurial outcomes.

With 'scientific entrepreneurship engineering', Philipp Magin and Harald F.O. von Kortzfleisch introduce methods and tools to support entrepreneurial activities in universities. They present results of an exploratory empirical study investigating all the existing scientific entrepreneurship methods and tools at universities in Germany, Switzerland, and Austria. William Allen and Rory O'Shea examine 'Promoting effective university commercialization' exploring the factors that can either stimulate or inhibit an academic's ability to commercialize a spin-off venture within a university environment. Uwe Obermeier *et al.* study co-authorship networks of scientists at a university as an archetypical example of complex evolving networks. Collaborative research has become the dominant and most promising way to produce high-quality output that leads to innovation.

Part III of the book is dedicated to modeling contributions regarding the systemic aspects of innovation. They complement what was said in the introduction about the role of agent-based modeling (ABM), especially as a tool to support innovation policymaking.

Nigel Gilbert *et al.* present some simulation experiments of their SKIN model (Simulating Knowledge dynamics in Innovation Networks) concerning learning in innovation networks. In their model, firms with different knowledge stocks attempt to improve their economic performance by engaging in radical or incremental innovation activities and through partnerships and networking with other firms. In trying to vary and/or to stabilize their knowledge stocks by organizational learning, they attempt to adapt to environmental requirements while the market strongly selects on the results. Marco Villani and Luca Ansaloni model the interplay between artifact innovation and attributions of functionality; where radical innovations are created by a process they call 'exaptation'. Tommaso Ciarli *et al.* address the role of interfaces in explaining the relation between product architecture and the organizational choices of firms involved in technological competition using a pseudo-NK simulation model. The simulation results show how the firms' organization vary with the number and the type of interfaces mediating the relationships between the components themselves. Interfaces are difficult to innovate, but provide the largest returns from innovation. Francesca Giardini and Federico Cecconi model the interplay between economic performance and social evaluation in industrial clusters investigating how social evaluations may affect the dynamics of innovation in a network of firms. Different types of social evaluations may affect the emergence of innovation and the network configuration of artificial firms working in an industrial cluster.

Flaminio Squazzoni and Riccardo Boero question the state of the art of policymaking from a complexity perspective. They provide an overview on how ABMs can change policymaking in a more complexity-friendly way. While standard policymaking is an attempt to reduce or eliminate complexity, ABMs allow us to understand and 'harness' complexity. Ramon Scholz *et al.* simulate various aspects of politically-induced innovation networks using an extended version of the SKIN model (Pyka *et al.* 2009) for describing the science landscape in the European Union. The model focuses on the network structures resulting from the research cooperations of the actors participating in the European Framework Programmes. They compare model results to empirical EU R&D collaboration networks, and conduct some policy-driven experiments. The book ends with some conclusions for innovation policy – consequences for the relation between investment in knowledge/technology/innovation and economic growth – put forward by the editor.

Notes

1 We do not discuss this body of literature here, but just hint at the fact that there is a huge amount of it concerning complexity issues in and around organizations and their management, and their leadership.
2 Connected to this issue, there are the sophisticated details of realism and constructivism, of determinism and voluntarism, of the role of the observer and of causality etc. in the various disciplines (discussed extensively in Ahrweiler 2001).
3 The chapters are introduced by their titles and some author statements, which illustrate the role of the single contributions for the whole volume.

References

Abrahamson, E. and Rosenkopf, L. (1997) 'Social Network Effects on the Extent of Innovation Diffusion: a Computer Simulation', *Organization Science*, 8 (3), pp. 289–309.
Ahrweiler, P. (1999) 'Complexity Theory and the Social Sciences: David Byrne', *Emergence*, 1 (2), pp. 101–5.
Ahrweiler, P. (2001) *Informationstechnik und Kommunikationsmanagement. Netzwerksimulation fuer die Wissenschafts- und Technikforschung*, Frankfurt: Campus, Information Technology and Communication Management.

Ahrweiler, P. and Gilbert, N. (2005) 'Caffè Nero: the Evaluation of Social Simulation', *Journal of Artificial Societies and Social Simulation*, 8 (4).

Ahrweiler, P., Gilbert, N. and Pyka, A. (2006)'Institutions Matter, but ... Organisational Alignment in Knowledge-based Industries', *Science, Technology and Innovation Studies*, 2 (1), pp. 3–18.

Ahrweiler, P., Pyka, A. and Gilbert, N. (2004) 'Simulating Knowledge Dynamics in Innovation Networks', in: R. Leombruni and M. Richiardi (eds) *Industry and Labor Dynamics: The Agent-based Computational Economics Approach*, Singapore: World Scientific Press.

Ahrweiler, P., Pyka, A. and Gilbert, N. (forthcoming) 'A New Model for University–Industry Links in Knowledge-based Economies', *Journal of Product Innovation Management*.

Ahuja, G. (2000) 'Collaboration Networks, Structural Holes, and Innovation', *Administrative Science Quarterly*, 45, pp. 425–55.

Albert, R. and Barabási, A.-L. (2002) 'Statistical Mechanics of Complex Networks', *Reviews of Modern Physics T4*, 1, pp. 47–97.

Anderson, P. (1999) 'Complexity Theory and Organization Science', *Organization Science*, 10 (3), pp. 216–32.

Arthur, B. (1989) 'Competing Technologies, Increasing Returns, and Lock-In by Historical Events', *Economic Journal*, 99, pp. 116–31.

Arthur, B. (1998) *Increasing Returns and Path Dependence in the Economy*, Ann Arbor: University of Michigan Press.

Barabási, A.-L. (2002) *Linked: The New Science of Networks*, Cambridge, MA: Perseus.

Barabási, A.-L. and Albert, R (1999) 'Emergence of Scaling in Random Networks', *Science*, 286, pp. 509–12.

Bar-Yam, Y. (1997) *Dynamics of Complex Systems*, Reading: Addison Wesley.

Bar-Yam, Y. (2004) *Making Things Work: Solving Complex Problems in a Complex World*, Cambridge, MA: Knowledge Press.

Beije, P. (1998) *Technological Change in the Modern Economy: Basic Topics and New Developments*, Cheltenham: Elgar Publishers.

Braha, D., Minai, A. and Bar-Yam, Y. (eds) (2008) *Complex Engineered Systems: Science Meets Technology*, New York: Springer.

Bridgeforth, B.W. (2005) 'Toward A General Theory of Social Systems', *International Journal of Sociology and Social Policy*, 25 (10/11), pp. 54–83.

Brown, S.L. and Eisenhardt, K.M. (1998) *Competing on the Edge: Strategy as Structured Chaos*, Boston: Harvard Business School Press.

Buijs, J. (2003) 'Modelling Product Innovation Processes, from Linear Logic to Circular Chaos', *Creativity and Innovation Management*, 12 (2), pp. 76–93.

Burt, R.S. (2004) 'Structural Holes and Good Ideas', *American Journal of Sociology*, 110 (2), pp. 349–99.

Burt, R.S. (1992) *Structural Holes*, Cambridge, MA: Harvard University Press.

Byrne, D. (1998) *Complexity Theory and the Social Sciences*, London: Routledge.

Casti, J. (1995) *Complexification: Explaining a Paradoxical World through the Science of Surprise*, New York: HarperCollins.

Chiva-Gomez, R. (2004) 'Repercussions of Complex Adaptive Systems on Product Design Management', *Technovation*, 24 (9), pp. 707–11.

Choi, T.Y., Dooley, K.J. and Rungtusanatham, M. (2001) 'Supply Networks and Complex Adaptive Systems: Control Versus Emergence', *Journal of Operations Management*, 19 (3), pp. 351–66.

Cowan, R. and Jonard, N. (2004) 'Network Structure and the Diffusion of Knowledge', *Journal of Economic Dynamics and Control*, 28, pp. 1557–75.

Cowan, R., Jonard, N. and Zimmermann, J.-B. (2007) 'Bilateral Collaboration and the Emergence of Innovation Networks', *Management Science*, 53, pp. 1051–67.

Cunha, M.P. and Comes, J.E.S. (2003) 'Order and Disorder in Product Innovation Models', *Creativity and Innovation Management*, 12 (3), pp. 174–87.

Dooley, K. and van de Ven, A. (1999) 'Explaining Complex Organizational Dynamics', *Organization Science*, 10 (3), pp. 358–72.

Eisenhardt, K.M. and Bhatia, M.M. (2002) 'Organizational Complexity and Computation', in: J.A.C. Baum (ed.) *Companion to Organizations*, Oxford: Blackwell.

Etzkowitz, H. and Leydesdorff, L. (eds) (1997) *Universities and the Global Knowledge Economy: A Triple Helix of University–Industry–Government Relations*, London and Washington: Pinter.

European Commission/DG Research (2002) *Benchmarking National Research Policies: The Impact of RTD on Competitiveness and Employment (IRCSE)*, Brussels: European Commission/DG Research.

Eve, R.A., Horsfall, S. and Lee, M.E. (1997) *Chaos, Complexity and Sociology: Myth. Models, and Theories*, Thousand Oaks: Sage.

Fagerberg, J., Mowery, D. and Nelson, R.R. (2006) *The Oxford Handbook of Innovation*, Oxford: Oxford University Press.

Fagerberg, J. (2003) 'Schumpeter and the Revival of Evolutionary Economics: An Appraisal of the Literature', *Journal of Evolutionary Economics*, 13, pp. 125–59.

Flake, G.W. (1999) *The Computational Beauty of Nature*, Cambridge, MA: MIT Press.

Foerster, H. von (1984) *Observing Systems*, Salinas, CA: Intersystems Publications.

Foerster, H. von and Zopf, J. (1962) 'Principles of Self-Organization', *Transactions of the University of Illinois Symposium on Self-Organization*, Robert Allerton Park, 8 and 9 June 1961.

Fornahl, F. and Brenner, T. (2003) *Cooperation, Networks and Institutions in Regional Innovation Systems*, Cheltenham, UK: Elgar Publishers.

Frenken, K. (2006) *Innovation, Evolution and Complexity Theory*, Cheltenham, UK: Elgar Publishers.

Gell-Mann, M. (1994) *The Quark and the Jaguar*, New York: Freeman & Co.

Gibbons, M., Limoges, C., Nowotny, H., Schwartzman, S., Scott, P. and Trow, M. (1994) *The New Production of Knowledge: The Dynamics of Science and Research in Contemporary Societies*, London: Sage.

Gilbert, N. (2007) *Agent-based Models*, London: Sage.

Gilbert, N. and Troitzsch, K. (2005) *Simulation for the Social Scientist*, Berkshire, UK: Open University Press.

Gilbert, N., Ahrweiler, P. and Pyka, A. (2007) 'Learning in Innovation Networks: Some Simulation Experiments', *Physica A-Statistical Mechanics and Its Applications*, 378 (1), pp. 667–93.

Gloor, P. (2006) *Swarm Creativity – Competitive Advantage through Collaborative Innovation Networks*, Oxford: Oxford University Press.

Granovetter, M. (1973) 'The Strength of Weak Ties', *American Journal of Sociology*, 78 (6), pp. 1360–80.

Hagedoorn, J. and Roijakkers, N. (2002) 'Small Entrepreneurial Firms and Large Companies in Inter-Firm R&D Networks – the International Biotechnology Industry', in: M. Hitt, D. Ireland, M. Camp and D. Sexton (eds) *Strategic Entrepreneurship*, Cambridge, MA: Blackwell Publishing.

Hejl, P.M. (1992) 'Selbstorganisation und Emergenz in sozialen Systemen [Self-Organization and Emergence in social Systems]', in: W. Krohn and G. Kueppers (eds) *Die Entstehung von Ordnung, Organisation und Bedeutung*, Frankfurt: Suhrkamp.

Holland, J.H. (1995) *Hidden Order: How Adaptation Builds Complexity*, Reading, MA: Addison-Wesley.

Imai, K., Baba, Y. (1991) 'Systemic Innovation and Cross-Border Networks, Transcending Markets and Hierachies to Create a New Techno-economic System', *Technology and Productivity: the Challenge for Economic Policy*, Paris: OECD.

Kaemper, E. and Schmidt, J. (2000) 'NetzwerkealsstrukturelleKopplung [Networks as structural Coupling]', in: J. Weyer (ed.) *Soziale Netzwerke*, Muenchen: Oldenbourg.

Kauffman, S.A. (1993) *The Origins of Order: Self-Organization and Selection in Evolution*, New York: Oxford University Press.

Kauffman, S.A. (1995) *At Home in the Universe: The Search for the Laws of Self-Organization and Complexity*, New York: Oxford University Press.

Klir, G.J. (ed.) (1978) *Applied Systems Research*, New York: Plenum Press.

Kowol, U. and Krohn, W. (1995)'Innovationsnetzwerke: Ein Modell der Technikgenese [Innovation Networks: A Model of the Genesis of Technology]', in: J. Halfmann, G. Bechmann and W. Rammert (eds) *Technik und Gesellschaft: Jahrbuch 8*, Frankfurt and New York: Campus, (own translation where cited).

Krohn, W. (1995) *Die Innovationschancen partizipatorischer Technikgestaltung und diskursiver Konfliktregelung [The Innovation Chances of participatory Technology Shaping and discursive Conflict Resolution]*, IWT-Paper 9/95, Bielefeld (own translation where cited).

Krohn, W. and Kueppers, G. (1989) *Die Selbstorganisation der Wissenschaft [The Self-Organisation of Science]*, Frankfurt: Suhrkamp.

Lane, D., van der Leeuw, S., Pumain, D. and West, G. (eds) (2009) *Complexity Perspectives in Innovation and Social Change*, Berlin and New York: Springer.

Luhmann, N. (1987) *Soziale Systeme: Grundriss einer allgemeinen Theorie [Social Systems: Foundation of a general Theory]*, Frankfurt: Suhrkamp (own translation where cited).

Luhmann, N. (1992) *Die Wissenschaft der Gesellschaft [The Science of the Society]*, Frankfurt: Suhrkamp (own translation where cited).

Luhmann, N. (1994) *Die Wirtschaft der Gesellschaft [The Economy of the Society]*, Frankfurt: Suhrkamp (own translation where cited).

Luhmann, N. (1997) *Die Gesellschaft der Gesellschaft [The Society of the Society]*, Frankfurt: Suhrkamp (own translation where cited).

Lundvall, B.-Å. (ed.) (1992) *National Innovation Systems: Towards a Theory of Innovation and Interactive Learning*, London: Pinter.

McKelvey, B. (1999) 'Self-Organization, Complexity, Catastrophe, and Microstate Models at the Edge of Chaos', in J.A.C. Baum and B. McKelvey (eds) *Variations in Organization Science – in Honor of Donald T. Campbell*, Thousand Oaks, CA: Sage Publications.

Malerba, F. (2002) 'Sectoral Systems of Innovation and Production', *Research Policy*, 31, pp. 247–64.

Malerba F., Nelson, R., Orsenigo, L. and Winter, S. (1999) 'History Friendly Models of Industry Evolution: the Computer Industry', *Industrial and Corporate Change*, 1, pp. 3–41.

Maturana, H. (1991) 'Gespraech mit Humberto R. Maturana [Interview with Humberto

Maturana]', in: V. Riegas and C. Vetter (eds) *ZurBiologiederKognition*, Frankfurt: Suhrkamp.

Nelson, R.R. (ed.) (1993) *National Innovation Systems: A Comparative Analysis*, Oxford: Oxford University Press.

Nelson, R.R. and Winter, S. (1982) *An Evolutionary Theory of Economic Change*, Cambridge, MA: Harvard University Press.

Newman, M. (2003) 'The Structure and Function of Complex Networks', *SIAM Review*, 45, pp. 167–256.

Oliver, A.L. and Liebeskind, J.P. (1998) 'Three Levels of Networking for Sourcing Intellectual Capital in Biotechnology: Implications for Studying Interorganizational Networks', *International Studies of Management and Organization*, 27, pp. 76–103.

Parsons, T. (1961) 'An Outline of the Social System', in: T. Parsons, E.A. Shills, K.D. Naegele and J.R. Pitts (eds) *Theories of Society: Foundations of Modern Sociological Theory*, New York: Free Press.

Porter, K.A., Bunker Whittington, K.C. and Powell, W.W. (2005) 'The Institutional Embeddedness of High-tech Regions: Relational Foundations of the Boston Biotechnology Community', in: S. Breschi and F. Malerba (eds) *Clusters, Networks, and Innovation*, Oxford: Oxford University Press.

Powell, W.W., White, D.R., Koput, K.W. and Owen-Smith, J. (2005) 'Network Dynamics and Field Evolution: The Growth of Inter-organizational Collaboration in the Life Sciences', *American Journal of Sociology*, 110 (4), pp. 1132–205.

Prewo, R. (1979) *Max Webers Wissenschaftsprogramm: Versuch einer methodischen Neuerschliessung [Max Weber's Programme of Science]*. Frankfurt: Suhrkamp (own translation where cited).

Prigogine, I. and Stengers, I. (1984) *Order out of Chaos*, New York: Bantam Books.

Pyka, A. and Kueppers, G. (eds) (2003) *Innovation Networks – Theory and Practice*, Cheltenham, UK: Elgar Publishers.

Pyka, A. and Saviotti, P.P. (2005) 'The Evolution of R&D Networking in the Biotech Industries', *International Journal of Entrepreneurship and Innovation Management*, 5 (1/2), pp. 49–68.

Pyka, A., Ahrweiler, P. and Gilbert, N. (2009) 'Agent-based Modelling of Innovation Networks: The Fairytale of Spillovers', in: A. Pyka and A. Scharnhorst (eds) *Innovation Networks: New Approaches in Modeling and Analyzing*, Berlin and New York: Springer.

Rechtin, E. (1991) *Systems Architecting: Creating And Building Complex Systems*, Englewood Cliffs, NJ: Prentice-Hall.

Schilling, M.A. and Phelps, C.C. (2005) 'Interfirm Collaboration Networks: the Impact of Small World Connectivity on Firm Innovation', *Management Science*, 53 (7), pp. 1113–26.

Schmid, M. and Haferkamp, H. (1987) 'Einleitung [Introduction for Sense, Communication and Social Differentiation.Contributions to Luhmanns Theory of social Systems]', in: M. Schmid and H. Haferkamp (eds) *Sinn, Kommunikation und soziale Differenzierung: Beitraege zu Luhmanns Theorie sozialer Systeme*, Frankfurt: Suhrkam. (own translation where cited).

Schumpeter, J. (1912) *The Theory of Economic Development*, Oxford: Oxford University Press.

Schwemmer, O. (1987) *Handlung und Struktur [Action and Structure]: Zur Wissenschaftstheorie der Kulturwissenschaften*, Frankfurt: Suhrkamp.

Siegel, D.S., Waldman, D., Atwater, L. and Link, A.N. (2003) 'Commercial Knowledge

Transfers from Universities to Firms: Improving the Effectiveness of University–Industry Collaboration', *Journal of High Technology Management Research*, 14, pp. 111–33.

Smith, H.L. and Ho, K. (2006). 'Measuring the Performance of Oxford University, Oxford Brookes University and the Government Laboratories' Spin-Off Companies', *Research Policy*, 35, pp. 1554–68.

Sorenson, O., Rivkin, J. and Fleming, L. (2006) 'Complexity, Networks and Knowledge Flow', *Research Policy*, 35 (7), pp. 994–1017.

Sterman, J.D. (1994) 'Learning in and about Complex Systems', *System Dynamics Review*, 10 (2–3), pp. 291–330.

Sterman, J.D. (2002) 'All Models are Wrong: Reflections on Becoming a System Scientist', *System Dynamics Review*, 18 (4), pp. 501–31.

Stewart, I. (1989) *Does God Play Dice? The Mathematics of Chaos*, Cambridge, MA: Blackwell.

Tesfatsion, L. and Judd, K. (2006) *Handbook of Computational Economics: Agent-based Computational Economics*, Missouri: Elsevier.

Thursby, J. and Kemp, S. (2002) 'Growth and Productive Efficiency of University Intellectual Property Licensing', *Research Policy*, 31, pp. 109–24.

Uzzi, B. (1997) 'Social Structure and Competition in Inter-firm Networks: The Paradox of Embeddedness', *Administrative Science Quarterly*, 42, pp. 35–67.

Valente, T.W. (1996) *Network Models of the Diffusion of Innovations*, Cresskill: Hampton Press.

Verspagen, B. and Duysters, G. (2004) 'The Small Worlds of Strategic Technology Alliances', *Technovation*, 24, pp. 563–71.

Waldrop, M.M. (1992) *Complexity: The Emerging Science at the Edge of Order and Chaos*, New York: Simon & Schuster.

Walker, G., Kogut, B. and Shan, W. (1997) 'Social Capital, Structural Holes and the Formation of an Industry Network', *Organization Science*, 8, pp. 108–25.

Wasserman, S. and Faust, K. (1994) *Social Network Analysis: Methods and Applications*, Cambridge, UK: Cambridge University Press.

Watts, D. (1999) *Small Worlds*, Princeton: Princeton University Press.

Watts, D. and Strogatz, S. (1998) 'Collective Dynamics of "Small-World" Networks', *Nature*, 393, pp. 440–2.

Weidlich, W. (1991) 'Physics and Social Science – The Approach of Synergetics', *Physics Reports*, 204 (1), pp. 1–163.

Part I

The systemic aspects of innovation (theory)

2 When normal isn't so normal anymore

Why life-changing technologies appear far more often than we think

John L. Casti

Onward and upward – then downward

Shortly after being acquitted of child molestation charges in California in the summer of 2005, the late pop star Michael Jackson mysteriously surfaced in the Gulf city of Dubai accompanied by the son of the king of Bahrain. Those sensitized to the bizarre ways of media publicity nowadays immediately sensed this as a signature event in Dubai's 'coming of age' in an age of mediocrity. During his royally guided tour of the city, Michael would not have had to look far to see what is intended to be the world's tallest building, Burj Dubai, coming out of the ground that sunny day.

History records many examples of societies seemingly being compelled to leave a visible testament for the ages to their increasingly positive view of the future by creating the world's tallest building. In each case, construction begins on these behemoths as the social mood starts accelerating upward. But skyscrapers don't appear overnight, and by the time construction is actually completed several years later, the positive mood has almost always given way to a deeply pessimistic one. Here is a picture of the three latest contenders in the skyscraper derby – Indonesia, Taiwan and Dubai – projected alongside a chart of the financial market averages in the respective countries. The current troubles of property developers in Dubai are not at all surprising, at least to those who understand this 'Skyscraper Curse'. Bad things tend to happen in countries – especially to 'little guys' who want to become 'big guys' – when they start trying to express their confidence in the future by erecting the world's tallest building.

This curse leads one to wonder about the future of Saudi Arabia, which in late 2008 announced its intention to outdo Dubai by building a skyscraper in Jeddah vastly taller than Burj Dubai. And this is not to mention the fortunes of that darling of evangelists of globalization everywhere, India, where a local architect in Delhi announced plans to build the world's tallest building, making the statement: 'It is about status. It is about glorification. It is high time that people started realizing that we too are a great nation.' *Sic transit gloria!*

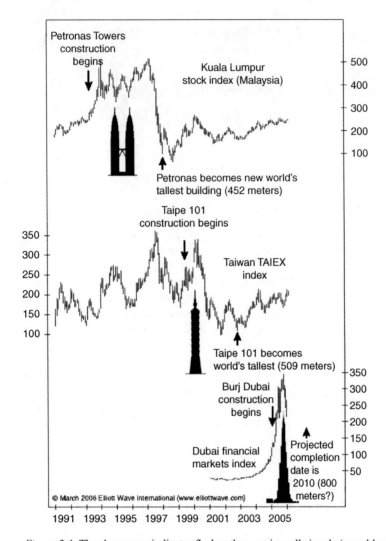

Figure 2.1 The skyscraper indicator flashes three major sell signals (monthly data).

What about technology?

From 1966 through late 1968, as public sentiment skyrocketed with the increasing stock market, technological wonder ran rampant. Futurists with visions of colonies on the moon, the sea floor and Mars were routinely quoted in the business press. Public fascination with such forecasts was reflected in a popular book, *The Year 2000*, by famed futurist Herman Kahn, which anticipated the conversion of seawater to drinking water and the use of artificial moons. One project that caught the imagination of reporters was 'Probe', a think tank of 27 top scientists established by TRW Inc. The team used existing technologies to forecast over 335 wonders yet to come. Not a single one came true. And six years later the Dow was down 45 percent with many of the most popular technology companies having gone bankrupt. A similar story can be told about the Internet boom in the late 1990s. My belief is that we will see the same pattern unfold from the mini-boom of 2003–7: Euphoria denied!

To test this hypothesis, have a look at Figure 2.2, below, showing major technological developments over the period 1920–2001 plotted against the Dow-Jones Industrial Average in that same period. Do you notice anything unusual about this chart?

What's interesting here are the periods 1920–38 and 1970–83. During these periods, very little 'breakthrough' technologies were introduced compared with the rest of the chart. And when are these times of low technological breakthrough? Just exactly when the financial markets are either in freefall, like the late 1920s to early 1930s, or going sideways with bursts of decline as in the early 1970s to mid-1980s. Given what we see in the world today and can reasonably project to the world of the next decade, it's not too difficult to make the call that life-changing technologies are going to be thin on the ground during this period.

It doesn't take an undiscovered genius to know that advances in technology go hand-in-hand with innovation. Without new products, or at least new ideas for how to use existing technology, we'd still be riding on horses and communicating by smoke signals. While it's difficult to measure 'innovation' directly, a

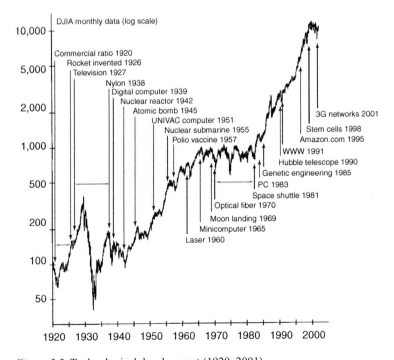

Figure 2.2 Technological development (1920–2001).

good surrogate is to simply look at how many patents are granted since patents necessarily involve something new and different. To get a feel for the way social mood impacts innovation, consider Figure 2.3 showing patents versus the DJIA for the twentieth century:

When we put this figure together with the fact that the social mood is always a *leading* indicator of social events, implying that we must shift the lower curve to the left by a few years to properly compare the two processes, the reader will have little trouble forecasting the future of innovation, hence wondrous new technologies, in the coming decade or more.

The proverbial perceptive reader will have noticed that these stories of sky-scrapers, technology and patents have all been set against movements in financial market indexes. Here's why.

Vox populi

On 19 March 2003, US forces rained 'shock and awe' down upon the hapless residents of Baghdad, thereby initiating the Iraq War, the first great folly of the

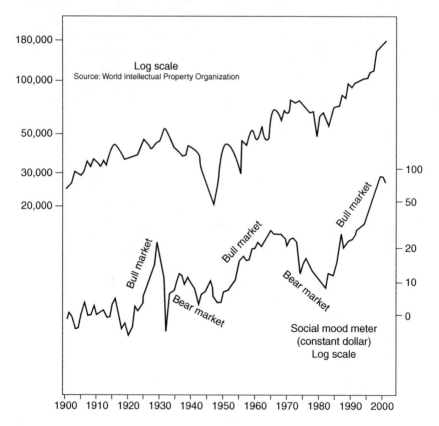

Figure 2.3 US patents granted and Dow Jones Industrial Average (1900–2001).

twenty-first century (but certainly not the last!). While there is, and will be, much to say about this sorry affair, it is enough here to note only that it illustrates perfectly the role played by the 'mood' of a population in creating a social climate, a kind of Zeitgeist, within which actions, behaviors and events of all types unfold. And the nature and texture of those events are dramatically impacted by whether that mood is optimistic or pessimistic.

The concept of the *mood* of a population as setting the tone for collective social events of all types, ranging from tastes in popular culture to shifts in political ideologies to the rise and fall of civilizations, is the root cause of my pessimism about what we are likely to see in the coming years and decades. Put simply, the social mood represents how the population feels about the future–on all timescales. And as with the Iraq War, the potential represented by the population's sense of the future, its mood, is realized in vastly different types of events, depending on whether the mood is waxing positive or waning negative on the timescale appropriate for the type of event in question. But to make the notion of the mood of a population useful for either explanation or prediction, we have to have an effective way to measure it.

'Sociometers'

Some years back, financial analyst Robert Prechter coined the term 'socionomics' for the way the social mood leads to social actions. He then proposed using the financial market averages as a way of measuring the mood. He called this measure a 'sociometer', as it serves much the same purpose for measuring social mood that a thermometer serves for measuring the overall motion of a collection of molecules. The underlying argument is that a market average like the Dow Jones Industrial Average (DJIA) reflects bets that people make about the future on all timescales from seconds to decades. The financial markets collect all these bets and process them into a single number: a change of price. That price change then serves as a very effective measure for how people feel about the future. If they are positive, they tend to buy and prices increase; if they're pessimistic, the tendency is to sell and prices go down. And the stronger the collective sentiment, the larger the bets.

But just as a thermometer doesn't measure what every single molecule is doing, the financial market averages do not represent the feelings of every single person in a population either. However, experience shows that an index like the DJIA serves as a much better characterization of the social mood than other types of measures of mood, such as opinion surveys, annual births, and the like. Moreover, accurate financial data is easy to find in every daily newspaper, and is available over quite long periods of time.

To illustrate this idea, consider the years 1930–2000. These seven decades divide into two completely different periods on a decade-to-decade basis: a 20-year span of negative global social mood from 1930–50, followed by 50 years of increasingly positive mood that ended in early 2000. During the pessimistic period, we saw events like the rise of dictatorships in Nazi Germany and the Soviet Union, the Holocaust, and the Great Depression, while in the postwar period the Berlin Wall came crashing down, apartheid ended in South Africa, and the European Union was formed. Note the qualitative difference in character between these events. Entirely different types of events tended to occur during the period of negative social mood than those taking place when the world, in general, was more optimistic about the future.

This difference is the crux of our argument for a rather more pessimistic view of what to expect over the coming decades. The global social mood started rolling over from positive to negative in about the year 2000. My contention is that it will accelerate in the downward direction for at least a decade or more before we hit the bottom. As a result, the types of events we can expect to see will be of a decidedly different nature than what has been the case over the last 50 years. The illustrations of skyscrapers and patents illustrate the overall situation we face today. Here is yet another big example to hammer home the point.

The decline and fall of globalization

Unlike skyscrapers, which are an inherently local phenomenon, physically confined to a particular geographical space, the once trendy idea of globalization, the view that the world is one gigantic marketplace perfectly structured to solve the ills of humankind, unfettered by the inconveniences of restrictions on the flow of capital, labor, materials or ideas, is another collective social phenomenon that is in the process of coming undone. Since the driving forces behind globalization are to a substantial degree American corporations, we look at the DJIA from 1970 as an indicator of the overall worldwide mood, since the New York Stock Exchange is still about the closest thing we have to a global financial market. Every single milestone in the path to globalization from the launching of the basic idea at Davos in 1975, to the formation of the World Trade Organization in 1996, to China's joining of the WTO in 2000 took place at a peak in social mood (see Figure 2.4).

Since 1975, the global social mood has been rosy enough to emit heat. In such times, the types of events we expect to see are ones that can be labeled with words like 'unifying', 'joining', and 'expanding'. Sad to say, the picture shows that this global mood is rolling over to begin a decades-long decline that is likely to lead to just the opposite types of social events. Globalization will be replaced by localization, unification will be replaced by fragmentation, and openness to strangers will be replaced by xenophobic behaviors. Distant early-warning signals of all of these types of behaviors are already apparent in the pages of your favorite daily newspaper or on the Internet.

Meanwhile, back in Europe

In many ways, the European Union represents a kind of 'dress rehearsal' for globalization, starting on its path to the form we see today with the Treaty of Rome in 1957. Numerous headlines over the past year have described European

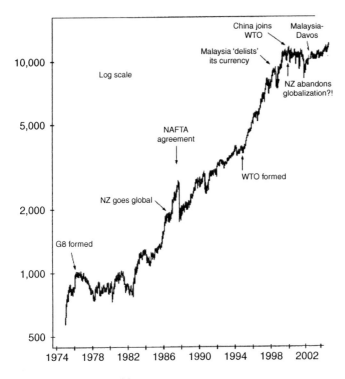

Figure 2.4 DJIA monthly.

concerns, such as 'Islamophobia' and other tensions over immigration, the right of 'free expression', Brussels-style, the Belgian separatist movement, expansion of the EU to 'non-European' nations and the like. We can summarize the social mood in Europe with the chart in Figure 2.5, showing the relevant financial index for European matters, the Dow Jones Stoxx 50, and milestone events in the EU. Here we see that the ups and downs of the EU follow in lockstep with the social mood: Nice things tend to happen when the mood is positive, not-so-nice things when the mood shifts to the negative. So if I were to go to sleep today and wake up in 20 years, one thing I would certainly not expect to see would be a European Union, or at least not one that bears any resemblance to what we have today.

The juggernaut of history

When I recently presented this rather downbeat vision of the next several years at an international symposium on 'the future', a member of the audience accused me of presenting a 'doomsday' scenario. I pointed out to that gentleman that the future I've presented here is very far from any kind of doomsday, as it's easy to

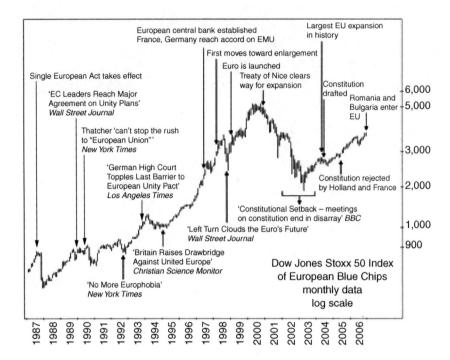

Figure 2.5 European Union waxes with the rise in stock prices.

imagine futures vastly worse than this. It is also not simply my imagination running amok either; this future has already happened – in the 1930s – and it can happen again, only this time rather worse.

The main point to bear in mind is that humanity survived the 1930s and it will survive the 2030s, as well. But as the old saying goes, the situation is desperate but it's not serious. And part of minimizing the pain of an unpleasant future is being prepared for it. So whether you're managing a family, a company, or a country, if you don't plan today for the possibility of a future you don't like, you'll be squashed flat by it when it arrives. A rerun of the 1930s may not be the future you want. But the juggernaut of history doesn't care what you want. It just rolls on.

And how is the flow rolling for innovation and technology? The outlook for the next several years is for a period of polishing existing apples to a brighter shine, not one for the introduction of major, life-changing technologies like practical fusion power or commercially workable supersonic flight.

Since the beginning of the global financial meltdown in 2007, much attention has been directed to the fact that extreme events, the 'once in a hundred year floods' happen a lot more frequently than one would like to believe. The fact is that the so-called 'normal' distribution for how events are distributed in both time and magnitude simply isn't so normal any more. We live in an age of extremes, and coming to terms with this fact will be one of the growth industries in intellectual life in the coming decades. Let's take a longer look at the dimensions of this issue in the context of technology and engineering.

The fat tail wagging the ordinary dog

Open-source software projects, like *Mozilla* and *Apache*, are modified by different programers who sometimes change more than one piece of software at a time. Stability of the software system is often measured by the number of changes per unit of time over a particular time interval. Researchers recently looked at the time between modifications to the code, as well as the number of files modified at one go. It turns out that both the distribution of modification times and the distribution of the number of files simultaneously modified obey a probability distribution that attaches much greater likelihood to extreme values of both quantities than what we'd expect from the old standby bell-shaped normal distribution.

Just as in evolutionary biology, where there are spurts of activity when a huge number of new species emerge almost overnight but when most of the time nothing is happening, so it is with open-source software development too. Most of the time no one is making any changes. Then, all of a sudden there is a burst of activity when the code is changed dramatically. Then everyone goes back to sleep. Here is another particularly graphic example in the same vein.

Following Hurricane Katrina's devastation of New Orleans in 2005, General Carl Strock of the US Army Corps of Engineers stated:

> When the project was designed ... we figured we had a 200 or 300 year level of protection. That means that the event we were protecting from might be exceeded every 200 to 300 years. That is a 0.5 per cent likelihood. So we had an assurance that 99.5 per cent of this would be okay. We, unfortunately, have had that 0.5 per cent activity here.

Stock's claim rests on the assumption that hurricanes of the size of Katrina occur with a frequency that can be described by the classical bell-shaped curve, the so-called normal distribution. Sad to say for New Orleans, statisticians have known for more than a century that the extreme events falling near the ends of a statistical distribution cannot usually be usefully described that way. Just as with the meltdown of the global financial system, the normal distribution dramatically underestimates the likelihood of unlikely events. Such events follow a different type of probability curve informally termed a 'fat-tailed' distribution. The difference is shown in Figure 2.6 below. Using this fat-tail law to describe the New Orleans situation, the 0.5 percent mentioned by General Strock would have been closer to 5 percent and the 300 years would have shrunk to about 60 years.

The key reason fat tails exist in financial market returns is that investor's decisions are not fully independent (a key assumption underlying the normal distribution). At extreme lows, investors are gripped with fear and they become

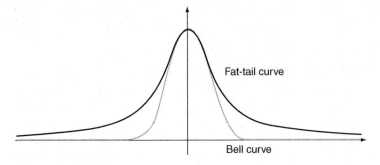

Figure 2.6 Fat-tail curve.

more risk-averse, while at extreme market highs investors become 'irrationally exuberant'. This type of interdependence then leads to herding behavior, which in turn causes investors to buy at ridiculous highs and sell at illogical lows. This behavior, coupled with random events from the outside world, pushes market averages to extremes much more frequently than models based on the normal distribution would have one believe.

A graphic illustration of this point is that the *casus casusorum* of the current global financial crisis is the almost universal use of the so-called Black–Scholes formula for pricing asset returns like options and other derivative securities. This rule, for which Myron Scholes and Robert Merton received the 1997 Nobel Prize in economics (Fisher Black having died in 1995) is to put it simply, just plain wrong. Why is it wrong? Because it is based on the normal distribution, which causes the formula to vastly underestimate the risk of the very types of events that actually occurred, thus setting off the chain reaction of bank failures and financial house collapses that continue to this day. Just another reason why there shouldn't be a Nobel Prize in economics IMHO!

Power(ful) laws

Conan Doyle's famous short story *Hound of the Baskervilles* consists of 59,498 words, of which 6,307 are different. This means that many words are repeated in this story. Not surprisingly, the most frequent word is 'the', which appears 3,328 times, followed by 'and' which occurs 1,628 times and 'to' which weighs-in at 1,429 appearances. Plotting the word frequency versus word rank on a logarithmic scale, we are led to the chart shown in Figure 2.7.

The straight-line relationship on the log–log scale between word rank and frequency is what is often termed a *power law* relationship (or, in this particular case of word frequency versus rank, it's usually called Zipf's Law). Power laws appear all over the place, ranging from the distribution of surnames in the United States to the relationship between the frequency and magnitude of earthquakes.

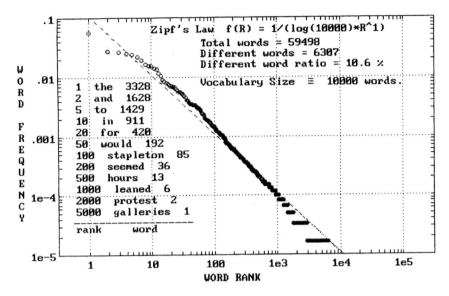

Figure 2.7 Word frequency *Hound of the Baskervilles*.

What's important about power laws in relation to fat-tailed distributions is the slope of the line. If the slope is less than −0.5, the extreme data (words of low rank) can be expected to appear with a higher frequency than one would expect from independent sampling of all words in the dictionary. So power laws are another way of characterizing the extreme events living on the fat tail.

The 98 percent rule

A common adage in business holds that 80 percent of a firm's sales come from 20 percent of the customers. Or, as is often the case in university departments, 80 percent of the papers are published by 20 percent of the professors. These are examples of what is called *Pareto's Principle*, and is closely related to fat-tailed distributions. There's really nothing special here about the number 80, as it could be anything between 50 and 100. And, in fact, the connection with fat tails is even more pronounced when instead of a conventional business with a High Street storefront and a shop window display one considers instead an Internet business like Amazon.

In a normal bookshop, there might be around 100,000 titles on the shelves of which 80 percent don't sell a single copy during the course of a month. So here the '80–20 Rule' prevails. Now consider Amazon, which has nearly four million titles available, or perhaps an online music site like *iTunes*, and ask how many of the titles they have available sell at least one copy the course of a month. The answer is not 20 percent or 50 percent or even 80 percent. Instead, it's a staggering 98 percent! So nearly every single title gets some action nearly every single month.

How can this possibly be? What makes Amazon different from the High Street shop? In his best-selling book, *The Long Tail*, Chris Anderson described the magic. What's needed, he says, is for there to be a functioning way to drive demand for those niche products out near the end of tail. First you need a 'head', consisting of a relatively small number of hits. Then comes a tail of many niche volumes, the kind only the author, his mother and a small band of fanatics and connoisseurs could ever love. So there must be not only a huge inventory of products, but also a way to direct prospective customers from the head to the tail by means of suggestions, background profiles from past purchases and all the other things a place like Amazon does to match readers with the books they really want. People need to start with something familiar, and then move via filters and suggestions to the unfamiliar. So you need a head to bring the customers in, a filtering mechanism to direct them to niche products, and an essentially unlimited shelf space filled with niche products to immediately service any customer's wishes. If any of these three ingredients is missing, then it's no sale!

Extreme behavior in technological processes

Events can be extreme in several different ways. The magnitude can be rare as with huge price changes in financial markets or with super volcanic eruptions. Or the event may be one whose time of occurrence is rare, as with bursts of speciation in the fossil record or the timing of major earthquakes or hurricanes. We have the same type of phenomena in technology as well, evidenced by periods of feverish inventive creativity, separated by long periods where the inventions are 'digested' prior to their reappearance as marketable commodities (sometimes!).

A couple of years ago, researchers Jerry Silverberg and Bart Verspagen examined a set of 247 technological innovations that occurred during the years 1764–1976, a period of over two centuries. They considered what fraction of the years saw 0, 1, 2, ... 7 innovations. The chart in Figure 2.8 below shows the result.

Nearly half the time, nothing is happening (zero innovations), while feverish activity (5 or 6 innovations) takes place less than 10 percent of the time. So yet again we have a distribution that exaggerates the tail events (0 or 6 innovations) far beyond what we'd expect to see if the innovations were distributed normally. It's worth mentioning in passing that these same researchers also found the fat-tail phenomenon in other data on patents and values of a wide range of technological innovations.

Emergence of a stochastic science

The take-home message from this quick tour of highly non-normal processes is that there is a lot more going on out on the 'fringe' than we ever imagined. Thus, fostering an environment that encourages exploration of 'long-shots' is a lot more likely to produce winners than by trying to 'design for success'. Whatever design there is rests in creating the environment. Evolution – and fat tails – will do the rest.

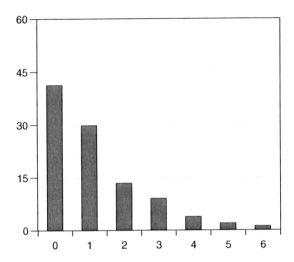

Figure 2.8 Silverberg and Verspagen's analysis of number of innovations during the period 1764–1976.

Credit

This chapter is an adaptation of material that first appeared in *Engineering and Technology*, published by the Institute of Engineering and Technology, UK. The author gratefully acknowledges permission to reprint this material in slightly modified form for this volume.

References

Albeverio, S., Jentsch, V. and Kantz, H., eds. (2006) *Extreme Events in Nature and Society*, Heidelberg: Springer.

Casti, J. (1994) *Complexification*, New York: HarperCollins.

Casti, J. (2010) *Mood Matters*, New York: Copernicus Books.

Silverberg, G. and Verspagen, B. (2003), 'Breaking the Waves: A Poisson Regression Approach to Schumpeterian Clustering of Basic Innovations', *Cambridge Journal of Economics*, vol. 27, pp. 671–93.

Silverberg, G. and Verspagen, B. (2007) 'The Size Distribution of Innovations Revisited: An Application of Extreme Value Statistics to Citation and Value Measures of Patent Significance', *Journal of Econometrics*, vol. 139, pp. 318–39.

Taleb, Nassim (2007) *The Black Swan*, New York: Random House.

3 A neo-Schumpeterian approach towards public sector economics

Horst Hanusch and Andreas Pyka

1 Introduction – a new research programme in economics: comprehensive neo-Schumpeterian economics

Since the 1980s questions of economic growth and economic development experienced a renaissance in economics after almost 25 years of silence. Seemingly unsolvable problems to overcome decreasing rates of marginal capital productivity and the methodological problems of an aggregate production function (e.g. Sraffa 1976) put economic growth theory after Solow's promising start in the 1950s (Solow 1956) into offside. In this discussion the major drivers of quantitative growth as well as – as some protagonists of neo-Schumpeterian economics claim – of qualitative growth (e.g. Saviotti 1996) are technological, organizational and institutional innovations. Economists widely agree on this.

Despite this general agreement on the important role of innovations two different schools of thought developed in economics:

- Neo-classical growth theory experienced a proper rejuvenation with the so-called new growth theory (among others Romer 1987, Lucas 1988). New growth theoretical approaches allow to get rid of the major problems of decreasing marginal capital productivity and convergence of growth rates by considering positive feedback effects (e.g. so-called technological spillovers or the explicit consideration of human capital) emerging in innovation processes. Although theoretical inconsistencies cannot be denied – in particular, the concept of positively interpreted spillovers on a macro-economic level cannot be complemented on a micro-economic level where technological spillovers are generally negatively interpreted (see Pyka *et al.* 2009) – new growth theory is considered to fit well into the theoretical framework of neo-classical theory which supported considerably its diffusion.
- The alternative approach in neo-Schumpeterian economics which started almost at the same time has chosen a different approach without only trying to integrate into the body of theory of neo-classical economics (e.g. Nelson and Winter 1982). Referring to the *Theory of Economic Development* by Joseph A. Schumpeter (1912) the contradictoriness of economic development driven by innovation with concepts like Olympic rationality and

economic equilibria is emphasized. Economic growth driven by innovation is compulsorily accompanied by structural change endogenously caused by the purposeful and sometimes erroneous actions and interactions of economic agents, i.e. knowledge generation and diffusion processes. Increasing efficiency on a sectoral level (i.e. process innovation) raises resources which are used for the explorative purposes (i.e. product innovation) which might lead to the emergence of new industries supporting long run economic growth by simultaneously triggering qualitative development (Saviotti and Pyka 2004). The dynamics are to be observed on a microeconomic level (entrepreneurship) and in the case of success manifest themselves on the sectoral level (industry life cycles). What is measured, however, on the macro-economic level as economic growth is only the average from structural dynamics on the meso-level of an economy and very likely does not tell anything on the causes of development.

Today neo-Schumpeterian economics has become an independent and also widely recognized research programme (e.g. Dosi *et al.* 1988, Fagerberg *et al.* 2005, Dopfer 2005 and Hanusch and Pyka 2007a) which influences considerably the design of innovation and technology policy in particular by international organizations such as the OECD (OECD 1991), the World Bank (World Bank 1999) and the European Commission. Since the mid-1990s, technology and innovation policy cannot be analysed without neo-Schumpeterian concepts such as technological clusters (e.g. Braunerhjelm and Feldman 2007), innovation networks (e.g. Pyka 2002) and entrepreneurship (e.g. Grebel *et al.* 2003).

Until most recently, however, the innovation-orientation of the neo-Schumpeterian approach is applied almost exclusively to manufacturing and service industries. In Hanusch and Pyka (2007b) we show that the innovation-orientation in the industrial sector is only one prerequisite for economic growth and development. The growth success of an economy similarly depends at least on the innovation-orientation or respectively future-orientation of financial markets as well as the public sector. Economic growth and development are carried by these three pillars of economic systems which are encompassed by the bracket of true uncertainty (Knight 1921) which is inseparably connected to all kinds of innovative development processes. This intrinsic uncertainty of innovation processes is the major cause why mainstream economic approaches can be applied neither for analysing industries or financial markets, nor for the activities of the public sector when it comes to true innovation. The concept of rationality applied in neo-classical economics is not applicable in uncertain situations. The Olympic rationality of neo-classical economics leads to a pathological pessimism concerning any kind of innovation. Without a general willingness to innovate, i.e. a willingness to deal with the *ex ante* non predictable possibility of failure and economic losses, any innovative behaviour becomes impossible.

From this, one can easily see that for an economic analysis of the potentials for growth and development of economies one cannot apply the idea of innovation to industrial sectors only. The innovation orientation has to be transferred to the financial markets and the public sector, as well as to the important mutual influences between these three realms of economic development. By this transfer of the innovation- respectively future-orientation and the accompanying uncertainty we develop Comprehensive Neo-Schumpeterian Economics (CNSE) (Hanusch and Pyka 2007c). Only in CNSE concepts used in the theory of financial markets like venture capital or in theory of public choice like political entrepreneurship receive their original innovative meaning.

The purpose of this chapter is to summarize our work (Hanusch and Pyka 2007a, 2007b, 2007c, 2007d) on the future orientation of the public sector in the CNSE approach. For this purpose we outline in Section 2 a neo-Schumpeterian theory of the public sector. In Section 3 we focus on the normative view on the public sector. Section 4 then looks at the co-evolutionary dimension. Section 5 finally applies empirically our theory to the public sectors of the European Union and their dynamics. Section 6 concludes.

2 The public sector in CNSE

Our considerations of a neo-Schumpeterian theory of the public sector focus on the justification of the state and encompass a normative perspective in the sense of defining tasks for public activities as well as a positive-empirical perspective supposed to explain real developments.

Justification of a public sector

The existence and necessity of a public sector can be explained within the neo-Schumpeterian approach by the persistence and inevitability of uncertainty accompanying every kind of innovation. Schumpeter's notion of creative destruction in his 1942 book *Capitalism, Socialism and Democracy* hints at the two sides of the innovation coin: in every innovation process, we find winners and losers. *Ex ante* it is impossible to know who will win and who will lose the innovation game. Accordingly, the uncertainty of innovation processes throws a veil of ignorance over the economic actors. In this sense, the ideas of John Rawls' Theory of Justice (1971) can be transferred to the neo-Schumpeterian context. An individual as a member of society can agree on a social contract to deal with the peculiarities and imponderables of innovation processes. This social contract then has to be executed by a state authority. In the neo-Schumpeterian context, sure enough the social contract also applies to firm actors and entails both support for uncertain innovation activities as well as social responsibilities in the case of innovative success (e.g. Acs 2007).

3 The normative view on the public sector

In the CNSE perspective the process dimension of innovation outweighs the artefact dimension prevailing in neo-classical economics. Innovation is considered to be the general strategy to deal with and to overcome problems in all spheres of economic life. Whereas in neo-classical economics innovation is considered to be the exogenous setting of new restrictions, in a CNSE perspective innovation is the proactive and therefore endogenous displacement and movement of restrictions. Accordingly, welfare – as the final goal for all kinds of policy initiatives – is no longer a static concept but also relates to the processes of innovation. In other words welfare strictly corresponds with the risk appetite and the appetite for experimentation in a society which are the prerequisites to develop the capabilities to design and to create a desired future.

This normative perspective of an economic theory of the state is supposed to guide the deviation and design of all public activities – encompassing public expenditures as well as public revenues – which in a neo-Schumpeterian context has to include the developmental potential of the economy. In this sense, basically all public interventions have to be scrutinized, as to whether they support or hinder the potential of economic development. Accordingly, for public activities, an orientation towards the future is postulated which directly is attached to the capabilities to design a desired future.

Two types of failure generally endanger this goal and can be considered the cardinal errors of economies: the first deals with the danger of discarding promising opportunities too early, whereas the second deals with the possibility of staying for too long on exhausted trajectories (Eliasson 2000). In both cases, resources for future development are wasted, which demands for policy intervention.

But why do economies and economic actors tend to these failures? The sources of potential failures are manifold, but again stem from the uncertainty underlying economic processes as well as the complex nature of novelties:

A first example is given by consumers' decisions concerning so-called merit goods as introduced by Richard Musgrave (1958) in public finance. Due to the future orientation and the complex character as well as the high probability of positive spillover effects of merit goods, individuals tend to undervalue strongly their consumption, e.g. as in education, or to underinvest in respective activities, e.g. as with respect to R&D. A future-oriented policy, therefore, has to consider these shortfalls, e.g. by improving the knowledge of economic actors concerning the benefits of the respective goods and activities and/or by supporting their consumption, use and production.

A second example deals with different and unbalanced speeds of development, which is symptomatic of dynamic innovation-driven processes. Creative destruction in a Schumpeterian sense is most often closely connected to the obsolescence of labour qualifications which might cause severe problems of mismatch unemployment on the labour markets – the new qualifications are not sufficiently available, whereas obsolete qualifications abound. From the perspective of neo-Schumpeterian economics this mismatch on labour markets demands not only an administrative design of labour policy, but also an active future-oriented design.

To complicate matters further, economic and social policies are not independent but are highly interdependent and affect one another. Not all policy packages are stable or beneficial. This is tantamount to saying that policies and the economy or society on which they are applied are components of a system. Changing one policy can in principle affect the whole system and the outcome of any policy change may not coincide with the one expected *ex ante*. In particular, there can be coherent or incoherent, or virtuous or vicious, policy packages. For example, an economic system which faces a massive obsolescence of competencies due to innovation accompanied by an increasing international labour division and which has:

- unemployment caused by an imbalance between competencies supplied and demanded;
- very high barriers to laying-off people;
- no, or inadequate, training programmes aimed at changing the competencies of the labour force, and
- a system of unemployment benefits in which benefits are paid without attempting to induce people to be retrained in order to change occupation,

has from a neo-Schumpeterian perspective an incoherent policy package. Unemployment is mostly created by inadequate competencies and the fact that people on unemployment benefits are paid to stay out of the labour market, thus further degrading their competencies. This combination creates a vicious circle in which people becoming unemployed can never re-enter the labour market and their number is likely to increase if the misalignment of supplied and demanded competencies persists. However, modifying only labour market policies is very likely not to be sufficient in this situation, but additional measures affecting education, conditions for entrepreneurship, etc. might be necessary. It is at the heart of Comprehensive Neo-Schumpeterian Economics that isolated modifications of one subsystem might cause not the desired effect, but causes via feedback effects detrimental effects in other subsystems.

With respect to recent labour market policy designs, the Danish model implemented since the 1990s is a good example of a future-oriented approach in a neo-Schumpeterian fashion which avoids the above outlined vicious circle by introducing a proactive dimension into labour market policies (see e.g. Lorenz and Lundvall 2006). On the contrary, German labour market policies aiming at the re-establishment of private demand very likely are responsible for a deepening of structural mismatch unemployment.

The positive-empirical view on the public sector

With respect to a positive-empirical approach of a neo-Schumpeterian theory of the state, which seeks to explain real developments, a promising starting point again comes from public finance and an empirical observation discussed more than 100 years under the heading of Wagner's Law (Wagner 1892). Adolph Wagner (1835–1917) formulated this law following empirical observations that the development of an industrialized economy is accompanied by an increasing absolute and relative share of public expenditures in GNP (gross national product). According to Wagner, the reasons for the income elasticity above unity towards public goods are to be seen in the increasing importance of law and power issues as well as culture and welfare issues in industrializing and developing economies. This way, public dynamics are narrowly connected to neo-Schumpeterian dynamics, which demand higher qualities of public goods such as infrastructure, education, basic research etc. as a condition sine qua non for economic development.

To avoid either an unbounded growth of public activities, which Schumpeter (1950) himself labelled the march into socialism, or an increasing privatization of public goods (e.g. in the health and education sector – which goes hand in hand with an increasing uneven distribution of services, itself an obstacle for economic development) a policy recommendation of neo-Schumpeterian economics has to focus on adding a qualitative dimension to Wagner's quantitative dimension. This can be achieved only by taking seriously the normative requirement in the design of all public activities of the neo-Schumpeterian approach, namely their orientation towards future development. In the case of potential insane Wagnerian dynamics, leading to an overall expansion of the public sector, a neo-Schumpeterian policy design will have to encompass a strengthening of the absorptive capacities of consumers towards superior merit goods.

4 The co-evolutionary dimension

Conceptually the co-evolutionary dimension is, by far, not new to economics and usually leads to a set of assumptions which are considered to frame economic processes and decision making. In pure economic theory, however, the co-evolutionary nature is more or less neglected by referring to the so-called *ceteris paribus* assumption. We claim that in order to investigate the relationship between the public sector and economic development this assumption is not applicable. Instead in addition to interfaces also the intersections between the public and private sectors have to be considered in order to capture interdependencies and co-evolutionary potentials.

A brief historical example describing the co-evolution of innovation policy and economic development may help to illustrate this point. This example furthermore sheds some light on the development from a manufacturing-based economy towards a knowledge-based economy which has triggered several changes in the role of the public sector as well as of the public–private interactions. Until the 1970s the most important task of the public sector with respect to the future-orientation of an economy was seen in the financing and coordination of the basic research sector as well as the institutional design of intellectual property rights to provide incentives for private actors of being engaged in R&D. Since these days crucial changes have taken place.

An increasing international competitive pressure (e.g. the rise of the Japanese economy and their successes in car manufacturing and consumer electronics) introduces a new field of activity for the public sector, namely technology transfer. The large successes of the basic research (and military research) system (e.g. American Apollo programme among others) should be transferred more quickly and effectively to the industrial application domain. New in the design of respective measures was the cooperative and pro-active approach between public and private agents which complements the framework setting role of the public sector with its current prevailing governance role.

In the 1980s large structural transformations in the industrial sectors were triggered by the advent of new general purpose technologies (see Lipsey *et al.* 2005) like information and communication technologies and biotechnologies. Besides the large national monopolies in telecommunication and the large pharmaceutical companies, small technology-oriented companies appear which start to play a decisive role for knowledge generation and diffusion processes in economies. (Very similar developments can be observed in the energy sector in the first decade of the twenty-first century with small companies being engaged in renewables challenging the business models of traditional huge energy firms.) Those economies which were able to provide a prolific environment for entrepreneurship (e.g. institutional settings focussing on low entry barriers, provision of public credits, flexible rules dealing with spin-offs from universities etc.) were able to create decisive advantages for their relative position in global economic development within one decade. They managed to get into worldwide leading positions in these new technologies. For those economies who were not among the first countries which introduced these new industries it turned out later to be extremely expensive to catch up (e.g. biotechnology in Germany) or even impossible to do so (e.g. ICT in Germany).

The cooperative dimension of the public sector is strengthened in this period. With respect to the ICT industries this can be seen in the importance of national (sometimes even international) institutions of standardization. Even more visible is this cooperative dimension in the modern organization R&D in pharmaceutical biotechnology: Public research laboratories and universities, private actors like dedicated biotechnology companies, and large diversified firms, are connected in innovation networks where knowledge is generated and diffused by interactive development processes. In many economies the cooperation dimension between the public and private sphere also includes financial relationships when public policy programmes were designed to fill the gap of missing venture capital for a certain new technology.

From this brief example one can see that intersections of the public sector comprise the industrial–public intersection as well as the public–financial intersection.

- The industrial–public intersection has an important manifestation in the design of modern innovation organization which in the literature is labelled as collective innovation processes (e.g. Pyka 1999). Private firms and public research institutes collaborate in knowledge creation and diffusion which includes besides inter-institutional collaborations between firms and public research institutes, the engagement of private firms in basic research e.g. among others in areas such as molecular biology and nanotechnologies, as well as pro-active technology transfer in public–private research partnerships.

Or consider the international and interregional competition for industrial settlement, its impact on future development of nations and regions, and the role the design of tax systems plays in this competition. A future oriented neo-Schumpeterian policy has to scrutinize whether the conditions generated by public activities allow for, or even open up, developmental potentials for the industrial sectors in the future.

- The public–financial intersection comprises policy activities to attract financial actors i.e. their international location decisions, and to provide for knowledge and information in high uncertain areas of innovation and industry development in order to support the decision making processes of financial actors. It also includes the cooperation of financial actors when it comes to the implementation and application of policy programmes to support innovation and entrepreneurship. In particular a long-term commitment based on sound technological forecasts is postulated to be an essential ingredient of a future orientation in financial markets, which, however, demands for joint efforts of and fine-tuned coordination between the public and financial sectors.

The above examples can show how many different interrelationships and intersections exist between the public sector and other economic domains and how relevant they are for a future-oriented concept of economic development.

The concept of a neo-Schumpeterian corridor

As we already saw, CNSE focusing on innovation driven, future-oriented development has to offer theoretical concepts to analyse the various issues of industry, financial markets, and the public sector and their encompassing qualitative interrelations. Innovation and, as a consequence thereof, uncertainty are ubiquitous phenomena characteristic of each economic domain and also of their intrinsically interwoven connectiveness. An improved understanding of the development processes going on in modern capitalistic economies can only be expected when these co-evolutionary dimensions are taken into account. This is illustrated with the concept of a neo-Schumpeterian corridor shown in Figure 3.1. The neo-Schumpeterian corridor is purely illustrative. The origin of the coordinate system represents the present stage. The corridor is widening in time because of the uncertainty shaping future development.

In a CNSE perspective, there exists only a narrow corridor for a prolific development of socioeconomic systems. Profound and comprehensive neo-Schumpeterian development takes place in a narrow corridor between the extremes of uncontrolled economic success (growth) and exploding bubbles, on the one hand, and stationarity, i.e. economic stagnancy, on the other hand. Consider for example the case of the financial sector, exaggerating the developments taking place in the real sector and leading to dangerous bubble effects, which might cause a breakdown of the whole economy. Or think of the case in which the public sector cannot cope with the overall economic development, and infrastructure, education, social security etc. become the bottlenecks of system development.

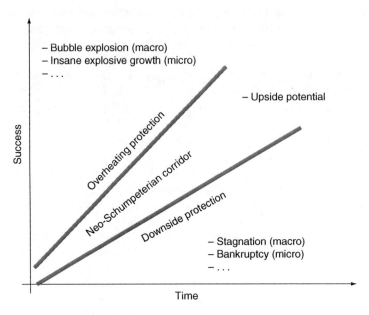

Figure 3.1 The neo-Schumpeterian corridor.

Economic policy in the sense of CNSE is supposed to keep the system in an upside potential including both overheating-protection, i.e. on the macro level bubble explosions and on the micro level insane explosive growth, and downside-protection, i.e. on the macro level stagnation and on the micro level bankruptcy.

5 The dynamic patterns of the future-orientation of the public sector in a CNSE-perspective

In this last section we apply the CNSE approach empirically in order to detect different designs of future-orientation of the public sectors in European economies. Promising possibilities to approach the future-orientation of the public sector from a positive-empirical dimension are so-called indicator-based models. International comparisons become possible by applying a comprehensive set of indicators describing the future-orientation of the public sector to a cluster analysis, which identifies economies with similar compositions and separates them from economies with different set-ups concerning their future-orientation. Without doubt, the future-orientation of CNSE entails also the dynamics as well as the potential for self-transformation of the public sector. In time new patterns of future-orientation of the public sectors are to be expected which allow conclusions concerning country-specific development profiles.

In Hanusch and Pyka (2007d) we investigate European economies and the future-orientation of their public sectors. The indicators applied in this model are listed in Table 3.1.

These indicators were applied to a cluster-analysis (for details concerning the method see Balzat and Pyka 2006) in order to detect commonalities and differences in the respective future-orientation as well as the underlying pattern dynamics.

The pattern of clusters in the future-orientation of public sectors (Figure 3.2) for the year 2000 is strongly geographically determined. This particular pattern corresponds with patterns identified in the varieties of capitalism approach (e.g. Amable 2003). We find three larger clusters, a central European, a Scandinavian and a Mediterranean public sector group. For the group of Scandinavian countries, a common alignment in one cluster clearly follows the idea of the Scandinavian welfare state which shapes the design of the public sector even visible with regard to the future orientation. This holds particularly for the education and science sector and the importance which is attached to a highly developed public infrastructure.

Obviously different enough to the Scandinavian strong welfare-orientation, the clustering algorithm identifies a Central European public sector group. Here the social responsibility of the public sector is also pronounced, but the particular public areas with a high future orientation (e.g. the education system and the knowledge infrastructure) seem to play a minor role.

Concerning the Mediterranean public sector group encompassing Spain, Greece and Italy, the public sector has a different influence on economic life compared to the Scandinavian and Central European cluster. One can assume a less dominant role in the social domain as well as in the domains of futurity. Of particular note, the education and knowledge system as well as the future-oriented public infrastructure seem to be less important.

Ireland, and surprisingly also Portugal, form their own country clusters and are therefore identified as structurally different to the other three European clusters.

Finally, the clusters of the public sectors' future-orientation of the European economies in the period from 2001 to 2005 are identified and displayed in Figure 3.3. Within this five-year period, strong changes can be identified: The large European economies – United Kingdom, France and Germany – make up one cluster, whereas the smaller economies – Denmark, the Netherlands, Belgium and Austria – are incorporated into another cluster. It seems as if the size of the economies influences the future orientation of the public sectors. A third larger

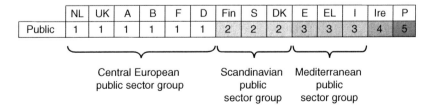

Figure 3.2 Country clusters of the public sector 2000.

Table 3.1 Indicators describing the future-orientation of the public sector

GOVERD (government expenditure on R&D) in per cent of GDP, 2000, 2005

GOVERD, average 1991 to 2000, average 2000 to 2005

GERD (gross domestic expenditure on R&D) in per cent of GDP, 2000, 2005

GERD, average 1991 to 2000, average 2000 to 2005

Tax burden for companies (corporate income tax, highest level, on non-distributed gains, reciprocal values), 2001, 2005

Tax burden for households (highest level of income tax, reciprocal values), 2001

Index of political stability, 2002, 2005

Index of regulatory quality (higher values indicating lower regulatory burden), 2002, 2005

Quality of internet access, broadband penetration rate, 2001, 2005

Number of personal computers per 100 inhabitants, 2001, 2005

Internet users per 100 inhabitants, 2001, 2005

Business internet penetration, number of internet hosts per 10,000 inhabitants, 2001, 2005

Number of secure internet servers per million inhabitants, July 2001, 2005

Employment rate of the population that has attained tertiary education and is aged 25–64, 1999, 2005

Perceived R&D subsidies, 2001, 2005

Perceived R&D tax credits, 2001, 2005

Tax treatment of R&D for large manufacturing firms, 1999–2000, 2004–5

Tax treatment of R&D for small manufacturing firms, 1999–2000, 2004–5

Number of scientific publications per million population, 1999, 2005

Percentage of scientific publications with a foreign co-author, 1995–7, 2003–5

Percentage of the population of 25- to 34-year-olds that has attained tertiary education, average 1993–2000, 2001–5

Total expenditure on non-tertiary education in per cent of GDP as of 2000, 2005

Total expenditure on tertiary education in per cent of GDP, 2000, 2005

HERD in per cent of GDP, 2000, 2005

Teaching staff per 1,000 students in primary and secondary educational establishments, 2001, 2005

Graduation rates at PhD level, 2001, 2005

Total public expenditure on education, all educational levels combined, 2000, 2005

Change in expenditure on educational institutions (1995, 2002) (2002, 2005) cluster is composed of the Mediterranean countries with Italy, Spain, Portugal and Greece. The other three clusters are single country clusters of Finland, Ireland and Sweden.

Only the Mediterranean cluster is characterized by certain stability. In a way, the consistency of this cluster is even strengthened as Portugal has now joined the group of the other Mediterranean economies with respect to the future orientation of its public sector. However, the other geographic consistency concerning the public sector which we detected in the 2000 pattern, namely the clustering of the Scandinavian countries is no longer visible. Finland and Sweden constitute single country clusters, and the future orientation of Denmark's public sector is identified as being similar to that of the smaller European countries. The focus on the pattern dynamics clearly is an advantage of the CNSE approach compared to the varieties of capitalism approach which has, due to its systemic nature, difficulties in depicting dynamics. A further advantage of this fine-resolution picture of varying public sectors' future orientation as well as their dynamics allows a correlation of the growth performance of the economies with their future-orientation make-up (Hanusch and Pyka 2007d) as well as an allocation of the economies into the neo-Schumpeterian corridor: Ireland, e.g. is an economy which very likely hits the ceiling and drops out the corridor. Germany instead might re-enter the corridor from below.

6 Conclusions

In economics there is a wide agreement on the importance of innovation for economic growth and development. Neo-Schumpeterian economics is best suited for the economic analysis of innovation processes because concepts were developed which allow dealing with important characteristics of innovation like true uncertainty, irreversibilities and bounded rationality. So far this strong innovation-orientation is applied almost exclusively in the analysis of industrial innovation. We argue that only a pronounced innovation-orientation which encompasses, besides industry, also financial markets as well as the public sector will allow a sound understanding of the (co-)evolutionary development processes typical for capitalistic organized economies. For this purpose the innovation principle is applied as a normative principle for a Comprehensive Neo-Schumpeterian Economics (CNSE) approach.

This chapter summarizes the implications of the CNSE approach for an economic theory of the state. This is done from a normative perspective highlighting the need for public interventions as well as from a positive-empirical perspective highlighting the empirical conditions for the increasing importance of public interventions into economic processes.

	NL	A	B	DK	F	D	UK	Fin	E	EL	I	P	Ire	S
Public	1	1	1	1	2	2	2	3	4	4	4	4	5	6

Figure 3.3 Country clusters of the public sector 2005.

It is shown that in particular the shift from the framework setting role towards the governance role of the public actor which simultaneously shapes the transition from a manufacturing-based economy to a knowledge-based economy requires the CNSE perspective for a comprehensive analysis of the underlying dynamics and co-evolutionary processes. The empirical investigation of the future-orientation of the public sectors in Europe shows that there is no optimal design, but different approaches with which European economies succeed and/or fail to stay within the neo-Schumpeterian corridor of prolific economic development.

The CNSE approach of the public sector is far from being fully developed. In particular, distribution issues and social justice are so far only moderately considered. Furthermore, the normative approach is in need of a Schumpeterian complement to the traditional welfare concepts which will allow for policy recommendations to improve the future-orientation of the economic system.

References

Acs, Z. (2007) ' "Schumpeterian Capitalism", in: Capitalist Development: Toward a Synthesis of Capitalist Development and the 'Economy as a Whole', in: Hanusch, H. and Pyka, A. (eds), *The Elgar Companion to Neo-Schumpeterian Economics*, Cheltenham, UK: Edward Elgar.

Amable, B. (2003) *The Diversity of Modern Capitalism*, Oxford: Oxford University Press.

Balzat, M. and Pyka, A. (2006) 'Mapping National Innovation Systems in the OECD Area', *International Journal of Technology and Globalisation*, 2 (1–2), pp. 158–76.

Braunerhjelm, P. and Feldman, M. (2007) *Cluster Genesis: Technology-based Industrial Development*, Oxford: Oxford University Press.

Dopfer, K. (ed.) (2005) *The Evolutionary Foundations of Economics*, Cambridge, UK: Cambridge University Press.

Dosi, G., Freeman, C., Nelson, R., Silverberg, G. and Soete, L. (eds) (1988) *Technical Change and Economic Theory*, London: Pinter Publishers.

Eliasson, G. (2000) *The Role of Knowledge in Economic Growth*, Royal Institute of Technology, Stockholm, Working Paper TRITA-IEO-R, 2000, p. 17.

Fagerberg, J., Mowery, D. and Nelson, R. (eds) (2005) *The Oxford Handbook of Innovation*, Oxford: Oxford University Press.

Grebel, T., Pyka, A. and Hanusch, H. (2003) *An Evolutionary Approach to the Theory of Entrepreneurship, Industry and Innovation*, 10 (4), pp. 493–514

Hanusch, H. and Pyka, A. (eds) (2007a) *Elgar Companion to Neo-Schumpeterian Economics*, Cheltenham, UK: Edward Elgar.

Hanusch, H. and Pyka, A. (2007b) 'The Principles of Neo-Schumpeterian Economics', *Cambridge Journal of Economics*, 30.

Hanusch, H. and Pyka, A. (2007c) 'Manifesto for Comprehensive Neo-Schumpeterian Economics', *History of Economic Ideas*, 15.

Hanusch, H. and Pyka, A. (2007d) 'Applying a Comprehensive Neo-Schumpeterian Approach to Europe and its Lisbon Agenda', in: Welfens, P. (ed.) *Europe's Innovation Performance*, Heidelberg: Springer.

Knight, F.H. (1965 [1921]) Risk, *Uncertainty and Profit*, New York: Harper and Row.

Lipsey, R., Carlaw, K. and Bekar, C. (2005) *Economic Transformations: General Purpose Technologies and Long Term Economic Growth*, Oxford: Oxford University Press.

Lorenz, E. and Lundvall, B.A. (eds) (2006) *How Europe's Economies learn – Coordinating Competing Models*, Oxford, UK: Oxford University Press.

Lucas, R.E. (1988) 'On the Mechanics of Economic Development', *Journal of Monetary Economics*, 22, pp. 3–42.

Musgrave, R.A. (1958) *The Theory of Public Finance*, New York: McGraw Hill.

Nelson, R.R. and Winter S.G. (1982) *An Evolutionary Theory of Technological Change*, Cambridge, MA: Belknap.

OECD (1991) (ed.) *Technology and Productivity: The Challenge for Economic Policy*, Paris: OECD.

Pyka, A. (1999) *Der kollektive Innovationsprozess*, Berlin: Duncker & Humblot.

Pyka, A. (2002), Innovation Networks in Economics – From the incentive-based to the knowledge-based Approaches, *European Journal of Innovation Management*, 5, (3), pp. 152–63.

Pyka, A., Gilbert, N. and Ahrweiler, P. (2009) 'Agent-Based Modelling of Innovation Networks – The Fairytale of Spillover', in: Pyka, A. and Scharnhorst, A. (eds), *Innovation Networks – New Approaches in Modelling and Analyzing*, Heidelberg, London and New York: Springer, 101–26.

Rawls, J. (1971) *A Theory of Justice*, New York: Oxford UP.

Romer, P. (1987) 'Growth Based on Increasing Returns due to Specialization', *American Economic Review*, 77, pp. 565–762.

Saviotti, P.P. (1996) *Technological Evolution, Variety and the Economy*, Cheltenham, UK: Edward Elgar Publisher.

Saviotti, P.P. and Pyka, A. (2004) 'Economic Development by the Creation of New Sectors', *Journal of Evolutionary Economics*, 14 (1), pp. 1–36.

Schumpeter, J.A. (1912) *Die Theorie der wirtschaftlichen Entwicklung*, Berlin: Duncker & Humblot.

Schumpeter, J.A. (1942) *Capitalism, Socialism, and Democracy*, New York: Harper and Bros.

Schumpeter, J.A. (1950) 'The March into Socialism', *American Economic Review*, 40, pp. 446–56.

Solow, R. (1956) A Contribution to the Theory of Economic Growth, *Quarterly Journal of Economics*, 70, pp. 65–94.

Sraffa, P. (1976) Warenproduktion mittels Waren, Deutsche Ausgabe von Sraffa.

Wagner, A. (1892) Grundlegung der politischen Ökonomie, 3rd edition. Leipzig: Winter.

World Bank (1999) *World Development Report*, Washington, DC: Oxford University Press.

4 Complexity, the co-evolution of technologies and institutions and the dynamics of socio-economic systems

Pier Paolo Saviotti

Introduction

The objectives of this chapter are:

- To show that economic systems are complex systems in which the different components interact with one another and can give rise to types of dynamics which are not necessarily easily intuitive but which can be better understood in a complexity based framework. The Tevecon model described here is an example of such a complexity based approach.
- To show on the basis of the Tevecon model that structural change is a determinant of economic development and that, as a consequence, the competencies a country uses need to change in order to adapt to the evolution of the external environment.
- To show that different combinations of interacting policies can be useful only if these policies are coherent or synergistic. Furthermore, it will be argued that the general principles underlying the social models of different European countries can affect the ability of these countries to create and adopt innovations in order to improve their economic performance. In particular, it will be shown that the reforms introduced into the Scandinavian countries in the 1990s and the resultant concept of flexicurity can be derived from the Tevecon model.

Complexity and co-evolution

Complexity

A complete treatment of complexity would be far outside the boundaries of the present chapter. Here a limited number of considerations about complexity will be used to introduce the subject. A complex system is constituted by its *components* and by their *interactions*. The set of components and interactions of a given system constitutes its *structure*.

One of the most important stylized facts about the evolution of biological and socio-economic systems is the existence in them of a structure and the fact that such a structure can undergo changes in the course of time. Some of these are qualitative changes since they involve the emergence of new entities and constitute discontinuities. For example, in the period following the industrial revolution many new types of artefacts and of human activities not comparable to those which existed before came into being. Examples of the former are cars, computers, and aeroplanes and of the latter research and development, venture capital firms and, of course, all the activities required to produce the new types of artefacts. These are important changes in the structure of the economic system, which we can call structural changes. The term structural change has in the past been used in a more restricted sense in economics, referring to the changing balance of different economic sectors. In the context of a complexity based approach the term can be used in a more general sense as describing any changes in the structure of the system. Examples of these changes are (*a*) the emergence of new components or the extinction of older ones, (*b*) the emergence of new interactions or the extinction of older ones, (*c*) the changing nature of any existing component, (*d*) a change in the interaction of existing components. Clearly such a definition of structural change encompasses the emergence of new institutions, of organizational forms and of new scientific disciplines. For example, the emergence of a new discipline constitutes an example of structural change in science, which can induce structural change in the economy by leading to the creation of new industrial sectors.

The long run evolution of capitalist societies has been characterized by important discontinuities. The transition from feudal to bourgeois society and the industrial revolution are just examples of such discontinuities. A theory of economic development must be able to account for these changes, that is, to explain why and how they took place. Complexity based approaches can in principle be the basis for an explanation since they can predict both the emergence and the changes in the structure of complex systems giving rise to phase transitions and to transformations involving discontinuities. The term complexity based approaches has been used here because, as already pointed out, there is no complete and consensually accepted treatment of complexity. In what follows some of the most crucial concepts of complexity will be outlined here and used in the subsequent analysis. Two related concepts involved in the emergence of complex behaviour are those of *feedback* and of *autocatalysis* (Allen 2007, Nicolis and Prigogine 1989). Feedback occurs when a component A of a system is both affected by and affecting another component B. In this case any change in A leads to a change in B which then leads to a further change in A. Autocatalysis occurs when the output of a given process is also one of its inputs. Nicolis and Prigogine (1977) showed that the feedback occurring in a chemical reaction called Brusselator can give rise to the emergence of structure in a previously homogeneous system. In principle feedback and autocatalysis can lead to the emergence and to changes of structure in any type of system, including biological and socio-economic ones. However, feedback does not operate according to the same mechanism in these different types of systems. While all molecules corresponding to a chemical formula can in principle be considered equivalent, the members of a biological or socio-economic population cannot. The essential difference consists in the presence of *microdiversity* in biological and socio-economic systems (Allen 2007). The competitive interactions between different biological species affect preferentially the weakest members of each population, thus influencing the fitness of each interacting population. A further change occurs when passing from biological to socio-economic systems. In this case not only inter-population interactions affect differentially the weakest members of each population but learning and strategy change can take place leading to changes in the nature of system components. This situation shows one of the basic features of complexity based approaches: there is a level of generality at which statements can be made about any system, but the detailed behaviour of any system can only be predicted by taking into account the specific features of the system. In summary, feedback and autocatalysis can lead to structural change in socio-economic systems. As new components emerge they have to self-organize themselves into relatively stable and coherent new structures.

Another important feature of socio-economic systems is that they never settle on any permanent structure, although some system structures can be relatively stable and persist for long periods. In other words, as Arthur (2007) points out, in a socio-economic system there cannot be any general equilibrium. Socio-economic systems have a tendency to transform themselves, where the term transformation implies the possibility of qualitative change (Saviotti 2007) represented by the emergence of new types of system components or by new types of interactions giving rise to new structures.

The co-evolution of technologies and institutions

Co-evolution is a phenomenon initially studied by biologists (Maynard Smith 1974, Roughgarden 1996) and recently also in the social sciences. In a general sense co-evolution occurs when two system components evolve by influencing each other. Thus, the evolution of a component A during a given period can affect that of a component B, which in turn will affect the evolution of A in a subsequent period. In general co-evolution tends to focus the feedback loops occurring in a pair of interacting components. In this sense it can be considered a simplified approach to complexity.

Many different types of co-evolution can be detected in a socio-economic system. For example, Nelson (2004) talks about the co-evolution of practice and understanding. However, the type of co-evolution which will be the object of this chapter is the co-evolution of technologies and institutions. During the development of innovation studies it became gradually evident that innovations cannot be created in an institutional vacuum. There is evidence that the institutions existing in a given society can either facilitate or hinder the process of economic development (North 1990, Helpman 2004). Furthermore, in addition to general institutions, such as property rights, technologies need specific institutions capable of creating standards, infrastructures or simply to lobby in favour of a new industry. The awareness of the important role played by institutions in the creation and utilization of innovations gave rise to the idea that appropriate institutions are required in order for innovations to emerge and to be adopted in socio-economic systems. Thus, Perez (1983) and Freeman and Perez (1988) introduced the concept of a techno-economic paradigm as the combination of a technological paradigm and of the appropriate institutions required to translate a technology into economic development. They argue that the creation of appropriate institutions is slower than that of innovations but that without the required institutions innovations would not lead to economic development. Nelson (1994) talks more explicitly about the co-evolution of technologies and institutions although he does not frame his analysis in terms of technological paradigms. In the literature about innovation systems the interaction between technologies and institutions occupies a central place. Starting with Freeman's (1987, 1998) study of the of national innovation system of Japan and continuing with the numerous other studies of national innovation systems (see for example Lundvall 1992 and 2007, Nelson 1992, Edquist 1997 and 2004) the system was defined as a system of interacting organizations and institutions which go far beyond R&D or higher education. What is especially important is that an innovation system involves not just the presence of institutions or organizations but their interaction with firms or research institutions. A very large number of studies of innovation systems has been carried out at different levels of aggregation – national (ibid.), regional (Cooke and Schall 2007), and sectoral (Malerba 2004a, 2004b). However, although the concept has become very widely used, it is still undertheorized (Edquist 2004). No list of the institutions which need to be taken into account to analyse an NIS exists. Scholars studying the NIS differ as to their conception, broad or narrow, of the concept. Authors favouring a narrow conception of the NIS focus mainly on scientific, technological or production institutions while those adopting a broad conception include a much wider range of institutions. Furthermore, most studies of innovation systems do not mention or do not use analytically co-evolution or complexity based approaches. In the view of the author of this chapter this is the path to be followed to improve our theoretical understanding of innovation systems.

Economic development and structural change

The composition of the economic system changes in the course of time because new types of output emerge and others become extinct. While this observation seems indisputable economists differ as to the role structural change plays in economic development. Some consider it only as an effect of economic growth. We considered it not only as an effect of previous but also as a determinant of future economic development. The resources allocated by governments to emergent high technology sectors show that politicians implicitly accept that the composition of the economic system is expected to have a positive influence on its growth potential. Emerging high technology sectors are expected to have higher rates of growth and of profit than mature ones, to contribute by supplying inputs and knowledge to other sectors etc. If new sectors, resulting for example from important innovations, had a rate of growth of demand, of output, of employment etc. considerably higher than the average for all the sectors in the economy, then the composition of the system at a given time would also be a determinant of its future growth path. The implication of this conclusion is that the composition of the economic system must be included as one of the variables in models of economic growth and development.

Economic development can be conceived as a process of *transformation*, (about this aspect see also Metcalfe *et al.* 2006, Lipsey *et al.* 2005) including both quantitative and qualitative change, which are expressed by two complementary long term trends or trajectories: growing efficiency; and growing variety. The first of these trends consists of performing more efficiently than in the past pre-existing activities, that is, of raising the ratio of outputs to inputs of given processes for a constant type of output. The second trend consists of creating new entities, qualitatively different from the pre-existing ones. This second trend is the result of *creativity*. We can then say that economic development requires both efficiency and creativity. In fact, these two trends can be considered complementary, as in the following two hypotheses (Saviotti 1996):

Hypothesis 1: The growth in variety is a necessary requirement for long-term economic development.

Hypothesis 2: Variety growth, leading to new sectors, and productivity growth in pre-existing sectors, is complementary and not independent aspect of economic development.

As a consequence of the above description an analytical representation of the composition of the economic system can be obtained by means of the concept of variety, defined as the number of actors, activities and objects required to describe the economic system. Tevecon is a model of economic development by the creation of new sectors (Saviotti and Pyka 2004a, 2004b, 2008). Given space constraints only a very short description will be given here.

The nature of the Tevecon model

The economic system in Tevecon consists of:

- An *endogenously* variable number of sectors, each created by a pervasive innovation and carrying out a set of activities resulting in a heterogeneous, highly differentiated, output. In principle such output could be either a material product or a service. Each sector is defined as the collection of firms producing a unique, even if highly differentiated, output. The number of sectors varies endogenously because the emergence of new sectors is *induced* by the dynamics of pre-existing ones.

  Depending on the vintage of the model, *production* is carried out by (*a*) human capital only (Saviotti and Pyka 2004a, 2004b), (*b*) by human and physical capital in more recent vintages

- The *demand* function contains the level of services performed by the product, the extent of product differentiation and product price
- A general analogue of R&D, *search activities* play a very important role in Tevecon. In the early vintages of the of the model search activities were only sectoral. They improved the services supplied by the products, increased the extent of product differentiation and reduced the price. In the course of time search effort (SE_i), the amount of resources allocated to search activities in sector *i*, increases with demand but at a less than linear rate
- In the model the most important source of economic development is constituted by the creation of new sectors. Each sector is created by an important innovation establishing an *adjustment gap*, an expression which indicates the *expected market size* of the potential market created by the innovation. The adjustment gap can be considered as a measure of the distance of the sector's market from saturation.

The basic structure of our model is very Schumpeterian. The first entrepreneur enters the market induced by the expectation of a temporary monopoly. If the innovation is successful imitators enter, gradually raising the intensity of competition in the sector. In this process the inducement for further entry falls until exit starts dominating entry. At this point what was initially an innovation has become part of the 'circular flow'. The sector then evolves towards a high degree of industrial concentration, eventually ending up in an oligopoly or in a monopoly. In the meantime, as production capacity and demand for the sector's output grows, the adjustment gap is gradually closed and in the end the sector becomes a saturated market. The joint dynamics of competition and demand give rise to an industry life cycle (Figure 4.1), in which the number of firms at first rises, then reaches a maximum and eventually starts declining. The decline of mature sectors induces entrepreneurs to look for new opportunities of temporary monopoly, to be found by exploiting new important innovations leading to new sectors (Saviotti and Pyka 2004a, 2004b, 2008).

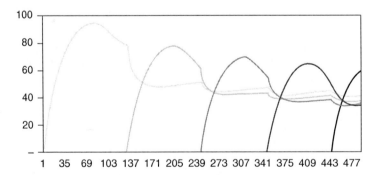

Figure 4.1 Number of firms in different sectors of the economic system.

Thus, in our model the process of economic development is closely linked to the creation of new sectors. Of course, the composition of the economic system changes during the process of economic development. Furthermore, such composition at a given time is a determinant of the rate of growth of the system in the following periods since the rates of growth of output, employment and demand are the highest in the early phases of the sector life cycle. An economic system which is particularly rich in new sectors is likely to grow at a faster pace than competing systems dependent on older and more mature sectors. Such conclusion is reinforced by a result we previously obtained from our model. Even if the rate of creation of employment within each sector falls, aggregate employment can still keep growing if there is an adequate inter-temporal coordination of the decline of mature sectors and of the creation of new ones (Saviotti and Pyka 2008). Thus, the composition of the economic system is both a consequence and a determinant of economic development.

A very important feature of Tevecon is its systemic nature. This is represented by the highly interactive nature of different subsets of the economic system and of the variables which represent them. The overall style of modelling was inspired by the literature on dynamical systems but relied heavily on inductive processes based on the observation of economic phenomena. Thus, the essential features of Tevecon which distinguish it from most existing economic models are: (*a*) the role played by structural change in economic development; and (*b*) the systemic nature given by the highly interactive nature of different subsets of the economic system. Amongst the consequences of this systemic nature there is the possibility to study the co-evolution of different subsets of the economic system. For example, in a study of the co-evolution of technologies and financial institutions we showed that excessive expectations of returns on their investment in innovating sectors by financial operators can lead to the formation of bubbles and to economic collapse (Saviotti and Pyka 2006). However, if financial operators learn to adapt their expectations to the evolution of the sectors the resultant fluctuations in economic performance can be considerably reduced (Saviotti and Pyka 2009).

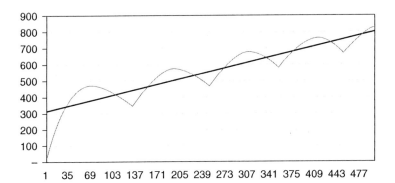

Figure 4.2 Aggregate employment.

Structural change, employment and innovation

In this section, following the complexity based approach described in this chapter, it will be pointed out that the interactions of different institutions and policies are as important as the institutions and policies themselves. Thus, not only does a country need to have an appropriate set of institutions to develop and use innovations, but it needs to coordinate these institutions to provide a coherent whole. It will be shown that different institutions and policies need to co-evolve synergistically in order to contribute to economic growth. In other words, it is the co-evolution of different institutions and not just the dynamics of one of them which needs to be taken into account. The discussion will include the relative importance of general institutions and of those which are specific to the innovations considered. It will be argued that some very general institutions constrain the elaboration and implementation of specific innovation policies and determine their success. In this context it will be shown that the social models of different EU countries can affect differentially their capacity to create and adopt innovations.

The labour market, especially for what concerns its flexibility, and the unemployment benefits of a country are often considered general institutions which can have a profound influence on the capacity of a country to grow. Here following the systems approach adopted in this chapter it will be argued that no isolated institution can determine the growth perspectives of an economic system. The interactions between institutions need to be taken into account. For example, we can expect a very rigid labour market with high barriers to hiring and firing workers to keep employment artificially high in mature sectors and to slow down the generation of resources required for the emergence of new sectors in violation of hypothesis 2, related to the complementarity of efficiency growth and variety growth. The presence of structural change implies that the distribution of competencies needs to be changed in the course of time at a rate depending on the rate of structural change itself. For relatively slow structural change it could be enough to rely on the succession of generations, with each new generation starting new competencies. However, when the rate of structural change grows most people will need to change their competencies one or more times during their working life. In general, structural change implies that the labour intensity of industrial processes falls as the corresponding sector matures while the fastest rate of growth of employment can be expected to occur in emerging sectors. Furthermore, emerging sectors are likely to provide higher rates of profit due to the limited competition existing in them. The ability of a country to change its output structure and to be first in emerging sectors is likely to provide it with higher rates of profit and with higher rates of growth of employment. Proponents of this view generally equate the concept of flexibility to the unconstrained ability of firms to hire and fire workers. However, for example, in a regime of fast structural change leaving things to the market could simply create a large pool of unemployed people who have the 'wrong' competencies and no way to improve them in order to re-enter the labour market. The creation of the 'right' competencies needs to be carried out in coordination with the displacement of the labour force from mature to emerging sectors. In other words, it is the co-evolution of different institutions and not just the dynamics of one of them which needs to be taken into account.

The capacity to change competencies in an economic system is linked to the social model of a country. The social model contains a set of assumptions about the rights and duties of the citizens of a country. Institutional reforms are likely to be more difficult to introduce in countries whose social models are incompatible with such reforms. For example, reducing employment in mature sectors can be expected to be more difficult in countries whose social models stress heavily the right of citizens to work. Although the EU is represented as a 'region' having adopted social models which tend to protect employment and slows down the creation and adoption of innovations, in reality there are at least four different EU social models (see Esping 1989). They differ for the degree of protection they provide for employment with the so called 'liberal' models providing very little protection and the more 'conservative' ones a much higher protection. However, although one might expect countries with liberal models to be able to change more quickly their competencies, an interesting experiment carried out in Scandinavian countries during the 1990s transformed their social models by making them compatible with a faster rate of change of competencies while preserving the high degree of social protection (health care etc.) that their citizens had traditionally enjoyed. The focal point of the reforms adopted in the Scandinavian countries consisted in the assumption that employment in high income countries could only be preserved in the presence of increasing competition by emerging countries by making their economic activities more knowledge intensive. In this way it would have been possible to create temporary monopolies in a number of economic activities and to preserve high employment in the presence of a positive wage differential with respect to emerging countries. Of course, Scandinavian countries are not identical. Thus, Sweden became the most R&D intensive country in the world while Denmark stressed learning even in medium or low technology activities (Lundvall 2002). However, the common feature of the models is the acceptance of the principle that the average worker cannot carry out the same activities based on the same competencies for all of his/her working life. As a consequence, job security is incompatible with employment security, a recognition which has given rise to the concept of flexicurity (Wilthagen and Tros 2004) according to which to attain employment security workers need to change their competencies once or more during their working lives. Of course, this implies that a high level of resources is required for retraining activities. Yet, if this allocation of resources allows to maintain high levels of employment it need not be considered as a cost but rather a *displacement* of resources from paying unemployed people to do nothing to retraining them. The typical approach one could expect in a 'conservative' social model is precisely to pay unemployed people to do nothing, which, to the extent that unemployment is due to maladjusted competencies, would gradually aggravate the problem. The good economic performance achieved by Scandinavian countries after the adoption of these reforms show that flexicurity is not only theoretically possible but that it works.

While the development of the reforms which rejuvenated the Scandinavian social model quite likely owe their origin to pragmatic if well informed policy considerations we can *ex post* interpret them as being fully coherent with, and in fact derivable from, the Tevecon model previously described. Thus, we can outline here a number of features of these reforms which can contribute to our understanding of economic development.

1   Structural change would require a change of competencies even if it occurred in a closed economy. Industrial sectors tend to require progressively less employment per unit of output as they mature. The higher the rate of innovation in the economy, the faster the rate of change of competencies needs to be. Competition from emerging countries, which typically enter low or medium technology sectors, tends to accelerate the flow of employment out of these sectors in post-industrial economies. The obvious solution for these economies is to create new high-tech sectors or knowledge intensive subsets of more traditional ones where they can preserve a degree of temporary monopoly and the corresponding wage differential. This implies to carry out the required investments.

2   These investments cannot be successful without the adequate coordination of different institutions and policies. For example, changes in R&D, training and education systems and in the labour market must be compatible and likely to lead to synergisms. Thus, the R&D system must generate opportunities for creation of new firms, which must be accompanied by establishing initiatives to facilitate entrepreneurship and to reduce the barrier for the creation of new firms. Furthermore, training must be easily available to allow workers to change their competencies as the economy requires. This must be done by a very high level of coordination between the institutions regulating the labour market and those providing training and education. A rigid labour market would not allow workers to move between different jobs, but dumping workers would probably worsen the problem if they were not given adequate competencies. Examples of coherent and synergistic and of incoherent policy combinations follow:

   • coherent and synergistic policy combination: low degree of job protection and high initial unemployment benefits to be rapidly replaced by training;
   • incoherent policy combination: high degree of job protection, high and persistent unemployment benefits lack of training.

   As previously pointed out, the increased investment in training required is not necessarily an extra cost since if it succeeds it is equivalent to paying people (previously unoccupied) to improve their competencies.

3   The economic system of a country can perform well only by adapting to its external environment. This includes the economic environment constituted by

all the countries of the world economic system. Amongst the possible changes in the external environment the ones of the greatest interest for this chapter are technical change and globalization. Technical change tends to reduce employment per unit of output in maturing sectors and this tendency is accelerated by globalization. In these circumstances post-industrial countries need to adapt by creating activities in which they have a temporary monopoly. This involves a change of competencies at a rate depending on that of structural change. The question which then arises is whether this adaptation can be carried out without renouncing the basic principles underlying the country's social model. To answer this question we need to distinguish between means and ends, the ends being the basic principles and the means the policies adopted. The answer which seems to come from the reforms carried out in the Scandinavian countries is that the ends can be preserved by adapting the means to the evolution of the external environment. In fact, not only changes in means/policies are possible but they are required to preserve the stability of the social model, which is viable only in presence of a good economic performance. The economic performance of EU countries is likely to be affected by the institutions compatible with their social models and in particular by the inability to separate policies from principles;when the external environment changes policies must be changed in order to preserve principles.

The previous discussion stresses the need for the co-evolution of the different policies of a country both to make them coherent and to improve the country's adaptation to the external environment. This highly interactive feature of real economies means that they are systems and that they should be studied as such on the basis of conceptual and analytical models which are based on theories of complexity. Tevecon is an example of this type of model and we believe it provides an adequate theoretical foundation for the relative success of the policies adopted by different EU countries regarding innovation, education, training and employment.

References

Allen, P. (2007) 'Self-organization in Economic Systems', in: Hanusch, H., Pyka, A. (eds) *Elgar Companion to Neo-Schumpeterian Economics*, Cheltenham: Edward Elgar, pp. 1111–48.

Arthur, W.B. (2007) 'Complexity and the Economy', in: Hanusch, H., Pyka, A. (eds) *Elgar Companion to Neo-Schumpeterian Economics*, Cheltenham: Edward Elgar, pp. 1102–10.

Cooke, P., Schall, N. (2007) 'Schumpeter and Varieties of Innovation: Lessons from the Rise of Regional Innovation Systems Research, in: Hanusch, H., Pyka, A. (eds) *Elgar Companion to Neo-Schumpeterian Economics*, Cheltenham: Edward Elgar.

Edquist, C. (1997) *Systems of Innovation: Technologies, Institutions and Organizations*, London: Pinter.

Edquist, C. (2004) 'Systems of Innovation, Perspectives and Challenges', in: Fagerberg, J., Mowery, D., Nelson, R.R. *The Oxford Handbook of Innovation*, Oxford: Oxford University Press, pp. 181–208.

Esping Andersen, G. (1989) *The Three Worlds of Welfare Capitalism*, Oxford: Blackwell.

Fagerberg, J., Shrolec, M. (2008) 'National Innovation Systems, Capabilities and Economic Development', *Research Policy*, 37, 1417–35.

Fagerberg, J., Verspagen, B. (2007) 'Innovation, Growth and Economic Development: Have the Conditions for Catch-up Changed', *International Journal of Technological Learning, Innovation and Development*, 1, (1), p. 13.

Freeman, C. (1987) *Technology Policy and Economic Performance*, London: Pinter.

Freeman, C. (1988) 'Japan: A New National System of Innovation?', in: Dosi, G., Freeman, C., Nelson, R., Silverberg, G., Soete, L. (eds) *Technical Change and Economic Theory*, London: Pinter, pp. 330–48.

Freeman, C., Perez, C. (1988) 'Structural Crises of Adjustment: Business Cycles and Investment Behaviour', in Dosi, G., Freeman, C., Nelson, R., Silverberg, G., Soete, L. (eds) *Technical Change and Economic Theory*, London: Pinter.

Helpmann, E. (2004) *The Mystery of Economic Growth*, Cambridge, MA: Harvard University Press.

Lipsey, R.G., Carlaw, K., Bekar, C.T. (2005) *Economic Transformations*, Oxford: Oxford University Press.

Lundvall, B.A., (2007) 'National Innovation Systems: From List to Freeman', in: Hanusch, H., Pyka, A. (eds) *Elgar Companion to Neo-Schumpeterian Economics*, Cheltenham: Edward Elgar, pp. 872–81.

Lundvall, B.A. (ed.) (1992) *National Systems of Innovation*, London: Pinter.

Lundvall, B.A. (2002) *Innovation, Growth and Social Cohesion, The Danish Model* Cheltenham: Edward Elgar.

Malerba, F. (2004a) 'Sectoral Systems, How and Why Innovation Differs across Sectors', in: Fagerberg, J., Mowery, D., Nelson, R.R. *The Oxford Handbook of Innovation*, Oxford: Oxford University Press, pp. 380–406.

Malerba, F. (2004b) *Sectoral Systems, Concepts, Issues and Analysis of Six Major Sectors in Europe*, Cambridge, UK: Cambridge University Press.

Maynard Smith, J. (1974) *Models in Ecology*, Cambridge, UK: Cambridge University Press.

Metcalfe, J.S., Foster, J., Ramlogan, R. (2006) 'Adaptive Economic Growth', *Cambridge Journal of Economics*, 30, pp. 7–32.

Nelson, R.R. (ed.) (1992) *National Innovation Systems*, Oxford: Oxford University Press.

Nelson, R.R. (1994) 'The Co-evolution of Technology, Industrial Structure and Supporting Institutions', *Industrial and Corporate Change*, 3, pp. 47–63.

Nelson, R.R. (2004) 'The Market Economy, and the Scientific Commons', *Research Policy*, 33, pp. 455–71.

Nicolis, G., Prigogine, I. (1977) *Self-organization in Nonequilibrium Systems*, New York: Wiley-Interscience.

Nicolis, G., Prigogine, I. (1989) *Exploring Complexity*, New York: Freeman.

North, D.C. (1990), *Institutions, Institutional Change and Economic Performance*, New York: Cambridge University Press.

Perez, C. (1983) 'Structural Change and the Assimilation of New Technologies in the Economic System', *Futures*, 15, pp. 357–75.

Roughgarden, J. (1996) *Theory of Population Genetics and Evolutionary Ecology: An Introduction*, Upper Saddle River, NJ: Prentice Hall.

Saviotti, P.P. (2007) 'Qualitative Change and Economic Development', in: Hanusch, H., Pyka, A. (eds) *Elgar Companion to Neo-Schumpeterian Economics*, Cheltenham: Edward Elgar, pp. 820–39.

Saviotti, P.P. (1996) *Technological Evolution, Variety and the Economy*, Aldershot: Edward Elgar.

Saviotti, P.P., Pyka, A. (2004a) 'Economic Development by the Creation of New Sectors', *Journal of Evolutionary Economics*, 14, (1), pp. 1–35.

Saviotti, P.P., Pyka, A. (2004b) 'Economic Development, Qualitative Change and Employment Creation', *Structural Change and Economic Dynamics*, 15, pp. 265–87.

Saviotti, P.P., Pyka, A. (2008) 'Product Variety, Competition and Economic Growth', *Journal of Evolutionary Economics*, 18, pp. 323–47.

Saviotti, P.P., Pyka, A. (2006)'On the Co-evolution of Technologies and Financial Institutions: Economic Evolution at the Edge of Chaos', presented at the Third Aix en Provence Complexity Workshop 'Complex Behaviour in Economics: Modeling, Computing, and Mastering Complexity' (COMPLEXITY2006) Aix en Provence (Marseilles), France, 17–21 May.

Saviotti, P.P., Pyka, A. (2009) 'Crises, Co-evolution and Economic Development', presented at the International Workshop on Coping with Crises in Complex Socio-Economic Systems held in ETH Zurich (Switzerland), 8–13 June.

Wilthagen, T., Tros, F. (2004) 'The Concept of "Flexicurity": A New Approach to Regulating Employment and Labour Markets'. Online, available at: http://ssrn.com/abstract=1133932.

Part II

The actors and networks of innovation (empirical research)

5 Does geographic clustering still benefit high tech new ventures?

The case of the Cambridge/Boston biotech cluster

Thomas J. Allen, Ornit Raz and Peter Gloor

Introduction

In this age of ubiquitous broadband connectivity, one might expect that the effect of separation distance on communication might have disappeared or at least be diminished.

If this is true, then one of its effects would be to disarm the arguments for similar firms, especially those newly formed and technology-based, to cluster geographically. Now there are many arguments for the benefits of geographic clustering, not the least of which is that a concentration of firms will attract resources, particularly of the human kind into an area. Still, the potential for synergy among like firms is considered a strong factor for locating in what often becomes a high rent district. It is widely believed that propinquity will stimulate communication and scientific exchange among firms, especially among small firms formed on the basis of a common technology. This is one of the basic premises supporting the argument for the geographic clustering of newly-formed high technology firms (Powell *et al.* 1996). Extensive research in recent years has demonstrated economic benefits for firms sharing a common technology within the same geographic cluster. Researchers identify different benefits to be derived from clustering. It will be easier to attract specialized staff, because the qualified pool of applicants is much larger. It is also easier to find venture capital, suppliers, and support services within a cluster (Saxenian 1994). Claims have also been made for the synergistic benefits of firms sharing scientific knowledge, especially if there are university laboratories near the cluster (Saxenian 1994).

Several studies have inferred inter-firm communication from the evidence of co-publishing and co-patenting across firms, (Schilling and Phelps 2005, Porter and Powell 2006, Porter *et al.* 2005, Powell *et al.* 1996). This is certainly a valid and effective way of detecting inter-firm communication, however a good amount of scientific exchange may occur that does not result in such products and does not therefore appear in such publicly accessible records. This less formal scientific exchange across firms, while resulting in a patent or paper, may still produce value for the communicating companies.

Of course, the arguments for communication being related to proximity, in other contexts have been around for a long time. Allen (Allen and Henn 2006), for example, has shown the probability of regular technical communication among engineers to decline as the inverse square of the distance between their work stations.

Of course, this decline was predicated on the need for face-to-face contact. So it does nothing to dispel the belief that modern media have diminished that need. This is in spite of the fact, that Allen and Hauptman (1987) showed face-to-face to be the preferred medium for complex or abstract messages, such as those typifying scientific communication. We are now in a new millennium and their work is more than 20 years old. Technology has advanced since their time and, probably more importantly, a new generation of scientists, more at home with modern media, has arrived on the scene. So today we may find less need for companies to cluster geographically. Scientists can potentially communicate effectively across firms through media other than face-to-face.

Many contemporary observers are now telling us that the day has arrived when we can forget about distance in its effect on communication. In fact, an eminent economist, Frances Cairncross has declared (in the title of her book) that distance is dead, 'new communications technologies are rapidly obliterating distance as a relevant factor in how we conduct our business and personal lives' (Cairncross 2001).

Does distance really no longer matter?

If Cairncross and others are really correct, there is no longer a rationale for the geographic clustering of new venture firms. Therefore one of our purposes in this research is to measure the effects of geographic proximity on communication among firms. Allen's work is both dated and based upon the study of communication among engineers and scientists all working within single organizations. A question remains whether Allen's observations carry through to communication among scientists in separate firms and living in a new millennium. The availability of an already existing biotechnology cluster in parts of Cambridge and Boston, Massachusetts provides a convenient opportunity to test this question.

This biotechnology cluster developed adjacent to MIT in Cambridge, Massachusetts. This so-called 'cluster' of mostly newly-formed biotechnology firms has come about over the past 20 to 30 years and continues to grow. Depending upon one's definition of what a biotech firm is, the number can range from 80 to over 200 firms. The Massachusetts Biotechnology Council, an industry trade association lists over 500 companies as members. We will be a little more conservative and restrict our selection on the basis of location and the nature of the firm's principal activity. We thus end up with fewer than 100 companies. Nevertheless, this is a sufficient size for a meaningful study.

Our basic hypothesis

When firms locate near one another a number of factors potentially influencing communication come into play. First of all, formal intentional communication is easier. Walking across the street is certainly easier than traveling a greater distance by car or plane. Informal communication is also more likely due to chance encounters. Finally, and influencing chance encounters, is the use of common facilities and locations such as restaurants, coffee shops, fitness centers, etc.

As far as we know, measurement similar to what Allen and his colleagues did for person-to-person communication has never been made for inter-firm communication. However, we see no reason that such communication would not be negatively affected by physical distance. We will in several ways test that hypothesis in this chapter.

First we will ask whether those firms located within the geographic bounds of the cluster communicate more themselves and show greater centrality in the communication network among organizations than do firms in the general region but outside the geographically defined boundaries. Then we will test whether the total amount of scientific communication reported by any organization with other organizations in the study will decline with the mean geographic distance between that organization and those other organizations.

Research method

The geographic cluster

The geographic extent of the cluster is defined through the use of postal zones (Zip codes). We choose these on the basis of the concentration of firms shown in the MIT Sloan School of Management Entrepreneurship Center's map of the location of firms (Figure 5.1).

Some of these postal zones cover part of Cambridge and others were in the city of Boston. They were generally in the vicinity of Harvard University, MIT, Boston University or Harvard Medical School. Organizations located outside of these regions will be employed as a control group.

Biotechnology companies

We were able to create an accurate listing of firms with the kind assistance of the Massachusetts Biotechnology Council and the MIT Entrepreneurship Center. We compiled a database listing of Boston's' biotechnology firms, pharmaceutical firms, hospitals and universities. Around 90 firms, hospitals and universities are located within the geographic cluster (Boston and Cambridge) while another 100 firms are located in a variety of suburbs. In order to focus our attention onto those firms working on human therapeutic applications of biologically-derived pharmaceuticals, we eliminated all companies that were in the agricultural, veterinary, and environmental products and services fields from the initial listings. We also eliminated those with a primary focus on diagnostics as opposed to therapeutics. This left us with a final sample of around 70 firms. Of these, we received data from 40 companies.[1] In each of the cooperating companies, we select a random listing of approximately 10 percent[2] of their bench-level scientists. The chosen scientists must have at least a PhD or MD level of education and be actively engaged in research.

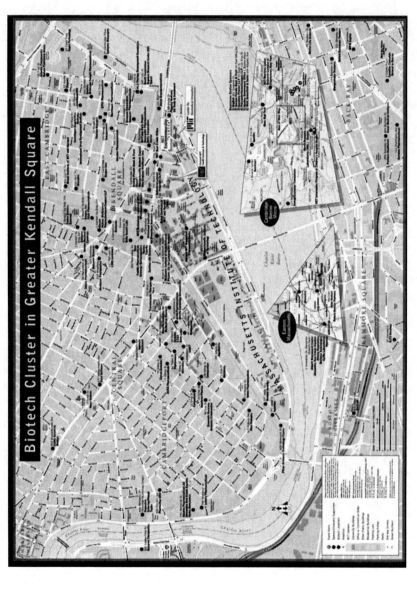

Figure 5.1 A street map of Cambridge and Boston, Massachusetts showing the locations of biotech companies (courtesy of MIT Sloan School of Management Entrepreneurship Center).

'Big Pharma'

A number of 'Big Pharma' or large broadly-based traditional pharmaceutical firms have recently located research operations in the Cambridge/Boston area. The goal of these larger firms is undoubtedly to tap into the scientific communication network that may exist among the smaller, newly-formed firms. Their longer-term goal is probably to acquire new technology and products through licensing from or acquisition of the firms owning the intellectual property. The large pharmaceutical firms are also included in our sample, to determine the degree to which they are successful in attaining this goal.[3]

A control group

Many of the firms, from which we collected data, are located outside of our selected postal zones. These more distant firms provide a convenient control group for comparison with the experimental group located within the selected postal zones.

Universities

There are five large research-based universities located within the selected postal zones. These are included in the study in a less direct way. Instead of asking scientists in the universities to report their communications, we rely on the reports of those in firms, who communicate with university scientists. Our reasoning for this is nothing more complex than ease of data collection. The initial phase of collecting data only on communication is followed by a web survey in which we seek further information on the exact university laboratories with which communication took place, the origins of the contact, etc. In looking at the universities, we hope to find the degree to which firms originating from these universities retain their connection with their 'mother' organizations, as well as the degree to which firms born elsewhere are able to develop relations with universities within the region.

Measuring the structure of communication networks; a web-based research method

As noted before, many current studies rely upon patent and publication databases and therefore are unable to detect communication that, while significant, may not result in such products. To capture this type of communication, we adapted a tool that we had previously used in the study of communication among individual scientists within organizations. In addition to being sensitive enough to capture much of this additional communication the web-based tool is also easy to use and is able to sense the dynamic aspect of the communication network as it evolves.

An email message with a link to a web page is sent once each week to each scientist. The web page contains a list of the names of the biotech/pharmaceutical organizations (including universities and hospitals) in the area. An example is shown in Figure 5.2. A scientist has merely to open the page and by mouse-click indicate which organizations (if any) that scientist had contact with on *that particular day*. This simple exercise, which can be completed in one minute or less, is then repeated on randomly chosen days, approximately once per week for a period of six months. The results collected with this tool provide a measure of the pattern of communication among the organizations and this can be related to measures of success.

The software tool we use to analyze the network is called Condor. Condor reveals the evolution of interaction patterns in social networks. It provides an environment for the visual identification and analysis of the dynamics of communication in social spaces (Gloor and Zhao 2006).

Data analysis

The first step in our data analysis is to plot the network of communication among scientists in the organizations in our sample (Figure 5.3). In this plot, each organization is represented as a network node. Pairs of nodes are connected if there is any communication reported between them. The length of each connection is inversely proportional to the amount of communication reported over the six-month period of the study.[4]

The network of Figure 5.3 exhibits an interesting characteristic. There is a set of organizations in the center, that have a higher than average level of communication,

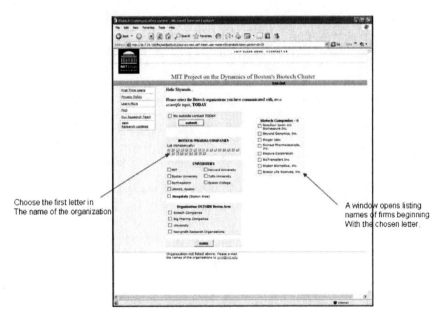

Figure 5.2 An example of the structure of the web page used in collecting communication data.

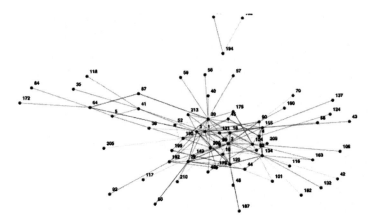

Figure 5.3 A view of the interorganizational scientific communication network as reported by scientists in a sample of biotech firms and organizations in the area of Eastern Massachusetts, USA.

among themselves.[5] Thus, there appears to be a sub-cluster, or perhaps a super cluster of high communicating organizations concentrated in the center of the overall network. It will be interesting to examine the membership of this super cluster (Figure 5.4).

However, before we get into the task of identifying membership,[6] let us look at one of our basic hypotheses. If the geographic propinquity of firms in a cluster enables more intense communication, this should be observable in our network. As a first step in testing this, we divided the entire set of organizations into two groupings, namely, the previously defined experimental and control groups.

Using the degree of centrality of each node within the network (Harary *et al.* 1965) as a metric for communication, we can compare the two groups (Figure 5.4). This reveals a significant difference in their centrality or embeddedness in the network. Those organizations within the geographically defined cluster region have nearly twice as many connections with other organizations

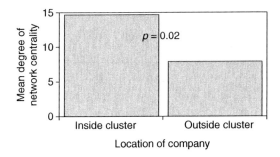

Figure 5.4 Mean centrality of organizations in the experimental and control groups.

(Figure 5.4). Since centrality indicates the number of other firms with which a given firm may be in contact but not the amount of contact, we also compare the experimental and control firms on the basis of the amount of communication reported (Figure 5.5). Here we find that scientists in firms that are within the geographic bounds of the cluster reported that along with distributing communication across a larger number of firms, they also simply communicated more. Thus we have the first elements of support for our basic hypothesis. Physical propinquity within a cluster may or may not be the cause of greater communication but it certainly correlates very strongly with the amount of inter-organizational scientific communication and assuredly makes it easier to attain greater levels of communication.

With some support for the communication benefits of clustering, let us learn a little more about the nature and makeup of the Cambridge/Boston cluster. We will do this graphically first by arbitrarily defining an area of concentration within the network of Figure 5.3. This will tell us the types of organization to be found within the region of intensive communication seen in the center of Figure 5.3. We will then turn to physical location and see what types of organization are actually located physically within the formally defined bounds of the cluster.

Categories of organization

Now to complicate things but make the analysis more interesting as well, not all of the organizations in the network of Figure 5.3 are newly-formed Biotech firms. As previously mentioned, there are also five universities, six large scale broad-based (traditional) pharmaceutical firms[7] and five larger, more established biotechnology firms. So let us look at each of these classes of organization separately.

Where are the universities?

A total of four of the universities can be readily identified within the network region that we have arbitrarily labeled a 'super cluster' (Figure 5.6). The remaining two are in the network but further from the center. The two most central uni-

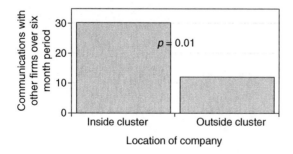

Figure 5.5 Mean level of communication of organizations in the experimental and control groups.

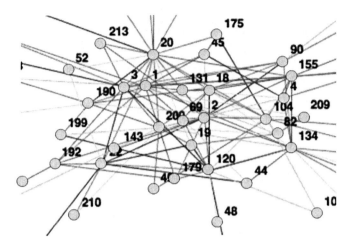

Figure 5.6 A view of the center of the network.

versities are Harvard (Medical School) and MIT. These were the two principal sources from which a majority of the new biotech firms originated. It is therefore not at all surprising to find their 'children' tightly connected to them. The parent locations are probably the chief reason why the newly formed firms located where they did. In fact, our choice of defining the geographic limits of the cluster by postal codes was based upon the locations of the 'parent' universities. We chose postal code zones that included the addresses of the principal universities. In addition, although the communication data cannot tell us this, interviews indicate that proximity to these universities was also a major factor underlying the location decisions of the major pharmaceutical firms.

What about 'Big Pharma'?

As stated above, there are five major pharmaceutical companies that have located in or near the experimental region. One of these has R&D activities at two sites about 33 km apart. So there are really six sites at which these large companies are active in R&D. Certainly, the principal reason for these large firms locating in the Boston area is the presence of so many newly-formed Biotech firms in that region. To put it simply, they want to become members of the scientific communication network. In addition to the university contact that their location enables, we would speculate that since most major pharmaceutical firms are working to develop new biologically based drugs, that they see membership in the network as an aid toward that goal. So how successful have they been? One does not have to look very long at Figure 5.6 to conclude that at least four of these firms have been very successful in at least gaining network membership. They are centrally located in the heart of the super cluster. Their scientists are in close communication with scientists in several smaller firms as well as in the universities.

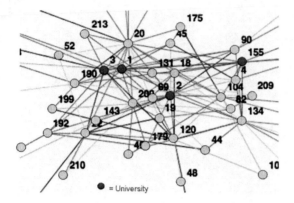

Figure 5.7 Center of the network, highlighting the universities.

'Big Bio'

We designated a subset of the biotechnology firms as 'Big Bio' simply on the basis of size and age. Such firms as Biogen, Amgen, Serono, Genzyme and Millennium are no longer new ventures. They are all large firms and have been in existence for more than a few years. Where are they in the network? Even a quick look at Figure 5.9 indicates that they are at least as central to the network as the larger pharmaceutical firms.

Comparisons on the basis of type of organization

We can see from the network that both 'Big Pharma' and 'Big Bio' are very well connected into the central core. The universities are as well, but that is no

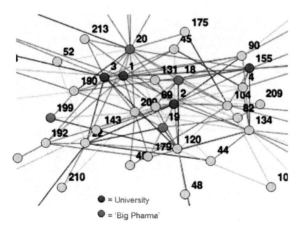

Figure 5.8 Center of the network highlighting the universities and the 'Big Pharma' firms.

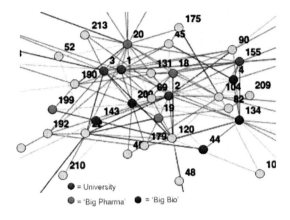

Figure 5.9 Center of the network highlighting the universities, 'Big Pharma' and 'Big Bio'.

surprise since they are the parents out of which most of the new ventures originated. We will now look and see in a more quantitative manner, just how embedded they are. At least in the case of the 'Big Pharma', this will be a test of how successful they have been in invading the network originally formed among the new venture firms. That they have been successful can be clearly seen in Figure 5.10.

Perhaps the most interesting aspect, though, is the low value of mean centrality for the small biotech firms. These are the firms who initially formed the network. Of course the universities were there from the beginning too, so it is no surprise to find them with a high degree of centrality. The larger firms were for the most part later arrivals and they are on average more deeply embedded than the startup firms.

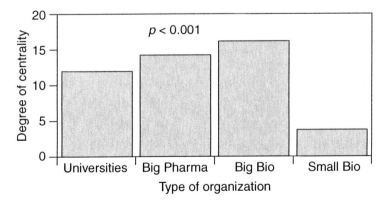

Figure 5.10 Mean network centrality for four categories of organization.

This certainly seems strange, but one possible explanation might be that these firms are not as old as we might initially assume. The set might be dominated by newly formed companies that haven't had sufficient time to embed themselves in the network. However, a closer look at the data shows this explanation to be invalid. Instead of the younger small companies having lower network central-ity, they actually have on average a higher degree of centrality than do the older small firms ($r=-0.45$; $p=0.05$). This significant negative correlation does not occur for the larger firms and universities.

Network centrality tells us how many organizations, on average, organiza-tions of a particular type communicate with. It does not reveal just how much communication actually occurs. Turning to that measure, we find the smaller firms in a much stronger position (Figure 5.11). Although the differences that appear in Figure 5.11 are not significant statistically, we can say that the smaller firms do not appear to be any less active in inter-organizational scientific com-munication than their larger neighbors.

Once again, however, what appear anomalous results for the smaller biotech firms. The older firms in this category report significantly less scientific ($r=-0.51$; $p=0.02$) communication outside of the firm than do the larger firms.

What we seem to be seeing here is that the newly formed firms work harder to establish scientific exchange with many neighboring organizations (not just with their university parent) and as they grow older, they narrow the number of targets for this activity and increase the amount of communication with this smaller number of organizations.

Mean distance and communication

As a further test of the basic clustering hypothesis, we measured the physical distances separating each pair of firms. Then for each firm, we computed the mean separation distance of that firm from all of the others. In Figure 5.12, we see that communication frequency is inversely related to inter-company distance. The inter-company distances are expressed in kilometers and were calculated

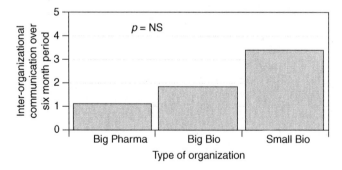

Figure 5.11 Interorganizational scientific communication reported by organizations in three categories.

using the coordinates of latitude and longitude for each company. Total communication was then plotted for each value of mean inter-company distance. As in our previous research, it is best expressed by a $1/x^2$ relationship as communication quickly decays as distance increases.

Each point on the graph in Figure 5.12 represents one company in our sample. The x-axis coordinate is the mean distance between that company and all of the other organizations. In other words, it is how far, on average, this particular company is from all of the other organizations in our sample. Because of the large concentration of companies in the experimental area, those with low average inter-company distances are in or very close to the geographic center of the biotech cluster. The vertical coordinate is the total amount of communication reported by that given company with all of the others.

This graph shows that companies in the physical center of the biotech cluster communicate more with other companies. As mean separation distance increases, the total communication with other companies decreases as $1/x^2$.

Since we did not ask scientists to report the medium used for each communication, this number could include email and telephone as well as face-to-face. Why then would it decay with distance? We do not have a definitive answer for that. However, there is evidence from a study by Allen and Hauptman (1987), that the use of different communication media is positively correlated. Regardless, however, of whether the communications were face-to-face, the important fact is the decline with distance. Those declaring the death of distance are at least in this instance patently wrong. Separation distance does matter in the twenty-first century, and these data once again support the idea of clustering high tech new ventures or at least biotechnology firms together geographically.

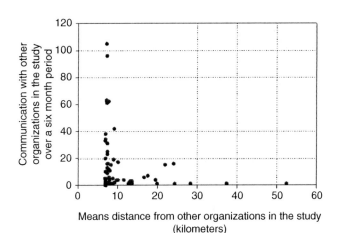

Figure 5.12 Interorganizational scientific communication reported as a function of mean separation distance from other organizations.

Acknowledgments

The authors would like to acknowledge the support, both financially and in terms of knowledge and support of the MIT Sloan School Entrepreneurship Center. Without their help and encouragement, this research would have been impossible to conduct.

We must also acknowledge and thank the many companies and scientists, who provided us with data. There are too many for individual acknowledgments, but thank you all!

Notes

1　The failure to reach all 70 companies is not due to low response from the companies. In fact, only one company declined to participate. It was instead due to our inability to reach all 70. This, in turn, was due to a lack of resources and time. There were only two of us working at that point in the study.
2　In companies with fewer than ten scientists, we sample all of those engaged in bench level research. Cooperation is, of course, voluntary and the overall research plan has been reviewed and approved by the MIT Committee on the Use of Human Experimental Subjects.
3　While the presence of the larger firms in many ways is a benefit to the startups, since they bring resources in the form of much needed money for licensing or even outright acquisition, this is not universally viewed in a positive light. One entrepreneur described the relationship of firms such as his with 'Big Pharma' as 'a bunch of monkeys playing with gorillas'. He was very concerned with protecting his intellectual property or at least getting what he considered a fair price for it.
4　Since it was impossible to visit and recruit all 40 collaborating firms on the same day, we used a sliding window. Data collection began on a slightly different day for each firm and continued for six months. So the six month period of data collection is slightly different (but overlapping) for each firm. This could have had serious implications for the study had something critical had affected the industry during the study. Fortunately for the investigators, no such untoward event occurred.
5　In terms of the number of scientific communications reported per unit of time.
6　No organizations, aside from universities will be specifically identified. Understanding what can result when people can identify high and low communicators, for example, we have had a standard policy, for many years of concealing identities. We will therefore identify only classes of organization.
7　Two of these have merged since the study. However, we will continue to treat them as separate in the study.

References

Allen, T.J. and O. Hauptman (1987) 'The Influence of Communication Technologies on Organizational Structure, a Conceptual Model for Future Research', *Communication Research*, 14 (5), pp. 575–8.
Allen, T.J. and G.W. Henn (2006) *Organization and Architecture for Innovative Product Development,* New York: Elsevier.
Cairncross, F. (2001) *The Death of Distance: How the Communications Industry is Changing Our Lives*, Boston, MA: Harvard Business School Press.
Gloor, P. and Y. Zhao (2004) *TeCFlow: A Temporal Communication Flow Visualizer for*

Social Networks Analysis, ACM CSCW Workshop on Social Networks. ACM CSCW Conference, Chicago.

Harary, F., R.Z. Norman and D. Cartwright (1965) *Structural Models: An Introduction to the Theory of Directed Graphs*, New York: Wiley.

Porter, K.A. and W.W. Powell (2006) 'Networks and Organizations', in: S. Clegg, C. Hardy, T. Lawrence and W. Nord (eds) *The Handbook of Organization Studies: Ten Years On*, Thousand Oaks, CA: Sage Publishing, pp. 776–99.

Porter, K.A., K.C. Bunker Whittington, and W.W. Powell (2005) 'The Institutional Embeddedness of High-tech Regions: Relational Foundations of the Boston Biotechnology Community', in: S. Breschi and F. Malerba (eds) *Clusters, Networks, and Innovation*, Oxford, UK: Oxford University Press, pp. 261–96.

Powell, W.W., K.W. Koput and L. Smith-Doerr (1996) 'Interorganizational Collaboration and the Locus of Innovation: Networks of Learning in Biotechnology', *Administrative Science Quarterly*, 41, (1), pp. 116–45.

Saxenian, AnnaLee (1994) *Regional Advantage: Culture and Competition in Silicon Valley and Route 128*, Cambridge, MA: Harvard University Press.

Schilling, M.A. and C.C. Phelps (2005) 'Interfirm Collaboration Networks: The Impact of Small World Connectivity on Firm Innovation', *Management Science*, 53 (7), pp. 1113–26.

6 Technological specialization and variety in regional innovation systems

A view on Austrian regions

Bernd Ebersberger and Florian M. Becke

Introduction

Recently regions have attracted increasing interest as the locus of innovation and as the geographical domain of innovation policy. The latter is remarkable as the discussion and the analysis of innovation policy has largely focused national measures (Fritsch and Stephan 2005). As a basis for suitable tailoring of policy measures and their profound discussion broad and multifaceted information has to be available about the content and the structure of the regional innovation activities (Laranja *et al.* 2008). In this sense learning about the regional innovation capabilities and the specialization profiles is an essential part of a strategic intelligence for innovation policy making (Kuhlmann 2005).

This chapter is based on the assumption that innovation is regional. It seeks to sketch out an approach how insight can be gained into the technological specialization profiles of regions. We first discuss the conceptual background and the research question guiding the analysis. Then we introduce the data set used and the measures derived. We discuss how we construct indicators from the measures to capture the regional specialization and variety. Subsequently we present the results and conclude with a brief discussion and an outlook into further research.

Conceptual background and research questions

This analysis builds on an evolutionary tradition, where learning, knowledge, competencies and their cumulative development are accepted concepts in the analysis of innovation activities on the micro and firm level as well as on the regional and national level (Malerba 2002). The analysis also builds on the notion of innovation systems as the key environment in which companies create new knowledge, combine and recombine existing and new knowledge and develop applications to eventually commercialize the results. The idea of innovation systems embraces the idea that linkages, collaboration and networks among a wide variety of actors are crucial in understanding the creation and diffusion of innovations (Edquist 1997). Based on the demarcation between elements that constitute the system and elements that do not, national systems (Lundvall 1992;

Nelson 1993; Freeman 1987), sectoral systems (Malerba 2002), technological systems (Carlsson 1995; Carlsson and Stankiewitz 1995; Callon 1992) and regional systems of innovation (Cooke *et al.* 1997) are distinguished.

Regional innovation systems

The latter has been appreciated as a way to better understand the innovation processes in the regional economy and to focus the attention to the success stories of regional development (Asheim and Gertler 2004) and the knowledge creation process within the region (Maskell and Malmberg 1999). The region can be regarded as the locus of innovation (Doloreux and Parto 2004). If innovation is a social process, then the region plays a particular role reflecting the institutional, political and social context (Rondé and Hussler 2005). This geographically bound process utilizes regionally shared knowledge bases (Asheim and Isaksen 1997; Maskell and Malmberg 1999) and social structures (Lam 2000). Porter argues that competitive advantage is generated by localized capabilities, competencies and interaction structures at the regional level (Porter 1998). Reflecting the notion of the innovation system the generation and diffusion of innovation is a social process. It is heavily embedded in informal social interactions, relationships and networks (Bathelt *et al.* 2004; Owen-Smith and Powell 2004).

Technological specialization

The interaction with a geographically bound region creates localized or regional spillovers which lead to regionally distinct profiles of capabilities and competencies (Storper 1997). These competencies and capabilities manifest themselves as the technological specialization of the region. The interdependencies between the infrastructure, the knowledge generation and the cumulativeness generating competencies might create local lock-ins or path dependencies and strong temporal persistence of specialization pattern. The localized externalities and the overall proximity of actors create a milieu where the odds are better for individual actors to pick up information which eventually turns out to be useful (Malmberg and Maskell 2002). Actors mutually stimulate knowledge creation and adaption (Malmberg 1997).

Eventually, the creation of innovation is not a target in itself (Howells 2005). Rather, is it the economic growth dynamics associated with innovation driven competitiveness of a region which drives policy interest in regional innovation systems (Fritsch and Stephan 2005). The discussion on growth has generally acknowledged that innovation and knowledge generating activities play a crucial role in determining the growth dynamics of national and regional economies. Generally, knowledge has either been interpreted as exogenous to the development and economic growth (Solow 1956), or interpreted as a one dimensional quantity (Temple 1999; Romer 1990). Only from the Schumpeterian or evolutionary perspective of economic growth variety and heterogeneity of actors, technologies and knowledge has been discussed (Pyka *et al.* 2000; Cantner *et al.* 2008; Frenken *et al.* 2007), where the latter introduce the notion of related and unrelated variety.

Variety

Related variety refers to the variety of knowledge bases, competencies or technologies which feed into the innovation process of an industrial sector. Related variety bases on the concept of Jacob's externalities (Jacobs 1969). It is assumed to have positive impact on the innovation activities within the sector. It also positively influences the regional growth potential through innovation. Frenken *et al.* find that related variety indeed exerts a positive effect on employment in a region (Frenken *et al.* 2007). Unrelated variety is the variety generated by different industrial sectors in a region, which are not related through a common knowledge base. Unrelated variety generates portfolio effects and immunizes the regional economy vis-à-vis exogenous shocks. Unrelated variety exerts negative effects on unemployment (Frenken *et al.* 2007).

Research questions

This chapter tries to explore the pattern and development of technological specialization and knowledge generation of regional innovation systems in Austria. In particular we are interested in (*a*) the *pattern of technological specialization* and (*b*) the *stability of this pattern* over time. As the nine regional innovation systems analyzed jointly form the Austrian national system we are interested in (*c*) the differences of the pattern between the regional systems, in particular whether regional systems are *complementary or substitutive*. Patterns of technological specialization are not analyzed for their own sake. Rather they are the foundation on which current and future economic development rests. Eventually we are interested in (*d*) the *sectoral structure which is supported by the variety of technological specialization* found in the regional innovation systems.

Data and methodological approach

The analysis here bases on the analysis of patent and utility model documents obtained from the Austrian Patent Office. The data contains the applications of patents and utility models covering the period from 1 January 1988 to 3 December 2008.

Patents and utility models

Patents and utility models are foremost legal instruments to protect intellectual property. Once granted they give temporary monopoly rights to their holders. This feature represents a valuable incentive mechanism to induce invention activities and application of inventions. For administrative purposes all the information contained in a patent or utility model application have to be prepared in a standardized way. However, filing for a patent or a utility model leads to the publication of the invention not later than 18 months after filing. All applications for patents or utility models, and the contained information, are published regardless of the success in the granting process. Some bibliographic data (e.g. inventor, applicant, title, IPC, date of filling) are published within one month from the date of the application at the Austrian Patent Office. The patent system enables individuals to receive temporary monopoly profits and society to gain from the dissemination of the published technologies.

In addition to being a legal instrument for IP protection, patent and utility model documents supply to the researcher a host of information about the act of the filing, the actors involved in the development of the knowledge codified in the document and about the knowledge itself. This information is used to construct the measures for the analysis below. The patent and utility model documents contain information about:

- date of filing
- the name and the location of the applicant
- classification of the technology (IPC).

This wealth of information and the reliability of the information make patent documents highly attractive as a data source for analysis of innovation systems at various levels (Breschi and Lissoni 2005; Moed *et al.* 2005; Porter and Newman 2005).

A utility model differs from a patent in certain aspects: Patents, like utility models are an instrument to protect technical inventions meaning solutions for technical problems. The requirements for the protection are: novelty, inventive step, industrial application. The main differences between patents and utility models are in the requirement of novelty and inventive step and in the granting procedure. The inventor (or the legal successor) has a six-month period of grace for the novelty requirement. That means a utility model is still achievable even if the invention was published or shown by the inventor or the legal successor no longer than six months before filing. Requirements for the inventive step are supposed to be smaller for the utility model than for the patent. Granting a patent means an examination procedure including the extensive appraisal of the state of the art. For the registration of a utility model only a formal examination of conformity with the law and a search report concerning the state of the art is required. Another difference concerns the duration of the intellectual property right, in that the maximal term of the utility model is only ten years in contrast to a term of 20 years for a patent. In Austria the range of innovations that can be protected by utility models is almost identical to that of patents and is not restricted to any technical fields.

Their cost advantage and flexibility make utility models a particularly interesting protection instrument for SMEs. As the Austrian economy and the Austrian innovation system are strongly based on SME activity – about 93 percent of the Austrian innovative companies have less than 250 employees

(Statistik Austria 2009) – focusing on patents only could create a bias towards larger enterprises. Hence we choose to collapse utility model and patent applications into a single data source for the analysis. Overall the data set contains 69,257 applications. The temporal distribution is reported in Figure 6.1.

Measures

As we are per se not interested in the protection behavior of firms in the nine Austrian regions, we use the application documents as to proxy the information required for the analysis.

Competencies To capture the technological specialization we require information about technological competencies. We use patent and utility model applications as an indicator for competencies. One or more international patent classification (IPC) code is assigned to each application. This allows us to approximate the technological content of the application and to capture the competencies of the involved actors by the assigned technological class(es).

Time and space For profiling the technological specialization pattern of regions we utilize the location of the applicant to assign competencies to certain regions. Investigating the dynamics of technological specialization we use the filing date to approximate the point of time where the competencies have been available in the region, where we equate the region with the Austrian federal states.

Indicators

The analysis of the specialization profile and the variety of the knowledge bases of the nine Austrian regions requires the construction of indices to capture specialization and the variety using the measures of competencies.

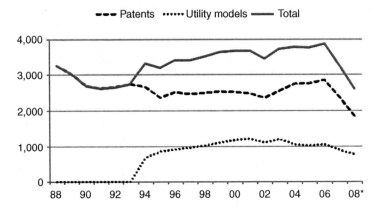

Figure 6.1 Patent and utility model applications in the data set.

Note
* 2008 covers 1 January to 3 December 2008. Even with correcting for this we observe a strong decline in patent application. Utility modes were established in Austria in 1994.

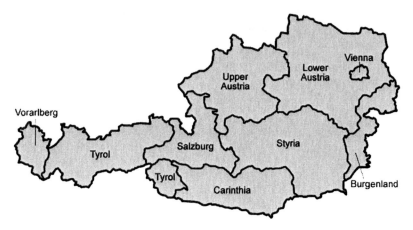

Figure 6.2 Austrian states.

Specialization

The structure and the pattern of technological specialization can be analyzed utilizing a number of indicators such as patent and utility model counts or patent and utility model growth. However, these indicators do not account for different propensities to file a patent or utility model across technological fields (Nesta and Patel 2005). The revealed technological advantage (*RTA*) index to depict the technological specialization of a region controls for these different propensities. The RTA_{ij} index is a transformation g of the ratio of regional p_{ij} and the national share p_{iA} of the technology i in region j where

$$p_{ij} = \frac{P_{ij}}{\Sigma_i P_{ij}}$$

$$p_{iA} = \frac{P_{iA}}{\Sigma_i P_{iA}}$$

P_{ij} is the number of region j's applications in a technological field i and P_{iA} is Austria's applications in the technological field i. Hence the RTA_{ij} is

$$RTA_j = g\left(\frac{\dfrac{P_{ij}}{\Sigma_i P_{ij}}}{\dfrac{P_{iA}}{\Sigma_i P_{iA}}}\right)$$

Where the transformation g normalizes the *RTA* to fall between 0 and 2

$$g(x) = \frac{x-1}{x+1} + 1$$

An *RTA* larger than 1 indicates relative technological advantage, whereas an *RTA* below 1 points towards a technological disadvantage. Based on the *RTA* index of technological advantage the coefficient of variation (*CV$_j$*) of the *RTA* is a measure of technological specialization of the region *j* (Nesta and Patel 2005).

$$CV_j = \frac{\sqrt{Var(RTA_j)}}{Mean(RTA_j)}$$

A high *CV* indicates that the region concentrates its competencies in narrow areas or niches, whereas a low *CV* points towards regions spreading out their competencies across a larger canvas of technologies.

Variety

Frenken *et al.* use 5-digit and 3-digit industrial classification to construct their indicator for related and unrelated variety (Frenken *et al.* 2007). In contrast to this procedure, we base our measure on the relationship between technological competencies and industrial sectors. We utilize a concordance between IPC 4-digit technologies and 21 related sectoral fields based on Schmoch *et al.* (2003) and Schmoch and Gauch (2004) as illustrated in Figure 6.3.

Given the mapping of technologies on sectors, indices for related and unrelated variety can be calculated. I_k denotes a set containing all technological classes supporting a given sectoral field $k \in \{1 \ldots K\}$. The technological share of this sector in region *j* is S_{kj}.

$$S_{kj} = \sum_{i \in I_k} p_{ij}$$

Unrelated variety *URV$_j$* in a region *j* is the entropy of sectoral fields in a region

$$URV_j = \sum_{k=1}^{K} S_{kj} \cdot lu\,(S_{kj})$$

IPC classes		Sectoral fields
G03H	Holographic processes or apparatus using light ...	
H04H	Broadcast communication	04–Audio-visual electronics (32.3)
H04N	Pictorial communication, e.g. television	
H04R	Loudspeakers, microphones, gramophone pick-ups ...	
H04S	Stereophonic systems	
...	...	

Figure 6.3 Illustration of the concordance between IPC and industrial sectors.

Related variety RV_j in a region j is the weighted sum of the entropy of within sector technologies.

$$RV_j = \sum_{k=1}^{K} S_{kj} H_{kj}$$

where

$$H_{kj} = \sum_{i \in I_k} \frac{P_{ij}}{S_{kj}} \log \left(\frac{S_{kj}}{P_{ij}} \right)$$

Results

Pattern of technological specialization

In a first step we analyze the technological specialization of the nine Austrian regions measured by the revealed technological advantage (*RTA*) based on the IPC-sections (Table 6.1).

The snapshot of the technological specialization of the nine Austrian regions reveals that the Austrian regions maintain a rather diverse specialization pattern, which can be further broken down into more detailed technologies. The more detailed break-down can be found in Table 6.6 in Appendix 6.1.

The overall breadth of the regional knowledge base can be assessed by the number of technological fields which are a technological advantage and hence have an *RTA* larger than 1 (cf. Table 6.2). Here it shows that Burgenland (BL) has the broadest knowledge base (41) closely followed by Vienna (VI), and Carinthia (CA) – both with 39 – and Lower Austria with 38 fields with a revealed technological advantage. On the lower end of the spectrum we find Vorarlberg (VA) with 17 technologies which represent a technological advantage.

The analysis of the technological specialization captured by the coefficient of variation of the *RTA* exhibits that two regions stand out markedly: Burgenland (BL) with a coefficient of variation of 70 percent; and Vorarlberg (VA) with a coefficient of variation of 60 percent. In the case of Burgenland, technological breadth meets with technological specialization, whereas in Vorarlberg low technological heterogeneity and low level of specialization coincide. Overall we observe three rather distinct groups of regions: highly specialized regions such as Burgenland and Vorarlberg; rather un-specialized regions exemplified by Upper Austria ($CV=26$ percent) and Lower Austria ($CV=31$ percent); and a group with medium specialization such as Tyrol ($CV=49$ percent) and Carinthia ($CV=42$). In an analysis of national innovation system Archibugi and Pianta find that specialization can partly be explained by the size of the innovation system (Archibugi and Pianta 1992). The results here also suggest that the specialization can partly be attributed to the differences in the size of the region. Measuring the size by the working population of a region we get a highly significant correlation of -0.81 ($p=0.009$) between specialization and size of the region. The smaller a region is the more it tends to be specialized.

Table 6.1 Technological specialization (2000–8) – IPC sections

	VA	TY	SZ	UA	ST	LA	BL	CA	VI
A – Human necessities	1.27	1.05	1.19	0.92	0.81	1.03	1.03	1.03	1.02
B – Performing op.; transport	0.81	0.95	0.94	1.18	1.01	0.94	0.78	0.99	0.86
C – Chemistry; metallurgy	0.86	1.08	0.72	1.08	0.98	0.92	0.85	0.81	0.99
D – Textiles; paper	0.46	0.82	0.38	1.20	1.16	0.63	1.08	0.49	0.54
E – Fixed constructions	1.22	0.94	1.07	0.95	0.95	1.11	0.98	1.22	0.84
F – Mechanical engineering	0.92	1.09	1.00	0.90	1.24	0.96	1.04	0.94	0.92
G – Physics	0.55	0.90	0.99	0.84	1.08	0.95	0.86	0.92	1.26
H – Electricity	0.74	0.94	0.75	0.81	0.79	1.15	1.45	0.88	1.26

Notes

RTA, based on the application data for patent and utility model obtained from the Austrian Patent Office, national applications only. Filing dates 1 January 2000 – 3 December 2008. Technological advantage, specialization is printed in italics.

Regions: VA = Vorarlberg, TY = Tyrol, SZ = Salzburg, UA = Upper Austria, ST = Styria, LA = Lower Austria, BL. = Burgenland, CA = Carinthia, VI = Vienna.

Table 6.2 Technological breadth and specialization

	VA	TY	SZ	UA	ST	LA	BL	CA	VI
Breadth (no. of fields with *RTA* > 1)	17	35	31	33	29	38	41	39	39
Specialization (CV)	0.60	0.43	0.49	0.26	0.34	0.31	0.70	0.42	0.36

Notes
Indices for breadth and specialization are based on the data documented in Table 6.6 and Table 6.7.
Regions: VA = Vorarlberg, TY = Tyrol, SZ = Salzburg, UA = Upper Austria, ST = Styria, LA = Lower Austria, BL = Burgenland, CA = Carinthia, VI = Vienna.

Stability of the technological specialization

Stability of the technological specialization means that the technological advantage persists over time: Areas which are characterized by technological advantage in the past tend to have technological advantage presently; disadvantaged technologies tend to remain disadvantaged. We investigate the stability of the specialization through the persistence of technological advantage over time. In particular we correlate the regional technological specialization pattern for the time 1988–92 with the pattern observed in 2004–8 Correlation coefficients are reported in Table 6.3.

The least degree of stability can be found in Burgenland (BL) where the current specialization is insignificantly correlated with the pattern observed in 1988–92 suggesting a dramatic change in the structure of technological competencies in the past two decades. Low levels of persistence of technological specialization can also be found in Vorarlberg (VA), Lower Austria (LA) and Carinthia (CA) where the correlation is low but significant. On the other hand a rather high level of persistence can be found in Vienna (VI), where previous technological specialization translates into current technological specialization.

Complementarity of specialization pattern

The specialization patterns of two regions are complementary if one region reveals a technological advantage where the other does not and vice versa. The specialization patterns are substitutive if the pattern of both regions reveal the same technological specialization. Hence the correlation between the specialization patterns of the regions allows us to identify whether they are complementary or substitutive. Strong positive correlation suggests similar e.g. substitutive pattern of specialization between two regions, negative correlation indicates complementarity between the specialization patterns. Table 6.4 gives the correlation between the regional pattern for the years 1988–92 and for the years 2000–8.

Generally for the current structure (2004–8) we find significantly positive correlation between most of the regional specialization pattern, suggesting a rather substitutive relationship between most of the regions. Only between Upper Austria, and both Vienna and Lower Austria, we find significantly negative correlation indicating complementary patterns of technological specialization. Overall, Upper Austria shows the most distinct pattern of technological specialization, only insignificant or significantly negative patterns emerge between Upper Austria and any one of the other regions. In stark contrast we observe the technological specialization pattern of Salzburg (SZ) which seems to be strongly substitutive with any other region – except Upper Austria. The difference between the correlations for the years 1988–92 and 2004–8 suggests that over the last two decades the pattern complementarity of the specialization pattern has not changed significantly for most of the regions. Only has Vorarlberg (VA) and Tyrol (TY) significantly changed the pattern; Vorarlberg more towards other

Table 6.3 Stability of the technological specialization (2000–8)

	VA	TY	SZ	UA	ST	LA	BL	CA	VI
Correlation	0.21*	0.45***	0.41***	0.39***	0.45***	0.26*	0.10	0.26*	0.61***

Notes
Correlation between the *RTA* (1988–92) and *RTA* (2004–8) for each region. *** (**, *) indicates significance at the 1 percent, (5 percent, 10 percent) level.
Regions: VA = Vorarlberg, TY = Tyrol, SZ = Salzburg, UA = Upper Austria, ST = Styria, LA = Lower Austria, BL = Burgenland, CA = Carinthia, VI = Vienna.

Table 6.4 Complementarity of specialization pattern

1988–92	VA	TY	SZ	UA	ST	LA	BL	CA	VI
VA									
TY	0.20**								
SZ	0.14	0.25***							
UA	0.12	0.04	0.14						
ST	−0.01	0.12	0.15	−0.16*					
LA	0.04	0.22**	0.17*	−0.01	0.16*				
BL	0.15	0.24***	0.24***	0.03	0.13	0.19**			
CA	0.08	0.23***	0.41***	0.13	0.09	0.09	0.29***		
VI	0.03	0.27***	0.14	−0.32***	−0.09	0.24***	0.13	0.13	
Mean Corr.	0.09***	0.20***	0.20***	0.00	0.05	0.13***	0.18***	0.18125	0.07

2004–8	VA	TY	SZ	UA	ST	LA	BL	CA	VI
VA									
TY	0.08								
SZ	0.36***	0.34***							
UA	0.08	−0.05	−0.04						
ST	0.21**	0.17*	0.24***	−0.01					
LA	0.17*	0.09	0.25***	−0.24***	0.05				
BL	0.32***	0.00	0.22**	0.04	0.00	0.24***			
CA	0.24***	0.17*	0.35***	−0.07	0.18*	0.27***	0.21**		
VI	0.15	0.17*	0.3***	−0.24***	−0.02	0.19**	0.19**	0.21**	
Mean Corr.	0.20***	0.12**	0.25***	−0.07	0.10**	0.13*	0.15***	0.20***	0.12*
Difference+	0.11**	−0.08*	0.05	−0.06	0.05	−0.01	−0.02	0.01	0.05

Notes

Correlation between the *RTA* (1988–92, 2004–8) of the regions *** (**, *) indicates significance at the 1 percent, (5 percent, 10 percent) level. + Mean of the difference between the *RTA* correlation (1988–92) and (2004–8), paired students t-test.

Regions: VA = Vorarlberg, TY = Tyrol, SZ = Salzburg, UA = Upper Austria, ST = Styria, LA = Lower Austria, BL = Burgenland, CA = Carinthia, VI = Vienna.

regions, Tyrol more into the direction of a distinct pattern. Vorarlberg hence developed a technological specialization profile which was rather unrelated to profiles of other regions into a profile which is more similar. Tyrol on the other hand decreased the similarity of the technological specialization pattern moving towards a more distinct and hence to a profile with is more complementary to other regions profiles. Stronger similarity is maintained relative to the profile of Salzburg, Carinthia and Vienna.

Technological variety supporting sectoral development

Having investigated the technological profile of the regions we now broaden the view to the sectoral fields which are supported by the technological profiles.

A concordance is used to map the patent and utility model applications based on their IPC code into sectoral related fields (Schmoch and Gauch 2004; Schmoch *et al.* 2003). This mapping allows for the construction of an indicator for related and unrelated variety reported in Table 6.5.

We observe that both Salzburg (SZ) and Upper Austria (UA) enjoy the highest level of related variety among the Austrian regions. The industries which these regions specialize in are supported by a broad knowledge base. This suggests that the innovation opportunities in these sectors are particularly favorable as they are fed by a broad and diverse knowledge base. On the contrary Vorarlberg (VA) and Burgenland (BL) exhibit the lowest related variety. The industrial sectors tend to be based on a rather narrow knowledge base, in a sense reducing the innovative opportunities in these regional innovation systems.

Both Vienna (VI) and Tyrol (TY) exhibit the highest level of unrelated variety. As higher levels of unrelated variety implies stronger portfolio effects these regional innovation systems seem to be less vulnerable to economic or technological shocks affecting only one or a small number of sectors. Vorarlberg (VA) and Salzburg (SZ) reveal to have the least unrelated among their industrial base suggesting that they are more exposed to potential shocks. We also find strong persistence of related and unrelated variety over time as we find strong correlation between the indicators for related (correlation=0.95, p=0.000) and for unrelated variety (correlation=0.84, p=0.005).

Conclusion

The analysis documented here set out to investigate the regional innovation systems in Austria based on the pattern of their technological specialization. Both the current specialization pattern as well as the development of the specialization has been investigated. We observe a rather distinct pattern of specialization for each of the regions which – to a large degree – are highly persistent over the last two decades. This suggests that technological specialization is a highly cumulative process resulting in path dependencies and restricting path for future development towards distinct technological specialization profiles. More detailed analysis into the development of Burgenland can offer insights into how

Table 6.5 Related and unrelated variety (2004–8)

	VA	TY	SZ	UA	ST	LA	BL	CA	VI
Unrelated variety	2.36	2.72	2.58	2.64	2.70	2.71	2.69	2.65	2.74
Related variety	2.20	2.89	3.05	3.24	2.93	2.97	2.38	3.00	2.92

Notes
Related and unrelated variety (Frenken *et al.* 2007) based on the applications 2004–8.
Regions: VA = Vorarlberg, TY = Tyrol, SZ = Salzburg, UA = Upper Austria, ST = Styria, LA = Lower Austria, BL = Burgenland, CA = Carinthia, VI = Vienna.

and why persistence of the technological specialization has been overcome and how broad technological heterogeneity has been developed together with high level of specialization.

What certainly remains a challenge in the overall system of Austrian regions is the high degree of similarity between the specialization profiles. Further analysis of the particularly distinct position of Upper Austria will shed light into the mechanism of achieving this complementarity. Comparison with other European regions can show whether this position is relative to the other Austrian regions or whether it holds also internationally. Only in the latter case can competitive advantage spring from this complementarity. Investigating the development of Tyrol in more detail can reveal the drivers and determinants for achieving a more distinct technological profile vis-à-vis other regions.

Appendix 6.1

Specialization pattern

Table 6.6 Pattern of specialization and technological advantage (2000–8)

	VA	TY	SZ	UA	ST	LA	BL	CA	VI
A01-Agriculture	0.83	0.89	0.93	1.20	0.93	1.06	1.16	1.29	0.67
A21-Baking	0.00	0.00	0.53	0.85	1.19	0.86	1.54	0.31	1.26
A23-Foods	0.36	0.52	1.30	0.72	0.98	1.16	1.27	1.17	1.05
A24-Tobacco	0.78	0.68	0.74	1.13	1.24	0.95	0.00	1.10	0.82
A41-Wearing apparel	1.11	0.88	1.34	1.00	0.77	1.10	1.21	0.57	1.10
A43-Footwear	0.85	1.41	1.42	1.03	0.64	0.77	1.04	0.72	0.95
A44-Jewellery	0.27	1.66	0.93	0.63	0.79	0.62	1.00	1.52	1.00
A45-Hand articles	1.03	1.08	0.89	0.97	0.89	1.15	1.00	1.05	1.11
A47-Furniture	1.65	0.78	1.15	0.95	0.73	1.03	0.76	0.95	0.81
A61-Medical or veterinary sci.	0.95	1.16	1.16	0.75	0.80	0.99	0.44	0.87	1.23
A62-Life-saving	0.70	1.08	0.91	1.02	0.57	0.86	1.66	1.38	1.08
A63-Sports	0.75	1.16	1.42	0.93	0.68	1.13	1.43	1.02	0.84
B01-Physical or chemical proc.	0.45	0.76	0.74	1.27	0.96	0.95	0.33	1.05	0.94
B02-Crushing, pulverizing	0.82	0.92	1.64	1.14	1.00	0.39	0.00	1.40	0.16
B03-Sep. of solid materials	0.00	0.70	1.09	0.61	1.25	1.24	0.00	0.00	0.98
B05-Spraying	0.96	1.20	0.92	0.91	0.80	1.13	1.52	1.49	0.85
B07-Separating solids	0.00	0.78	1.04	0.88	1.43	0.91	0.00	1.20	0.73
B22-Casting	0.71	1.08	0.34	1.50	0.89	0.38	0.00	0.35	0.32
B23-Machine tools	0.61	1.25	0.36	1.43	1.00	0.46	0.00	0.18	0.59
B24-Grinding	0.53	1.54	0.29	1.10	0.89	0.68	0.77	0.69	0.65
B25-Hand tools	1.16	1.16	0.92	1.26	0.91	1.02	1.03	1.38	0.46
B26-Hand cutting tools	0.77	0.78	0.93	0.98	1.28	0.76	0.70	1.54	0.42
B27-Working wood	0.98	1.22	0.94	1.27	0.96	0.81	0.00	1.38	0.21
B28-Working cement,	0.95	1.25	1.49	0.80	0.93	1.02	0.00	1.27	0.28
B29-Working of plastics	0.24	0.33	0.32	1.53	0.52	0.62	0.64	0.49	0.48

B32-Layered products	0.70	1.21	1.15	1.13	0.72	1.21	1.16	1.02	0.60
B41-Printing	0.58	1.28	0.00	1.44	0.36	0.45	0.00	0.57	0.93
B42-Bookbinding	0.31	0.74	0.67	0.99	0.63	0.62	1.11	0.93	1.36
B44-Decorative arts	0.73	0.83	0.69	1.02	0.91	0.68	1.60	1.47	1.06
B60-Vehicles in general	0.52	0.69	1.15	0.91	1.29	1.17	0.60	0.83	0.84
B61-Railways	1.58	0.47	0.66	0.39	0.91	0.86	0.71	0.72	1.27
B63-Ships	0.55	0.76	1.35	0.92	1.04	1.00	0.79	1.29	1.14
B64-Aircraft	0.32	1.22	0.52	0.95	1.07	1.04	1.33	0.96	1.24
B65-Conveying	0.78	0.83	0.87	0.95	0.97	1.09	1.08	1.14	0.94
B66-Hoisting	1.03	0.93	1.44	1.20	0.83	0.78	0.00	0.76	0.79
B67-Opening bottles	0.37	1.12	1.39	0.67	1.07	0.98	0.87	0.36	1.09
C01-Inorganic chemistry	0.00	0.00	0.86	0.72	1.08	1.19	1.17	1.63	0.95
C02-Treatment of water	1.01	1.15	0.81	1.01	1.05	1.15	0.98	1.23	0.81
C03-Glass	0.00	1.44	1.37	0.62	1.07	1.38	1.16	0.55	0.45
C04-Cements	0.70	0.99	1.15	0.62	0.78	1.10	1.16	1.02	0.81
C07-Organic chem.	0.00	1.12	0.29	1.18	0.78	0.33	0.60	0.00	1.21
C08-Organic macromolec.	0.71	0.72	0.22	1.34	0.52	0.99	0.00	0.57	1.05
C09-Dyes	0.56	0.48	0.53	1.20	1.33	0.98	1.15	0.55	0.64
C10-Petroleum, gas	1.04	0.83	0.75	0.96	0.95	1.10	1.20	0.91	0.97
C11-Animal or veg. oils	0.90	0.50	1.06	0.72	1.08	1.19	1.17	0.57	0.83
C12-Biochemistry	0.59	0.76	1.03	0.72	0.76	0.95	1.07	0.39	1.42
C21-Metallurgy of iron	1.61	0.00	0.00	1.39	0.89	0.09	0.00	0.00	0.28
C22-Metallurgy of iron	0.95	1.47	0.84	1.20	1.15	0.71	0.00	0.99	0.46
C23-Coating metallic mat.	0.00	1.55	0.00	0.95	1.36	0.88	0.00	0.94	0.66
C25-Electrolytic processes	0.00	1.01	0.00	1.28	0.97	0.74	1.49	0.89	0.96
D06-Treatment of textiles	0.69	1.60	0.65	0.97	0.29	1.05	1.57	0.86	1.01
D21-Paper-making	0.00	0.00	0.16	0.52	1.47	0.47	0.00	0.55	0.29
E02-Hydraulic engineering	1.09	1.03	1.32	1.10	0.89	1.08	0.00	1.30	0.78
E05-Locks	1.71	0.65	0.74	0.62	0.66	1.18	0.78	0.81	0.99
E06-Doors, windows,	0.81	0.83	1.33	1.12	0.71	1.03	1.35	1.23	0.50
E21-Earth or rock drilling	0.00	0.77	0.83	0.89	1.51	0.68	0.00	1.20	0.52
F01-Machines or engines	1.03	1.10	0.96	0.80	1.53	0.82	1.04	0.71	0.50
F02-Combustion engines	0.55	1.39	0.96	0.58	1.59	0.56	0.53	0.18	0.35
F03-Engines for liquids	1.01	0.55	1.13	0.76	0.80	1.33	1.36	1.26	1.19

continued

Table 6.6 Continued

	VA	TY	SZ	UA	ST	LA	BL	CA	VI
F04-Positive-displacement	0.58	0.37	0.86	0.99	1.36	0.84	0.00	1.18	1.05
F15-Fluid-pressure actuators	0.42	0.36	0.00	1.22	0.97	0.00	0.96	0.00	0.41
F16-Engineering elements	1.14	0.95	0.91	1.03	1.15	0.90	1.15	0.79	0.74
F21-Lighting	1.26	1.55	0.75	0.58	0.89	1.23	1.17	0.71	0.76
F23-Combustion apparatus	0.92	0.58	1.24	1.06	0.89	0.77	0.93	1.06	1.21
F24-Heating	0.76	1.16	1.13	0.95	0.73	0.94	0.42	1.12	1.21
F25-Refrigeration or cooling	0.00	0.85	1.08	0.87	1.05	1.21	1.43	1.22	1.14
F26-Drying	1.18	0.74	0.00	1.19	0.89	1.01	1.11	1.35	1.02
F28-Heat exchange	0.00	0.97	1.11	0.88	1.05	0.81	1.14	0.72	1.23
F41-Weapons	0.41	0.78	1.09	0.45	0.54	1.28	0.00	1.43	1.09
F42-Ammunition	0.65	0.00	0.00	1.13	0.27	1.41	1.67	1.40	0.52
G01-Measuring	0.44	0.88	0.92	0.77	1.32	0.88	0.49	0.60	1.18
G02-Optics	0.86	1.27	0.90	1.12	0.72	0.97	1.13	0.99	1.12
G03-Photography	0.58	0.79	0.00	0.71	0.00	1.30	1.16	1.08	1.30
G04-Horology	0.69	1.44	0.99	0.75	0.29	1.30	0.00	1.01	1.22
G05-Controlling	0.62	0.76	0.91	1.14	0.95	0.98	0.00	0.93	1.24
G06-Computing	0.61	0.64	0.93	0.82	0.82	1.02	0.96	0.83	1.38
G07-Checking-devices	0.47	0.59	1.26	0.92	1.04	0.79	1.03	0.82	1.30
G08-Signalling	0.49	0.77	1.06	0.80	0.79	1.01	1.28	1.30	1.28
G09-Educating	0.70	1.07	1.18	0.78	0.64	1.11	0.82	1.22	1.26
G10-Musical instruments	0.28	1.04	0.87	0.86	1.36	0.58	0.73	0.98	1.17
G11-Information storage	0.67	0.90	0.96	0.95	0.00	0.30	1.27	1.19	1.40
H01-Basic Electr. elements	0.76	1.13	0.47	0.86	0.69	1.28	1.63	1.06	0.99
H02-Electric power	0.66	0.57	0.72	0.92	0.72	1.23	1.29	0.84	1.28
H03-Basic electronic circuitry	0.51	0.44	0.48	0.38	1.30	1.18	0.00	0.79	1.35
H04-Electric communication	0.50	0.79	1.02	0.68	0.75	0.86	1.00	0.69	1.45
H05-Electric techniques	1.16	1.14	0.83	0.72	1.08	0.86	1.49	0.58	1.23
No. of fields with *RTA* > 1	17	35	31	33	29	38	41	39	39
CV (*RTA*)	0.60	0.43	0.49	0.26	0.34	0.31	0.70	0.42	0.36

References

Archibugi, D. and Pianta, M. (1992) *The Technological Specialization of Advanced Countries*, Dordrecht: Kluwer Academic Publishers.

Asheim, B. and Gertler, M. (2004) 'Understanding Regional Innovation Systems', in: Jan Fagerberg, David Mowery and Richard Nelson *Handbook of Innovation*, Oxford: Oxford University Press, pp. 77–86.

Asheim, B. and Isaksen, A. (1997) 'Location, Agglomeration and Innovation', *Towards Regional Innovation Systems in Norway?* European Planning Studies, 5 (3), pp. 299–330.

Bathelt, H., Malmberg, A. and Maskell, P. (2004) 'Clusters and Knowledge: Local Buzz, Global Pipelines and the Process of Knowledge Creation', *Progress in Human Geography*, 28 (1), pp. 31–56.

Breschi, S. and Lissoni, F. (2005) 'Knowledge Networks from Patent Data', in: H.F. Moed, W. Glänzel and U. Schmoch (eds) *Handbook of Quantitative Science and Technology Research: The Use of Publication and Patent Statistics in Studies of S&T Systems*, Dordrecht, Boston, London: Kluwer Academic Publishers, pp. 613–44.

Callon, M. (1992) 'The Dynamics of Techno-economic Networks', in: R. Coombs, P.P. Saviotti and V. Walsh (eds) *Economic and Social Analysis of Technology: Technological Change and Company Strategies: Economic and Sociological Perspectives.* London: Academic Press.

Cantner, U., Gaffard, J.-L. and Nesta, L. (2008) 'Schumpeterian Perspectives on Innovation, Competition and Growth', *Journal of Evolutionary Economics*, 18 (3–4), pp. 291–3.

Carlsson, B. (ed.) (1995) *Technological Systems and Economic Performance: The Case of Factory Automation: Economics of Science, Technology and Innovation, vol. 5.* Dordrecht: Kluwer Academic Publishers.

Carlsson, B. and Stankiewitz, R. (1995) 'On the Nature, Function and Composition of Technological Systems', in: B. Carlsson (ed.) *Technological Systems and Economic Performance: The Case of Factory Automation: Economics of Science, Technology and Innovation, vol. 5*, Dordrecht: Kluwer Academic Publishers.

Cooke, P., Gomez Uranga, M. and Etxebarria, G. (1997) 'Regional Innovation Systems: Institutional and Organisational Dimensions', *Research Policy*, 26 (4/5), pp. 475–91.

Doloreux, D. and Parto, S. (2004) *Regional Innovation Systems: A Critical Review*, Canada: University of Québec.

Edquist, C. (1997) *Systems of Innovation: Technologies, Institutions and Organizations. Science, Technology and the International Political Economy Series*, London: Pinter.

Freeman, C. (1987) *Technology Policy and Economic Performance: Lessons from Japan*, London: Pinter.

Frenken, K., van Oort, F. and Verburg, T. (2007) 'Related Variety, Unrelated Variety and Regional Economic Growth', *Regional Studies: The Journal of the Regional Studies Association*, 41, pp. 685–97.

Fritsch, M. and Stephan, A. (2005) 'Regionalization of Innovation Policy: Introduction to the Special Issue: Regionalization of Innovation Policy', *Research Policy*, 34 (8), pp. 1123–7.

Howells, J. (2005) 'Innovation and Regional Economic Development: A Matter of Perspective?: Regionalization of Innovation Policy', *Research Policy*, 34 (8), pp. 1220–34.

Jacobs, J. (1969) *The Economy of Cities*, New York: Vintage Books.

Kuhlmann, S. (2005) *Strategic Intelligence for Research Policy*, Manchester, UK: Manchester Business School.

Lam, A. (2000) 'Tacit Knowledge, Organizational Learning and Innovation: A Societal Perspective', *Organization Studies*, 21 (3), pp. 487–513.

Laranja, M., Uyarra, E. and Flanagan, K. (2008) 'Policies for Science, Technology and Innovation: Translating Rationales into Regional Policies in a Multi-level Setting', *Research Policy*, 37 (5), pp. 823–35.

Lundvall, B.-Å. (1992) *National Systems of Innovation: Towards a Theory of Innovation and Interactive Learning*, London: Pinter.

Malerba, F. (2002) 'Sectoral Systems of Innovation and Production', *Research Policy*, 31 (2), pp. 247–64.

Malmberg, A. (1997) 'Industrial Geography: Location and Learning', *Progress in Human Geography*, 21 (4), pp. 553–8.

Malmberg, A. and Maskell, P. (2002) 'The Elusive Concept of Localization Economies: Towards a Knowledge-based Theory of Spatial Clustering', *Environment and Planning A*, 34, pp. 429–49.

Maskell, P. and Malmberg, A. (1999) 'Localized Learning and Industrial Competitiveness', *Cambridge Journal of Economics*, 23, pp. 167–85.

Moed, H.F., Glänzel, W. and Schmoch, U. (eds) (2005) *Handbook of Quantitative Science and Technology Research: The Use of Publication and Patent Statistics in Studies of S&T Systems*, Dordrecht, Boston, London: Kluwer Academic Publishers.

Nelson, R.R. (1993) *National Innovation Systems: A Comparative Analysis*, New York: Oxford University Press.

Nesta, L. and Patel, P. (2005) 'National Patterns of Technology Accumulation: Use of Patent Statistics', in H.F. Moed, W. Glänzel and U. Schmoch (eds), *Handbook of Quantitative Science and Technology Research: The Use of Publication and Patent Statistics in Studies of S&T Systems*, Dordrecht, Boston, London: Kluwer Academic Publishers, pp. 531–52.

Owen-Smith, J. and Powell, W.W. (2004) 'Knowledge Networks as Channels and Conduits: The Effects of Spillovers in the Boston Biotechnology Community', *Organization Science*, 15 (1), pp. 5–21.

Porter, A.L. and Newman, N.C. (2005) 'Patent Profiling for Competitive Advantage', in: H.F. Moed, W. Glänzel and U. Schmoch (eds), *Handbook of Quantitative Science and Technology Research: The Use of Publication and Patent Statistics in Studies of S&T Systems*, Dordrecht, Boston, London: Kluwer Academic Publishers, pp. 587–612.

Porter, M. (1998) 'Clusters and the New Economics of Competition', *Harvard Business Review*, pp. 77–90.

Pyka, A., Cantner, U. and Hanusch, H. (2000) 'Horizontal Heterogeneity, Technological Progress and Sectoral Development', in: U. Cantner, H. Hanusch and S. Klepper (eds) *Economic Evolution, Learning, and* Complexity, Heidelberg: Physica-Verlag, pp. 73–96.

Romer, P.M. (1990) 'Endogenous Technological Change', *Journal of Political Economy*, 98, pp. 71–102.

Rondé, P. and Hussler, C. (2005) 'Innovation in Regions: What does Really Matter?: Regionalization of Innovation Policy', *Research Policy*, 34 (8), pp. 1150–72.

Schmoch, U. and Gauch, S. (2004) *Innovationsstandort Schweiz: Eine Untersuchung mit Hilfe von Patent- und Markenindikatoren*, Fraunhofer ISI.

Schmoch, U., Laville, F., Patel, P. and Frietsch, R. (2003) *Linking Technology Areas to Industrial Sectors* (Report to the European Commission), Karlsruhe: Fraunhofer ISI.

Online, available at: http://vg00.met.vgwort.de/na/39496c84eddf070704e9?l=http://publica.fraunhofer.de/eprints/urn:nbn:de:0011-n-205714.pdf.

Solow, R.M. (1956) 'A Contribution to the Theory of Economic Growth', *Quarterly Journal of Economics*, 70, pp. 65–94.

Statistik Austria (2009) *Statistik Austria – Innovation im Unternehmenssektor.* Online, available at: www.statistik.at/web_de/statistiken/forschung_und_innovation/innovation_im_unternehmenssektor/index.html accessed 23 July 2009.

Storper, M. (1997) *The Regional World*, New York: Guilford Press.

Temple, J. (1999) 'The New Growth Evidence', *Journal of Economic Evidence*, 37, pp. 112–56.

7   Innovation is small

SMEs as knowledge explorers and exploiters

Bjørn Asheim

Introduction

Small and Medium Enterprises (SMEs) are a heterogeneous group of firms. The majority is to be found in traditional, less knowledge intensive sectors such as furniture and textile as well as in more skilled based engineering industries producing machinery and tools. These firms are in best cases exploiters of knowledge coming from external R&D institutes and universities, but most commonly their production is either based on knowledge from the client firms (in the case of capacity subcontractors) or on their skilled workforce (if they are specialized suppliers or if they produce for the final market). However, one also finds SMEs which are both knowledge explorers and exploiters. These firms are new emerging firms in knowledge intensive sectors such as biotech and nanotech which base their production on newly created knowledge from universities in the case of university spin-offs, or in collaboration with universities in case of knowledge based entrepreneurship. Finally, one finds SMEs in businesses such as fashion, design and media, where knowledge for innovation is explored and exploited in quite different ways compared to the other types of SMEs. Thus, the activities of these different types of SMEs reflect their dependence on different knowledge bases, which is important to recognize to understand the way they carry out knowledge creation and innovation. In this chapter we distinguish between analytical, synthetic and symbolic knowledge bases (Asheim and Gertler 2005, Asheim and Coenen 2005, Asheim *et al.* 2007). Such a heterogeneous group of firms represents different challenges with respect to what kind of innovation support is adequate and efficient for promoting innovativeness and competitiveness which, consequently, require a fine-tuned innovation policy directed to the needs of the respective types of SMEs.

Regional innovation policy towards SMEs

What kind of regional policy should be implemented to promote innovation in SMEs? Research carried out in the SMEPOL project – SME policy and the regional dimension of innovation (Asheim *et al.* 2003) – demonstrated the need for a more system-oriented as well as a more pro-active innovation-based regional policy. In the project, SME innovation policy tools were classified in two dimensions, resulting in a four-quadrant table (Table 7.1). The table distinguishes between two main aims of the support tools. Some tools aim at giving firms access to resources that they lack to carry out innovation projects, i.e. to increase the innovation capacity of firms by making the necessary resource inputs available, such as financial support for product development, help to contact relevant knowledge organizations or assistance in solving specific technological problems, where the absorptive capacity of the firm is critical. The other type of instruments have a larger focus on learning, trying to change behavioural aspects, such as the innovation strategy, management, mentality or the level of awareness in firms, where the skill levels of the workforce are a major determining factor of the outcome.

An appropriate way to design and implement an instrument aimed at assigning lacking resources to firms is, thus, to do it according to a learning-to-innovate framework. In line with this perspective the objective of policy instruments is not solely to provide scarce resources (such as financial assistance) to innovating firms per se but also to promote learning about R&D and innovation and the acquisition of new routines within firms, where highly skilled people and adequate skill provision in the regions are critical resources in order to increase the absorptive capacity. Lack of demand is often a bottleneck for financial incentives to innovation activity, i.e. that firms initially do not see the need to innovate, or alternatively, that firms do not have the capability to articulate their need for innovation. Some policy instruments should, therefore, also attempt to enhance demand for initial innovation activity of firms (i.e. apply a learning perspective), and, thus, must include an explicit behavioral aspect with an ultimate policy target of promoting the innovation activity of enterprises.

The other dimension includes the target group of instruments. Some tools focus on innovation and learning within firms, to lower the innovation barriers of firms, such as lack of capital or technological competence. Other instruments to a larger extent have regional production and innovation systems as their target group, aiming at achieving externalities or synergies from complementarities within the regions. The barriers may for example be lack of user–producer interaction or lack of relevant competence in the regional knowledge organizations to support innovation projects.

Table 7.1 Regional innovation policy: a typology

	Support: financial and technical	*Behavioral change: learning to innovate*
Firm-focused	Financial support Brokers	Mobility schemes
System-focused	Technology centers	Regional innovation systems

Source: Asheim *et al.* 2003.

Obviously, all tools are not as relevant for all types of SMEs mentioned in the introduction. Mobility schemes would be most advantageous for traditional firms which need to increase their absorptive capacity, as would brokers to link them up with R&D institutes and technology centers to carry out 'real services' which the firms do not have the capacity to do at their own premises. Financial support is of course welcome to all types of SMEs, but they will need it in different forms. While new emerging firms would need risk capital in the form of venture capital and capital from business angels or from similar public agencies, as would also be the case for SMEs in media and fashion industries, traditional SMEs would be looking for more patient capital to finance investments in new production facilities from private banks or public funding institutions for SMEs. What is a relevant tool for all types of SMEs is being integrated in a regional innovation system (RIS), but also here the RIS have to be differentiated with respect to which type of SMEs needs innovation support. While new, emerging industries would be part of an RIS narrowly defined, the other types of SMEs will in addition also need to be surrounded by RIS broadly defined (see forthcoming sections). Once again, this reflects the different knowledge bases the types of SMEs depend on; the new, emerging firms depend on analytical, science based knowledge, the traditional on synthetic, engineering knowledge, and the fashion and media firms on symbolic knowledge (see next section).

Differentiated knowledge bases

Knowledge creation and innovation processes have become increasingly complex, diverse and interdependent in recent years. Therefore, the binary argument of whether knowledge is codified or tacit can be criticized for a restrictively narrow understanding of knowledge, learning and innovation (Johnson *et al.* 2002) and a need to go beyond this simple dichotomy can, thus, be identified. One way of doing this is to study the basic types of knowledge used as input in knowledge creation and innovation processes where a distinction can be made between 'synthetic', 'analytical', and 'symbolic' types of knowledge bases.

As a starting point distinction can be identified between two more or less independent and parallel forms of knowledge creation, 'natural science' and 'engineering science' (Laestadius 2000). Johnson *et al.* (2002: 250) refer to the Aristotelian distinction between on the one hand '*epistēmē*: knowledge that is universal and theoretical', and '*technè*: knowledge that is instrumental, context specific and practice related'. The former corresponds with the rationale for 'analysis' referring to understanding and explaining features of the (natural) world (natural science/know-why), and the latter with 'synthesis' (or integrative knowledge creation) referring to designing or constructing something to attain functional goals (engineering science/know-how) (Simon 1969). A main rationale of activities drawing on symbolic knowledge is creation of alternative realities and expression of cultural meaning by provoking reactions in the minds of consumers through transmission in an affecting sensuous medium (Table 7.2).

Table 7.2 Differentiated knowledge bases: a typology

Analytical (science based)	Synthetic (engineering based)	Symbolic (arts based)
Developing new knowledge about natural systems by applying scientific laws; *know why*	Applying or combining existing knowledge in new ways; *know how*	Creating meaning, desire, aesthetic qualities, affect, intangibles, symbols, images; *know who*
Scientific knowledge, models, deductive	Problem-solving, custom production, inductive	Creative process
Collaboration within and between research units	Interactive learning with customers and suppliers	Learning-by-doing, in studio, project teams
Strong codified knowledge content, highly abstract, universal	Partially codified knowledge, strong tacit component, more context-specific	Importance of interpretation, creativity, cultural knowledge, sign values; implies strong context specificity
Meaning relatively constant between places	Meaning varies substantially between places	Meaning highly variable between place, class and gender
Drug development	Mechanical engineering	Cultural production, design, brands

Source: Asheim and Gertler 2005; Asheim *et al.* 2007; Gertler 2008.

The distinction between these different knowledge bases takes specific account of the rationale of knowledge creation, the way knowledge is developed and used, the criteria for successful outcomes, and the strategies of turning knowledge into innovation to promote competitiveness, as well as the interplay between actors in the processes of creating, transmitting and absorbing knowledge. The knowledge bases contain different mixes of tacit and codified knowledge, codification possibilities and limits, qualifications and skills required by organizations and institutions involved as well as specific innovation challenges and pressures, which in turn help in explaining their different sensitivity to geographical distance and, accordingly, the importance of spatial proximity for knowledge creation (Asheim *et al.* 2009). Thus, the dominance of one mode arguably has different spatial implications for the knowledge interplay between actors than another mode of knowledge creation. Analytical knowledge creation tends to be less sensitive to distance-decay facilitating global knowledge networks as well as dense local collaboration. Synthetic and symbolic knowledge creation, on the other hand, has a tendency to be relatively more sensitive to proximity effects between the actors involved, thus favoring local collaboration (Moodysson *et al.* 2008).

As this threefold distinction refers to ideal-types, most activities are in practice comprised of more than one knowledge base. The degree to which certain knowledge bases dominate, however, varies and is contingent on the characteristics of the firms and industries as well as between different type of activities (e.g. research and manufacturing). According to Laestadius (2007) this approach also makes it unnecessary to classify some types of knowledge as more advanced, complex, and sophisticated than other knowledge, or to consider science based (analytical) knowledge as more important for innovation and competitiveness of firms, industries and regions than engineering based (synthetic) knowledge or artistic based (symbolic) knowledge.

Modes of innovation and regional innovation systems

This view corresponds with the ideas of Lorenz and Lundvall (2006) about different but complementary 'modes of innovation'. One the one hand we can talk about a broad definition of the mode of innovation as D(oing), U(sing) and I(nteracting) relying on informal processes of learning and experience-based know-how. The DUI mode is a market or demand driven model based more on competence building and organizational innovations and synthetic and symbolic knowledge producing mostly incremental innovations. Such a mode is typically found in non-R&D based economies (e.g. Denmark). On the other hand one finds a narrower definition of the mode of innovation as S(cience), T(echnology) and I(nnovation) based on the use of codified analytical, scientific knowledge, which is a science push/supply driven high-tech strategy able to produce radical innovations (e.g. found in Finland and Sweden).

These two modes of innovation will also be differently manifested with regard to regional innovation systems and clustering. Regional innovation systems can be defined as narrow or broad (Asheim and Gertler 2005). A regional innovation system broadly defined includes the wider setting of organizations and institutions affecting and supporting learning and innovation in a region. This type of system is less systemic than the narrowly defined types of innovation systems. Firms mainly base their innovation activity on interactive, localized learning processes stimulated by geographical, social and cultural/institutional proximity, without much formal contact with knowledge creating organizations (i.e. R&D institutes and universities) (Asheim and Gertler 2005). A narrow definition of innovation systems on the other hand primarily incorporates the R&D functions of universities, public and private research institutes and corporations, reflecting a top-down model of science and technology policies. The narrowly defined innovation system corresponds to the STI mode of innovation mentioned above, while the more broadly defined system is more easily accommodated by the DUI mode (Lundvall 1992).

This distinction is helpful in order to avoid a too one-sided focus on promoting science-based innovation of high-tech firms at the expense of the role of learning and experience-based, user-driven innovation. However, it also indicates limits of such innovation strategies in a longer term perspective and, thus, emphasizes the need for firms in traditional manufacturing sectors and services more generally to link up with sources of codified knowledge in distributed knowledge networks (Berg Jensen *et al.* 2007).

An example of this could be SMEs which may have to supplement their informal knowledge, characterized by a high tacit component (i.e. the DUI mode), with competence arising from more systematic research and development (i.e. the STI mode) in order to avoid being locked-in in price squeezing, low road competition from low cost countries. Thus, in the long run, it will be problematic for most firms to rely exclusively on informal localized learning. They must also gain access to wider pools of both scientific and engineering knowledge on a national and global scale (Asheim *et al.* 2003). However, the DUI-based type of innovations will remain the key to their competitive advantage, as strong tacit, context specific knowledge components, found in e.g. engineering knowledge dominating the DUI mode, are difficult to copy by other firms in different contexts, and, thus will be the basis for sustaining the firms' and regions' competitive advantage also in the long run (Porter 1998).

New research confirms that combining the two modes of innovation seems to be most efficient with regard to improving economic performance and competitiveness, i.e. firms that have used the STI-mode intensively may benefit from paying more attention to the DUI-mode and vice versa (Berg Jensen *et al.* 2007). The ability of firms to search and combine knowledge from different sources seems to be stronger associated with innovativeness than either interfacing predominantly with customers or suppliers applying a DUI mode of innovation, or with research system actors in STI oriented processes. (Laursen and Salter 2006) Thus, on the firm level these two modes of innovation are coexisting, but they will be applied in different combinations depending on the dominating knowledge base(s) of the regional industry as well as the absorptive capacity and cognitive distance between actors on the firm and system levels. The unanswered question is, however, how the capacity of combining the two modes of innovation can be further diffused to, and implemented, in less innovative firms as well as on the regional level.

SMEs and regional innovation systems

An explicit conceptual clarification of the linkage between on the one hand clusters of different types of SMEs and on the other regional innovation systems has so far received relatively little attention in the literature. By distinguishing between the firms' knowledge base and forms of linkages with the regional innovation system, the different industrial development paths of clusters where regional innovation systems are built in order to support innovation in already established industries, and the establishment of relations between clusters and regional innovation system from the formation of the cluster in order to promote emerging industries based on new knowledge, could be explained in a more systematic way. Thus, there are different logics behind constructing regional innovation systems contingent on the knowledge base of the industry it addresses as well as on the regional knowledge infrastructure which is accessible (Asheim 2007). In a territorially embedded regional innovation system, the emphasis lies on the localized, path-dependent inter-firm learning processes often involving innovation based on synthetic knowledge. The role of the regional knowledge infrastructure is therefore mainly directed to industry-specific, hands-on services and concrete, short-term problem solving. In a regionalized national innovation system, R&D and scientific research take a much more prominent position. Innovation builds primarily on analytic knowledge. Linkages between existing local industry and the knowledge infrastructure are however weakly developed. Instead they hold the potential to promote new industries at the start of their industrial and technological life cycle. In this, the role of the regional(ized) knowledge infrastructure is a very central one as it provides the cornerstone for cluster development (through the precarious task of commercializing science). Similar to the regionalized national innovation system, in the regionally networked innovation system the knowledge infrastructure plays an indispensible role, however more territorially embedded. But in contrast to it, the cluster is not wholly science-driven but represents a combination of a science and market-driven model. In comparison to the territorially embedded regional innovation system, the networked RIS often involves more advanced technologies combining analytic and synthetic knowledge as well as having better developed and more systemic linkages between universities and local industry. While territorially embedded RIS are often found in mature industries and regionalized national innovation systems found in emergent industries, networked regional innovation systems can typically support various types of industries in different life cycle phases. Firms and knowledge infrastructure form a dynamic ensemble, combining *ex-post* support for incremental problem-solving and *ex ante* support to counter technological and cognitive lock-ins. Table 7.3 shows combinations of different types of regional innovation systems and knowledge bases.

Cooke (2001) has, based on studies of the biotechnology industry in the United Kingdom, the United States and Germany, introduced a distinction between the traditional RIS (which he refers to as the institutional regional innovation system – IRIS) and the new entrepreneurial regional innovation

Table 7.3 Types of regional innovation systems and knowledge bases

	Analytical/scientific knowledge	Synthetic/engineering knowledge	Symbolic/creative knowledge
Embedded (grassroots RIS		IDs in Emilia-Romagna (machinery)	'Advertising village' – Soho (London)
Networked (network RIS)	Regional clusters – regional university (wireless in Aalborg)	Regional clusters – regional technical university (mechanical in Baden-Württemberg)	Barcelona as the design city
Regionalized national (dirigiste RIS)	Science parks/ technopolis (biotech, IT)	Large industrial complex (Norwegian oil and gas related industry)	

system (ERIS). The traditional IRIS (more typical of German regions or regions in the Nordic countries whose leading industries draw primarily from synthetic knowledge bases) is characterized by the positive effects of systemic relationships between the production structure and the knowledge infrastructure embedded in networking governance structures regionally and supporting regulatory and institutional frameworks on the national level. In contrast ERIS (found in the United States, United Kingdom and other Anglo-American economies) lacks these strong systemic elements, and instead gets its dynamism from local venture capital, entrepreneurs, scientists, market demand and incubators to support innovation that draws primarily from an analytical knowledge base. Thus, Cooke calls this a 'venture capital driven' system. Such a system will of course be more flexible and adjustable and, thus, will not run the same risk of ending up in 'lock-in' situations as traditional regional innovation systems caught in path-dependency on old technological trajectories. On the other hand, new economy innovation systems do not seem to have the same long-term stability and systemic support for historical technological trajectories, raising important questions about their long-term economic sustainability.

How to combine the DUI and STI modes of innovation

As was mentioned in a previous section combining the two modes of innovation seems to be the most efficient strategy for firms and regions to improve their innovativeness and economic performance. Firms that have used the STI mode intensively may benefit from paying more attention to the DUI mode and vice versa (Lorenz and Lundvall 2006, Berg Jensen *et al.* 2007). In this way, on the firm levels these two modes of innovation can (and should) co-exist, but they will be applied in different combinations depending on the dominating knowledge base(s) of the regional industry.

Here the perspective of cognitive distance becomes crucial (Nooteboom 2000). If the cognitive distance between the two modes of innovation is perceived by key actors to be too wide, then it will not be possible to combine them. They will be seen as incompatible alternatives rather than complementary modes. There will be a lack of absorptive capacity within firms and regions to

Table 7.4 ERIS and IRIS types of regional innovation systems

ERIS (entrepreneurial RIS)	*IRIS (institutional RIS)*
Venture capital driven	R&D driven/interactive learning
Serial start-ups	Networked
Market focused	Technology/production focused
Radical innovations	Incremental innovations
Incubators (university–industry relations)	Clusters
Initial public offerings	Bank borrowing
Knowledge-based economy	Learning economy

Source: Cooke 2001.

acknowledge and appreciate the potential gains of the other mode of innovation as well as to access and acquire the necessary competence for combining them. There are, however, two key 'bridging mechanisms' which could assist in achieving an optimal cognitive distance as a necessary condition for combining the two modes. The first of these deals with understanding that the STI mode is not only limited to an analytical knowledge base, but also includes synthetic and symbolic knowledge bases. In the case of the synthetic knowledge base this can be illustrated by reference to applied research undertaken at (technical) universities, which clearly must be part of the STI mode, but operates on the basis of synthetic (engineering) knowledge (drawing on basic research at science departments of universities creating new analytical knowledge); while the case of symbolic knowledge can partly be substantiated by the new tendency of changing design education from being artisan based to be placed at universities with research based teaching, and partly by the steadily increasing research in game software and new media, which in some countries is located at new, specialized universities. This broadening of what constitute the STI mode of innovation shows that also activities based on synthetic and symbolic knowledge bases need to undertake new knowledge creation and innovation in accordance with an STI mode, and, thus, need systemic relations with universities or other types of R&D institutes (e.g. in a regional innovation system context). The other bridging mechanism is the recognition that partly learning is not only reproductive but can also be developmental, and partly the innovative potential that a learning work organization can display in being the operative context for such learning. Even the most science based company will obviously benefit from organizing its work in such a way that learning dynamics are created by giving their employees autonomy in their work.

The role of universities in promoting new and emerging firms

Universities play a strategic role in promoting the formation of new, emerging firms, and could in many ways be seen as drivers of the new regional knowledge economy. Universities have at least three roles relevant for the promotion of spin-offs and knowledge-based entrepreneurship:

1   As a provider of human capital. The educational part of universities' tasks of producing highly skilled people is by far their most important task. With respect to SMEs a discussion can be raised of how general or dedicated to the needs of a region the educational programs should be organized and it is important to find an optimal combination of dedicated, short terms needs and more long term, general educational requirements to avoid negative lock-in situations as well as to secure sufficient absorptive capacity to be able to access new knowledge.

2   As a key node in the knowledge exploration subsystem of regional innovation systems (RIS). This role stems from the so called 'third task' of the universities, i.e. to interact directly and serve society with its knowledge, skills

and competencies in addition to carrying out teaching and research. This third task ranges from creating high-tech firms based on newly creating knowledge at university departments and research centers, consulting for local industry, delivering advice for politicians and informing general public debates. This role of universities in the knowledge exploration subsystem could be called a *deepening* contribution from a provider of human capital to an orchestrator of regional innovation support (in the context of RIS) (Benneworth *et al.* 2009).

3   As a technological transfer agency bringing exogenous knowledge into the region. Universities are increasingly becoming of strategic importance for regional development in the knowledge economy by often being the only actor able to transfer global state-of-the-art science and technology into the region. In Finland universities have explicitly taken on this role, e.g. when participating in EU's Framework Programmes. Likewise universities play a key role in the general internationalization of research and education, e.g. for the latter in the context of the Bologna process. This role could be called a *widening* involvement of universities using their national and international networks to extend the knowledge, learning and innovation networks of regional actors (Benneworth *et al.* 2009).

The increased interest for and importance of the third task of universities could be described as a change from mainly taking on 'generative' to more and more engaging in 'developmental' roles Gunasekara (2006). Generative roles refer primarily to the provision of limited, discrete knowledge outputs such as scientific and technological information, equipment and instrumentation, skills or human capital, networks of scientific and technological capabilities and prototypes for new products and processes) in response to business or public sector demands (Benneworth *et al.* 2009).

In taking on developmental roles universities in contrast constructively interact with broader regional governance structures shaping 'the development of regional institutional and social capacities' (Gunasekara 2006: 730) and, thus, more directly promote regional economic development. Universities typically become involved in strengthening as well as creating new systemic connections within RISs, hopefully resulting in positive, long-lasting impacts on regional economic growth (Benneworth *et al.* 2009).

In addition to universities there are other HEIs (higher educational institutions) in a region. Of specific relevance for especially more traditional types of SMEs are institutions such as polytechnics (Fachhochschule in German). They train highly skilled workers and provide necessary intermediate technical skills and competence of importance for these workers to engaging in user-producer interactive learning and undertake developmental learning in learning work organizations. These kinds of skills and competencies are of key importance for having a working RIS broadly defined as a complement to a narrowly defined RIS where universities play the leading role in the knowledge exploration subsystem as described above.

120   B. Asheim

Conclusion

According to the World Economic Forum Growth Competitiveness Report Finland, Sweden and Denmark for five years (2004–9) have consistently been among the six highest ranking nations with Finland and Sweden, most years among the top three. This impressive performance of the Nordic countries is achieved with very different innovation policies and strategies. Finland has pursued a science-driven, high-tech oriented strategy focusing on radical product innovations, with especially good results in the ICT sector, and Sweden a technology-based strategy of process innovations and complex product improvements, with both countries ranking as the top two nations with respect to R&D investments (Sweden 4 percent and Finland 3.8 percent). Denmark has implemented a demand driven, market based strategy characterized by mostly non-R&D, incremental innovations heavily oriented towards consumer goods sectors (e.g. furniture), sometimes with a design orientation, but not as a general rule such as in 'made in Italy' products. While Sweden is dominated by large firms in high-tech (analytical knowledge) and mature sectors (synthetic knowledge), Finland in a combination of similar types of large firms and small high-tech firms, Denmark is dominated by SMEs mostly in traditional sectors (synthetic knowledge) but with a growing number of SMEs using symbolic knowledge. These empirical facts and theoretical perspectives have a very important policy implication in that there is no 'one size fits all' policy formula, i.e. no optimal or best way with respect to innovation policy promoting competitiveness and innovation in various industries in different regions and nations in a globalizing knowledge economy (Tödtling and Trippl 2005). Instead, innovation policies must be fine tuned to take into account actual differences in industrial structures and social and institutional environments. This is not least the case with respect to the types of regional innovation policy to be developed and implemented to promote and support innovation in SMEs.

References

Asheim, B.T. (2007) 'Differentiated Knowledge Bases and Varieties of Regional Innovation Systems', *Innovation – The European Journal of Social Science Research*, 20 (3), pp. 223–41.
Asheim, B.T. and Coenen, L. (2005) 'Knowledge Bases and Regional Innovation Systems: Comparing Nordic Clusters', *Research Policy*, 34, pp. 1173–90.
Asheim, B.T. and Gertler, M.S. (2005) 'The Geography of Innovation: Regional Innovation Systems', in Fagerberg, J., Mowery, D. and Nelson, R. (eds) *The Oxford Handbook of Innovation*, Oxford: Oxford University Press, pp. 291–317.
Asheim, B.T., Isaksen, A., Nauwelaers, C., and Tödtling, F. (eds) (2003) *Regional Innovation Policy for Small–Medium Enterprises*, Cheltenham: Edward Elgar.
Asheim, B.T., Coenen, L., Moodysson, J., and Vang, J. (2007) 'Constructing Knowledge-based Regional Advantage: Implications for Regional Innovation policy', *International Journal of Entrepreneurship and Innovation Management*, 7 (2/3/4/5), pp. 140–55.
Asheim, B.T., Ejermo, O., and Rickne, A. (eds) (2009) 'Introduction: When is Regional "Beautiful"? Implications for Knowledge Flows, Entrepreneurship and Innovation', *Industry and Innovation*, 16 (1), pp. 1–9.

Benneworth, P., Coenen, L., Moodysson, J., and Asheim, B.T. (2009) 'Exploring the Multiple Roles of Lund University in Strengthening the Scania Regional Innovation System: Towards Institutional Learning?', *European Planning Studies*, 17 (11), pp.1645–64.

Berg Jensen, M., Johnson, B., Lorenz, E., and Lundvall, B.-Å (2007) 'Forms of Knowledge and Modes of Innovation', *Research Policy*, 36, pp. 680–93.

Cooke, P. (2001) 'Regional Innovation Systems, Clusters, and the Knowledge Economy', *Industrial and Corporate Change*, 10 (4), pp. 945–74.

Gertler, M. (2008) 'Buzz without Being There? Communities of Practice in Context', in Amin, A. and Roberts, J. (eds) *Community, Economic Creativity and Organization*, Oxford: Oxford University Press.

Gunasekara, C. (2006) 'Reframing the Role of Universities in the Development of Regional Innovation Systems', *Journal of Technology Transfer* 31 (1), pp. 101–11.

Johnson, B., Lorenz, E., and Lundvall, B.-Å. (2002) 'Why All This Fuss About Codified and Tacit Knowledge?', *Industrial and Corporate Change*, 11, pp. 245–62.

Laestadius, S. (2000) 'Biotechnology and the Potential for a Radical Shift of Technology in Forest Industry', *Technology Analysis and Strategic Management*, 12, pp. 193–212.

Laestadius, S. (2007) 'Vinnväxtprogrammets teoretiska fundament', in Laestadius, S. Nuur, C., and Ylinenpää, H. (eds) *Regional växtkraft i en global ekonomi: Det svenska Vinnväxtprogrammet*, Stockholm: Santerus Academic Press, pp. 27–56.

Laursen, K. and Salter, A. (2006) 'Open for Innovation: The Role of Openness in Explaining Innovation Performance among UK Manufacturing Firms', *Strategic Management Journal*, 27, pp. 131–50.

Lorenz, E. and Lundvall, B.-Å. (eds) (2006) *How Europe's Economies Learn: Coordinating Competing Models*, Oxford: Oxford University Press.

Lundvall, B.-Å. (ed.) (1992) *National Systems of Innovation: Towards a Theory of Innovation and Interactive Learning*, London: Pinter.

Moodysson, J., Coenen, L., and Asheim, B. (2008) 'Explaining Spatial Patterns of Innovation: Analytical and Synthetic Modes of Knowledge Creation in the Medicon Valley Life Science Cluster', *Environment and Planning A*, 40 (5), pp. 1040–56.

Nooteboom, B. (2000) *Learning and Innovation in Organizations and Economies*, Oxford: Oxford University Press.

Porter, M. (1998) 'Clusters and the New Economics of Competition', *Harvard Business Review*, November–December, pp. 77–90.

Simon, H. (1969) *The Sciences of the Artificial*, Cambridge, MA: MIT Press.

Tödtling, F. and Trippl, M. (2005) 'One Size Fits All? Towards a Differentiated Regional Innovation Policy Approach', *Research Policy*, 34 (8), pp. 1203–19.

8 Reverse knowledge transfer and its implications for European policy

Rajneesh Narula and Julie Michel

Introduction

The internationalization of R&D is by no means a new phenomenon. The growing intensity and spread of R&D activities by MNEs in a systematic way dates back to the post-Second World War period, and reflects – with a lag – how the MNE as a whole has evolved towards complex and interdependent organizational forms to undertake international business. Thus, until the early 1990s, the trend towards more intensive and complex R&D activities abroad was more of an exception, and limited to relatively few large and organizationally sophisticated firms. This has now started to change. Indeed, the growing spread and intensity of R&D is regarded as one of the central and most dynamic elements of the process of globalization, and is now much more commonplace. According to UNCTAD (2005), R&D expenditures of foreign affiliates worldwide more than doubled from 29 billion dollars to 67 billion dollars between 1993 and 2002. Between 1995 and 2003, R&D expenditures of foreign-controlled affiliates increased twice as fast as their turnover (OECD 2008). It remains the case that much of the R&D undertaken by MNEs abroad tends to be of a relatively low intensity, and primarily aimed at adapting technology developed in their home country for application by their foreign affiliates in response to local conditions and market needs – anecdotal evidence suggests perhaps 70–80 per cent of all overseas R&D expenditures is of such 'asset-exploiting nature'. Nonetheless, it is no longer uncommon that even relatively small MNEs engage in overseas R&D, and it is increasingly common that even resource-constrained or traditionally ethnocentric firms now seek opportunities to engage in 'asset-augmenting' activities, whether on their own, or in collaboration with other actors in the host economy.

Managing and organizing such a complex web of activities represents a managerial challenge to firms, and has been the subject of a growing literature (see e.g. Gassmann and von Zedtwitz 1999, Cantwell and Janne 1999, Zanfei 2000, Foss and Pedersen 2002, Criscuolo and Narula 2007, Yang *et al.* 2008). Much less has been said about how governments need to respond to these circumstances, although the dangers of the 'hollowing out' of the innovation capacity of home countries has been a matter of some concern even as early as the late

1970s (Lall 1979, Mansfield *et al.* 1979). More recently, R&D international-ization has been seen to be a signal of weakness of the technological competit-iveness of the home country, implying that the domestic innovation system does not meet the technological needs of firms in certain industries (Narula and Zanfei 2004).

In this chapter we argue that although asset-augmenting activities is seen as primarily benefitting the MNE, home country innovation systems can also benefit from reverse knowledge transfer. Policy makers need to promote these linkages and flows, rather than seeing R&D internationalization as a threat to the home economy. New knowledge developed abroad by firms can and should be encouraged to be transferred to the rest of the firm and to the local environment of the home country.

The importance of new knowledge sources for MNEs

Few MNEs can sustain their innovative capabilities by depending exclusively upon the innovation system of their home countries. Firms need access to know-ledge abroad, as the home country cannot develop all the technologies needed by the firm. This stems from the fact that innovation systems and technological specialization of nations evolve only very gradually, especially in new sectors (Narula and Zanfei 2004). Besides, cognitive limits to resources in smaller coun-tries exist that limit the breadth of technological expertise possible. These systems change more slowly than the needs of firms and, as a result, companies must seek to acquire the knowledge they need in foreign locations. MNEs can take advantage of local capacities of host countries in terms of technology stocks, research programmes and trajectories and creative human capital (Pearce *et al.* 2008). Indeed, despite the economic and technological convergence associ-ated with international production (through imitation and diffusion of knowledge and practices), there are distinct patterns of specialization among countries which MNEs are able to take advantage of, and build interactions with and between their subsidiaries. The kinds of specializations of a specific host loca-tion can be explained by the fact that the innovative potential of a region depends on relatively immobile factors, such as the highly skilled workforce, the poten-tial for spillovers, niche markets, research institutions and regulation (Hotz-Hart 2000). Despite globalization, a perfect convergence cannot be expected between the needs of firms and the technological resources associated with a given loca-tion. Therefore, there remains a variety and diversity among locations and an enduring pattern of regional specialization. Thus MNEs – which need access to a variety of technological areas – seek to tap into and integrate their R&D activ-ities with these location-specific assets if they are to sustain their innovative capability.

However, the desire to acquire new knowledge sources has to be tempered with the benefits of centralization. *Ceteris paribus*, firms prefer to concentrate their R&D activities at home. In doing so, MNEs benefit from economies of scale for these costly activities, and maintain strategic control of their R&D

investments by being close to headquarters. Centralizing R&D also decreases the costs of communication and coordination, which are non-trivial (Criscuolo and Narula 2007). The complexities of innovation and the complexities of building relationships with a large variety of external actors also results in considerable systemic and institutional inertia which makes it hard for firms to easily relocate such activities (Narula 2002). Firms tend to be risk averse, and the strategic importance of R&D means that firms are hesitant to take risks by relocating their R&D. Thus, the high costs of integration in innovation systems of 'new' host countries, compared to the relatively low marginal costs of staying in the innovation system of the country of origin creates an inertia that makes companies hesitant to internationalize their R&D (Narula 2002).

The embedding of MNEs and the stickiness of knowledge

Despite the advantages of concentrating their R&D activities in a few locations, firms increasingly need to be physically present abroad – whether in response to pressures to adapt their products and services to specific markets, or to access locationally bound foreign knowledge. Although some aspects of knowledge can indeed be acquired without physically establishing abroad, MNEs have greater access to the local knowledge in systems of innovation by doing so (Song and Shin 2008). 'By going directly to the places with more expertise in a given technological field the firm is able to penetrate at a lower cost such networks' (van Pottelsberghe de la Potterie and Guellec 2001). Knowledge is not always 'transportable' and requires a physical presence to be more easily accessible (see Criscuolo and Verspagen 2008 for a review). Even if ideas and innovative solutions are spreading rapidly through space, the tacit knowledge that leads to these ideas cannot be distributed across large distances without moving the people who possess such knowledge (Inkpen 2008). Tacit knowledge is often better transferred between people through face-to-face contacts or other person-based communication mechanisms, than codified communication (Piscitello and Rabbiosi 2007). Therefore, as MNEs need to be innovative and acquire constantly new knowledge, they have to be present where this knowledge circulates. Tacit knowledge is much more difficult to exchange or trade, and thus tends to be sticky and geographically less mobile. In industries where the tacit aspect is considerable, *ceteris paribus*, the propensity to geographically concentrate is higher (Iammarino and McCann 2006) than in sectors where the knowledge being exchanged is codifiable.

However, merely establishing a subsidiary in a host country is not sufficient to acquire the knowledge. Foreign companies may be unable to get access to tacit knowledge embedded in the regional interpersonal networks (Singh 2007). The subsidiary has to create links and relationships with other economic agents, and become part of the economic milieu of the host location, as knowledge is disseminated more quickly and easily when firms are embedded. Embedding a subsidiary is neither instantaneous nor costless, because of the effort required to acquire 'membership' of the relevant networks. The effect of interpersonal similarity, also known as '*homophily*' in the literature (which may be defined as a tendency to interact with similar others) facilitates the sharing of knowledge between individuals and within clusters (Makela *et al.* 2007). Furthermore, firms need to have something to share which other members of the agglomeration need. All these elements imply that MNEs should be present and must invest in these environments to be able to capture knowledge. The company must also communicate its ideas to be accepted by the local scientific community and to obtain the desired information (Porter 1993). This helps explain why the agglomeration of R&D activities in a few locations changes very slowly.

Of course, not all knowledge flows are intentional. Knowledge 'leaks' unintentionally, for instance when employees move from one firm to another, and these leakages are obviously greater when firms are collocated. The argument in favour of locating in close spatial proximity presumes that firms wish to benefit from and promote knowledge transfers. This is not always the case for two reasons. First, while all firms in principle seek to have positive inflows of knowledge and desire to benefit from unintended spillovers, few firms wish to be the source of unintended knowledge outflows. Although in the case of R&D (compared to sales or manufacturing) there is a greater active interest in seeking spillovers, this tendency will reflect the capabilities of the firm. Alcácer (2006) found that although R&D tends to be more concentrated relative to manufacturing and sales, firms with superior technologies are less keen of being physically proximate to other firms in the same industry (compared to less technologically competitive firms), regardless of the activity. In other words, firms may seek to *avoid* collocation of R&D to minimize leakages of valuable assets.

The importance of spillovers for MNEs

Thus, MNEs that seek to augment their existing competences tend to establish subsidiaries in regions where there are clusters of suppliers, clients, competitors, research institutions, universities or industry associations. As a large body of literature shows (for a competent review, see Iammarino and McCann 2006), firms agglomerate to take advantage of three types of spillovers: intra-industry, inter-industry and external sources of knowledge.

First, intra-industry spillovers are associated with the presence of many technologically active firms in a given sector and concentrated in an agglomeration. This concentration of firms in the same area creates externalities of specialization (as opposed to externalities of diversities due to the concentration of various industries). The companies are in cooperation and competition simultaneously, which can produce stable mechanisms of accumulation of collective knowledge. The competitive advantage of a system is created and maintained through a certain optimal level of rivalry between firms. The spatial concentration of firms can stimulate the intensity of the exchanges and collaboration between agents, thus creating a common attitude towards innovation. In addition, MNEs can monitor technological and competitive environment in a particular place (Doh *et al.* 2005).

Second, companies can install R&D activities in a location to benefit from inter-industry spillovers (see Jacobs 1970, 1986). These effects concern the externalities of diversity that promote the creation of new ideas *across* sectors and under the right circumstances create an increase in the scope of research.

Third, the efforts of firms to improve technology are supported by external sources of knowledge, often associated with location-specific knowledge infrastructure that provide quasi-public goods such as public research organizations, universities and industry associations (Asheim and Gertler 2005). These often tend to be spatially concentrated (Almeida and Kogut 1997, Saxenian 1994), and create opportunities for scientific and technological spillovers. However, the work on clusters emphasizes that although these opportunities are in principle available to all firms that are part of the spatial agglomeration, having access requires knowledge of informal institutions and time invested in being collocated with these actors, and are *not* automatically available to all firms (Tallman *et al.* 2004). A study by Schrader (1991) showed that the frequency of interactions between R&D employees of several firms has a positive impact on the frequency of innovations in these firms. Companies that do not interact with others risk missing opportunities and, as a result, the productivity of their innovation declines. Engineers from Sony, for example, had effectively isolated themselves in the 1990s, in the belief that ideas from outside the company were not good enough. They thus developed products such as cameras which were not compatible with various forms of memory used by clients (Hansen and Birkinshaw 2007). Firms tend to learn not just from their own experiences and employees, but also from their suppliers, partners, investors, scientists, inventors, customers, competitors and companies that are not necessarily in the same field. This exchange will allow the company to raise or deploy its own knowledge effectively.

The challenges of knowledge transfer

However, acquiring and internalizing knowledge derived from interactions with the host location's innovation system is just the first step (see Figure 8.1). Once these new competences are integrated inside the subsidiary, they need to be transferred to the rest of the MNE's operations. Furthermore, this cross-border integration of knowledge may also influence and upgrade the knowledge base of systems of innovation in other locations where the MNEs operate. Indeed, even though the home country of the MNE has tended to play a significant (and major) role as a net (and dominant) source of new technological competences for the MNE as a whole, there has been a considerable shift in the relative importance of the home country. That is, conventional knowledge transfer has tended to be from the home country to subsidiaries, with the subsidiaries acting as net exploiters of assets originally developed in the home base. However, a number of studies have highlighted the growing phenomenon of reverse knowledge transfer, with certain subsidiaries transferring knowledge in the reverse direction (Frost 1998, Håkanson and Nobel 2001, Criscuolo *et al.* 2005, Frost and Zhou

2005, Ambos *et al.* 2006, Rabbiosi 2008). There are three possible steps in the reverse transfer of knowledge: from the local environment of the host country to the MNE's subsidiary; from the MNE subsidiary to the parent company; and from the parent company to the local environment of the country of origin[1] (see Figure 8.1).

However, achieving successful knowledge transfer remains a challenge, regardless of the direction of knowledge flows. On 97 subsidiaries in 13 Swedish MNEs, most subsidiaries had a very low level of integration of knowledge with the rest of the MNE (Andersson and Forsgren 2000). Similar results were found in a survey of 255 foreign affiliates of German MNEs (Kutscker and Schurig 2002). Indeed, although the transfer of knowledge is supposed to be one of the defining attributes of MNEs (Casson 1979), it is surprising that the *lack* of knowledge transfer between units and individuals appears to be more common than its presence (Forsgren 2008). In response to the challenges of promoting efficient knowledge transfers as there is a growing geographical spread of MNE's centres of excellence, new R&D organizational structures are being utilized by MNEs that acknowledge that foreign subsidiaries can contribute as much as the home location of the MNE to the creation of new technological assets (e.g. Chiesa 1996, Gassmann and von Zedtwitz 1999, von Zedtwitz and Gassmann 2002, Criscuolo and Narula 2007). As such, MNEs are moving away from a 'centralized hub' to a multi-hub 'integrated network'.

It is clearly one thing to implement a dispersed R&D structure; it is quite another to achieve successful and efficient coordination, since personnel and management do not always adapt to these new structures readily due to organizational inertia. There are a number of barriers to the internal knowledge diffusion process connected to inter-unit geographical, organizational and technological distance and also to the motivational disposition of both the sender and the receiver units (see Gupta and Govindarajan 2000, Kogut and Zander 1993, Szulanski 2003). Thus if firms want to reap the benefits of a geographically dispersed R&D organization, they must ensure that knowledge generated in different units of the network is transferred to the rest of the organization and this requires the adoption of intricate and sophisticated mechanisms for the dissemination and integration of both explicit and tacit knowledge. In addition to the problems of transferring tacit knowledge across distances, there is often *technological distance* between subsidiaries, where the recipient subsidiary does not have the needed absorptive capacity to utilize the information being made available. Transfers are more efficient if the receivers of knowledge have appropriate levels of absorptive capacity (Narula 2003) allowing agents to internalize and use the knowledge made available to them. This means that agents need to properly understand, implement and assess the value of knowledge (Ambos *et al.* 2006). In addition, other subsidiaries may be reluctant to utilize knowledge developed elsewhere. Many of these individual subsidiaries have often had little experience of cooperating with each other, and in many instances have been engaged in inter-unit rivalry under a centralized hub model. Indeed, in the case of newly acquired operations, they may have been *de jure* competitors (Criscuolo and Narula 2007). Achieving a harmony of

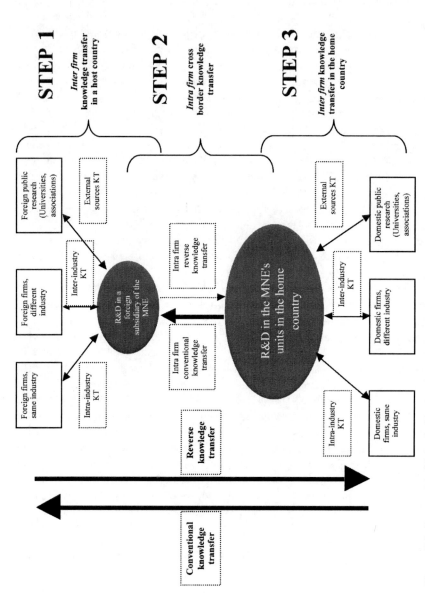

Figure 8.1 Conventional knowledge transfer and reverse knowledge transfer.

inter-facility division of labour is all the more difficult because of these inter-unit rivalries. MNEs whose subsidiaries have the appropriate skills and show some willingness to absorb and share knowledge are able to achieve better results in the transfer of knowledge (Veugelers and Cassiman 2004). Thus, the possession of knowledge and practices, and an effective way to manage communications and interactions among subsidiaries, are essential in the process of sharing knowledge (Adenfelt and Lagerström 2008).

Differences in cognitive knowledge, specialization, language, social norms and identities of individuals also create difficulties in communication (Buckley and Carter 2004, Welch and Welch 2008). It is thus essential to increase connectivity within the MNEs in one way or another to improve the internal transfer of knowledge (Makela *et al.* 2007). In addition, a gap between the vision adopted by management and the beliefs of subsidiaries may result in inconsistencies and conflicts can hinder knowledge transfer. The evidence suggests that while socialization mechanisms help to overcome some of these bottlenecks, there remain a number of obstacles in optimizing knowledge flows in physically and technologically dispersed R&D facilities (Criscuolo and Narula 2007).

Indeed, the promotion of reverse knowledge transfer presents remains one of the most vexing features of the modern dispersed and multi-hub R&D MNE structure. According to a recent questionnaire survey involving 35 major Swiss MNEs,[2] the transfer of knowledge from parent company to foreign subsidiaries is higher than from foreign subsidiaries to parent company. On a 4-point scale, the degree of importance for the conventional knowledge transfer had a mean of 3.11 whereas for reverse knowledge transfer was considerably lower, at 2.22. Further, when asked, 'Do your foreign R&D subsidiaries use knowledge from your parent company?', 80 per cent responded that it was crucially or very important (with an average of 3.52). When asked: 'Do your Swiss units use knowledge from your foreign R&D subsidiaries?', only 48 per cent indicated it was as important (with an average of 3.49). These knowledge transfers are associated with countries in which these MNEs have been most embedded. As Table 8.1 shows, the two most important home countries for reverse knowledge transfer are the United States and Germany, which are also the Switzerland's two most important trade partners, accounting for 29.3 per cent of Swiss exports. These two countries also account for 61.2 per cent of the foreign patents developed by these firms.

In general, the survey indicates that in the case of Swiss firms, their R&D centres are interconnected and there is a strong knowledge transfer between subsidiaries and the country of origin. However, the direction of knowledge transfer remains biased towards a conventional flow (Figure 8.1): subsidiaries benefit more from the knowledge created in Switzerland than the other way round. The transfer from foreign subsidiaries to the country of origin is lower than the one from the country of origin to foreign affiliates.

Given that the sharing of knowledge can become a competitive advantage, MNEs must ensure that new insights for the entire organization flow efficiently. According to Yang *et al.* (2008: 5), reverse transfer is much more difficult than conventional transfer because both are based on different transfer logics. While

Table 8.1 Location of subsidiaries engaged in reverse knowledge transfer to Switzerland

Countries	Frequency of responses	Percentage
United States	16	28.57
Germany	9	16.07
China	7	12.50
United Kingdom	6	10.71
France	5	8.93
Singapore	3	5.36
Italy	2	3.57
Japan	2	3.57
Austria	1	1.79
Canada	1	1.79
European Union	1	1.79
Finland	1	1.79
Mexico	1	1.79
Sweden	1	1.79
Total	56	100.00

Source: Authors' calculations based on a 2008 survey of 35 Swiss multinationals (see Michel 2009).

conventional transfer is more of a 'teaching' process, reverse transfer is a 'persuading' process (Yang *et al.* 2008). Indeed, in conventional knowledge transfer, the subsidiary is often obliged to replicate knowledge from the parent through the use of control mechanisms (Yang *et al.* 2008). On the other hand, subsidiaries are motivated to transfer knowledge to their parent firm because it could strengthen their strategic position in the whole organization (Gupta and Govindarajan 2000, Mudambi and Navarra 2004), and they have to persuade the parent firm that its knowledge can fit the parent's needs (Yang *et al.* 2008). The study of Swiss MNEs highlights the difficulties of reverse transfer of knowledge. The difficulties of intra-firm reverse knowledge transfer concern first the *high specificity* of foreign knowledge, second, its relevance to the parent company is not always immediately apparent and third, it may be regarded as inferior to those already available to the parent. The weakness of the transfer is thus not always related to the difficulties inherent in the transfer.

Policy implications

In general, the industrial and technological specialization of countries changes only very gradually, and – especially in newer, rapidly evolving sectors – much more slowly than the technological needs of firms. As a result, firms must seek either to import and acquire the technology they need from abroad, or venture abroad and seek to internalize aspects of other countries' innovation systems. Thus, in addition to proximity to markets and production units, firms also venture abroad to seek new sources of knowledge, which are associated with the innovation system of the host region. When firms do so, their R&D strategy is about actively tapping into foreign knowledge bases. It is important to emphasize that not many firms engage exclusively in either asset augmenting or asset-exploiting, rather they most often engage in both simultaneously (Criscuolo *et al.* 2005). When firms engage in R&D in a foreign location to avail themselves of complementary assets that are location-specific (and include those that are firm-specific), they are essentially aiming to explicitly internalize aspects of the systems of innovation of the host location. However, developing and maintaining strong linkages with external networks of local counterparts is expensive and time consuming, and is tempered by a high level of integration with the innovation system in the home location. Such linkages are both formal and informal, and will probably have taken years – if not decades – to create and sustain.

The process of engaging in reverse knowledge transfer efficiently – even by firms which seek to utilize it – is still not fully understood. However, is a growing phenomenon and of considerable importance to MNEs in an interdependent and competitive economic milieu. It is thus essential for policy makers to fine-tune R&D policy, if economic agents are to benefit from this new trend in R&D (Guimón 2009). It is not only about attracting R&D by foreign firms and promoting their embeddedness, but also about promoting their own national firms to venture abroad, and then to encourage them to share the benefits from their improved competitiveness with their home innovation system. Domestic companies are often the largest contributors to home country R&D activities. For instance, in the EU-15, firms under European control account for 85 per cent of aggregate industrial R&D outlays (OECD 2008). In Japan, firms under Japanese control account for 97 per cent of the total. In the United States, parent companies perform about 70 per cent of industrial R&D. These companies are now increasingly doing R&D abroad.

Most countries that have been historically inward-looking have also always regarded the need to import technologies as a sign of national weakness, and have – not coincidentally – a tradition of techno-nationalism. That is, they have sought to maintain in-country competences at whatever the cost. This problem is aggravated by the trend towards multi-technologies even in mature industries. The strategy of technological self-sufficiency is particularly untenable in economies that have limited resources (such as small countries). They must either spread their resources thinly across many technological competences or concentrate on a few. It is one thing to propose changing policies that have previously championed self-sufficiency, and quite another to change the attitudes of policy makers and organizations that implement policy. Institutions (in the sense of routines and procedures) create the milieu within which economic activity is undertaken and establishes the ground rules for interaction between the various economic actors. Nonetheless, systems do change, because the costs of supporting inefficient institutions may far outweigh the benefits of change. Systems and institutions are also evolutionary processes which require imitation, experimentation, learning and forgetting, and this most often means that change is gradual, slow and cumbersome.

Some policy makers feel that national champions should not venture abroad, but should seek complementary assets by arms-length mechanisms such as licensing and outsourcing. However, just as there are limits to the firm's use of non-internal sources of knowledge, there are limits to how much countries can rely on such arms-length means. Innovation based largely on improving and modifying external sources of technology acquired through arms-length means is an option only available *as long as there is something to imitate*. As countries approach the technological frontier, there are two problems. First, it may not be possible to buy cutting-edge technologies, since firms that own these technologies are reluctant to license or sell them. The reluctance has to do with the nature of technology (in that a price cannot be put on an unproven knowledge base for which no market exists) and the fact that firms will seek to maximize the rent from their inventions as long as they are in a monopoly position. To sell their new technologies would be to create a competitor. Second, imitation is not possible *at* the frontier, since it is difficult to predict *ex ante* which technology (of several competing nascent technologies) will become paradigmatic. This explains the popularity of strategic technology partnering at and around the frontier, because firms seek to collaborate when it helps to reduce uncertainty and reduce the innovation time span. They therefore seek partners who can improve the probability of 'winning', and these firms are those that have complementary resources to offer (Narula 2003). As they approach the frontier, countries must have the capacity not just to absorb and imitate technological development created by others, but also the ability to generate inventions of their own. This requires technological capabilities that are non-imitative. In other words, learning-by-doing and learning-by-using have decreasing returns as one approaches the frontier, and in-house learning by asset-augmenting R&D internationalization is probably the only efficient option.

There are three areas where government policy plays a significant role. First, there are the generic aspects of promoting an efficient innovation system. This concerns investment in R&D whether foreign-owned or domestic (e.g. helping the domestic actors to adopt foreign innovations, attracting foreign talent, promoting collaboration between domestic and foreign players, investing in public research infrastructure, establishing effective intellectual property rights regimes, etc. see Guimón 2009 for a review). Perhaps most importantly, it is the promotion of effective and efficient means by which firms and organizations within an innovation system communicate with each other, and this reflects the balance between the cooperation and rivalry within the milieu. These are 'generic' aspects of a knowledge infrastructure, and are quasi-public goods in that they are potentially available to all firms, and need to evolve to meet the needs of firms. Policies can sustain this reverse technology transfer, for instance, by encouraging the international mobility of skilled manpower or by encouraging the internationalization expansion of public R&D centres and universities. Knowledge transfer may also be encouraged by fostering contacts between research institutions, associations, universities and businesses.

Second, it is important for policy makers to distinguish between asset-exploiting activities and asset-augmenting activities. Asset-exploiting relates to foreign R&D which improve home products in adapting them to local markets. Asset-augmenting activities relate to foreign R&D that tap into new sources of knowledge abroad. Augmenting activities have a more innovative function than the exploiting type. Indeed, exploiting activities use initial firm-specific knowledge developed at home in order to adapt products of processes to local conditions. In this context, core activities are concentrated in the home country, and foreign activities enhance the technologies developed at home. Exploring activities develop core innovations in host countries. In this case, a new important source of competitive advantage is the capacity of foreign subsidiaries to create innovations based on host countries' technological competences. Domestic R&D activities are thus not the only sources of knowledge that MNEs exploit. They can also access foreign sources of knowledge to complement their R&D activities at home, or to acquire or create new unique intangible assets.

Policy makers also need to distinguish between firms that internationalize as an 'exit' because of the poor fit between the needs of the firm and the knowledge infrastructure, and those that internationalize because they need to augment their existing assets and those available to them at home. This requires a clearer understanding of the strategies of individual MNEs and their technology portfolios, rather than a one-size-fits-all approach, tailoring their policy tools to specific needs.

Third, policy makers can address means to promote MNEs to actively help them upgrade the home country innovation system through reverse knowledge transfer. Specifically, while Step 1 and Step 2 (see Figure 8.1) of the reverse knowledge transfer process have a direct bearing on the technological competitiveness of the MNE. Step 3 of the process, on the other hand, has a direct bearing on the quality of the location advantages of the home country, but may have few immediate benefits for the MNE. The system of innovation of home country can profit from the exploring activities. Indeed, this kind of R&D leads to the reverse knowledge transfer.

Notes

1 A questionnaire was sent in May 2008 to the 71 most innovative Swiss firms according to patent applications (see Michel 2009). In August 2008, 35 firms responded. High-technology industries represent 40 per cent of the respondent firms, against 37.1 per cent for high-medium technology industries, 11.4 per cent for medium-low technology industries and 5.7 per cent for low technology industries.

2 In fact, knowledge transfer may take place through MNE in at least five different forms (e.g. Gupta and Govindarajan 1991; Piscitello and Rabbiosi 2007): (*a*) flows from parent company to subsidiaries, (*b*) flows from subsidiaries to parent company, (*c*) flows from local environment to subsidiary, (*d*) flows from subsidiary to local environment, (*e*) flows to peer subsidiaries. However, in the subject of the reverse knowledge transfer, the three flows mentioned in the text are the most relevant.

References

Adenfelt, M. and K. Lagerström (2008) 'The Development and Sharing of Knowledge by Centres of Excellence and Transnational Teams: A Conceptual Framework', *Management International Review*, 48 (3), pp. 319–38.

Alcácer, J. (2006) 'Location Choices across the Value Chain: How Activity and Capability Influence Collocation', *Management Science*, 52 (10), pp. 1457–71.

Almeida, P. and B. Kogut (1997) 'The Exploration of Technological Diversity and the Geographic Localization of Innovation', *Small Business Economics*, 9, pp. 21–31.

Ambos, T., B. Ambos and B.B. Schlegelmilch (2006) 'Learning from Foreign Subsidiaries: An Empirical Investigation of Headquarters' Benefits from Reverse Knowledge Transfers', *International Business Review*, 15 (3), pp. 294–312.

Andersson, U. and M. Forsgren (2000) 'In Search of Centres of Excellence: Network Embeddedness and Subsidiary Roles in Multinational Corporations', *Management International Review*, 40(4), pp. 329–50.

Asheim, B. and M. Gertler (2005) 'The Geography of Innovation', in: J. Fagerberg, D. Mowery and R. Nelson (eds) *Handbook of Innovation*, Oxford: Oxford University Press.

Buckley, P.J. and J.M. Carter (2004) 'A Formal Analysis of Knowledge Combination in Multinational Enterprises', *Journal of International Business Studies*, 35 (5), pp. 371–84.

Cantwell, J. and O. Janne (1999) 'Technological Globalisation and Innovative Centers: the Role of Corporate Technological Leadership and Locational Hierarchy', *Research Policy*, 28, pp. 119–44.

Casson, M. (1979) *Alternatives to the Multinational Enterprise*, London: Macmillan.

Chiesa, V. (1996) 'Managing the Internationalization of R&D Activities', *IEEE Transactions on Engineering Management*, 43 (1), pp. 7–23.

Criscuolo, P. and R. Narula (2007) 'Using Multi-hub Structures for International R&D: Organisational Inertia and the Challenges of Implementation', *Management International Review*, 47 (5), pp. 639–60.

Criscuolo, P. and B. Verspagen (2008) 'Does it Matter where Patent Citations come from? Inventor versus Examiner Citations in European Patents', *Research Policy*.

Criscuolo, P., R. Narula and B. Verspagen (2005) 'Measuring Knowledge Flows among European and American Multinationals: A Patent Citation Analysis', *Economics of Innovation and New Technologies*, 14, pp. 417–33.

Doh, J.P., G.K. Jones, H. Teegen and R. Mudambi (2005) 'Foreign Research and Development and Host Country Environment: An Empirical Examination of US International R&D', *Management International Review*, 25 (2), pp. 121–54.

Forsgren, M. (2008) 'A Critical Review of the Evolutionary Theory of the MNC', in: Dunning, J. and P. Gugler (eds) *Foreign Direct Investment, Location and Competitiveness: Progress in International Business Research 2*, New York: Elsevier.

Foss, N.J. and T. Pedersen (2002) 'Transferring Knowledge in MNCs: The Role of Sources of Subsidiary Knowledge and Organisational Context', *Journal of International Management*, 8 (1), pp. 49–67.

Frost, T. (1998) *The Geographic Sources of Innovation in the Multinational Enterprise: US Subsidiaries and Host Country Spillovers 1980–1990*, Boston: Massachusetts Institute of Technology.

Frost, T. and C. Zhou (2005) 'R&D Co-practice and 'Reverse' Knowledge Integration in Multinational Firms', *Journal of International Business Studies*, 36 (6), pp. 676–87.

Gassmann, O. and M. von Zedtwitz (1999) 'New Concepts and Trends in International R&D Organization', *Research Policy*, 28 (2–3), pp. 231–50.

Guimón, J. (2009) 'Government Strategies to Attract R&D-Intensive FDI', *Journal of Technology Transfer*, 34 (4), pp. 364–79.

Gupta, A.K. and V. Govindarajan (1991) 'Knowledge Flows and the Structure of Control within Multinational Corporations', *Academy of Management Review*, 16 (4), pp. 768–92.

Gupta, A.K. and V. Govindarajan (2000) 'Knowledge flows within Multinational Corporations', *Strategic Management Journal*, 2, pp. 473–96.

Håkanson, L. and R. Nobel (1993) 'Determinants of Foreign R&D in Swedish MNCs', *Research Policy*, 22, pp. 397–411.

Håkanson, L. and R. Nobel (2001) 'Organization Characteristics and Reverse Technology Transfer', *Management International Review*, Special Issue, 41 (4), pp. 392–420.

Hansen, M.T. and J. Birkinshaw (2007) 'The Innovation Value Chain', *Harvard Business Review*, pp. 121–30.

Hotz-Hart, B. (2000) 'Innovation Networks, Regions, and Globalization', in: G.L. Clark, M.P. Feldman, and M.S. Gertler (eds) *The Oxford Handbook of Economic Geography*, Oxford: Oxford University Press, pp. 432–50.

Iammarino, S. and P. McCann (2006) 'The Structure and Evolution of Industrial Clusters: Transactions, Technology and Knowledge Spillovers', *Research Policy*, 35 (7), pp. 1018–36.

Inkpen, A.C. (2008) 'Managing Knowledge Transfer in International Alliances', *Thunderbird International Business Review*, 50 (2), pp. 77–90.

Jacobs, J. (1970) *The Economy of Cities*, New York and Toronto: Random House.

Jacobs, J. (1986) *Cities and the Wealth of Nations*, Harmondsworth: Penguin.

Kogut, B. and U. Zander (1993) 'Knowledge of the Firm and the Evolutionary Theory of the Multinational Corporation', *Journal of International Business Studies*, 24 (4), pp. 625–45.

Kutscker, M. and A. Schurig (2002) 'Embeddedness of Subsidiaries in Internal and External Networks: A Prerequisite for Technological Change', in: V. Havila, M. Forsgren, and H. Håkansson (eds) *Critical Perspectives on Internationalization*, Amsterdam: Pergamon.

Lall, S. (1979) 'The International Allocation of Research Activity by US Multinationals', *Oxford Bulletin of Economics and Statistics*, 41 (4), pp. 313–33.

Makela, K., H.K. Kalla, and R. Piekkari (2007) 'Interpersonal Similarity as a Driver of Knowledge Sharing within Multinational Corporations', *International Business Review*, 16 (1), pp. 1–22.

Mansfield, E., D. Teece, and A. Romeo (1979) 'Overseas Research and Development on Foreign Sales Performance', *Economica*, 46, 182, pp. 187–96.

Michel, J. (2009) *Investissements directs à l'étranger dans les activités de R&D: théorie et application aux entreprises suisses*, Berne: Peter Lang.

Mudambi, R. and P. Navarra, (2004) 'Is Knowledge Power? Knowledge Flows, Subsidiary Power and Rent-seeking within MNCs', *Journal of International Business Studies*, 35 (5), pp. 385–406.

Narula, R. (2002) 'Innovation Systems and "Inertia" in R&D Location: Norwegian Firms and the Role of Systemic Lock-in', *Research Policy*, 31, pp. 795–816.

Narula, R. (2003) *Globalization and Technology*, Cambridge, UK: Polity Press.

Narula, R. and A. Zanfei (2004) 'Globalisation of Innovation: The Role of Multinational Enterprises', in: J. Fagerberg, D. Mowery and R.R. Nelson (eds) *Handbook of Innovation*, Oxford: Oxford University Press.

OECD (2008) *The Location of Investment of Multinationals linked to Innovation*, Paris: OECD.

Pearce, R., D. Dimitropoulou, and M. Papanastassiou (2008) 'Locational Determinants of FDI in the European Union: The Influence of Multinational Strategy', in: J. Dunning and P. Gugler (eds) *Foreign Direct Investment, Location and Competitiveness: Progress in International Business Research 2*, Missouri: Elsevier.

Piscitello, L. and L. Rabbiosi (2007) 'The Impact of Knowledge Transfer on MNEs' Parent Companies: Evidence from the Italian Case', in: L. Piscitello and G. Santangelo (eds) *Do Multinationals Feed Local Development and Growth?*, Missouri: Elsevier, pp. 169–94.

Porter, M. (1993) *L'avantage concurrentiel des nations*, Paris: InterEditions.

Pottelsberghe de la Potterie, B. van and D. Guellec (2001) 'The Internationalisation of Technology Analysed with Patent Data', *Research Policy*, 30 (8), pp. 1256–66.

Rabbiosi, L. (2008) 'The Impact of Subsidiary Autonomy on MNE Knowledge Transfer: Resolving the Debate', *SMG Working Paper*, 16/2008.

Saxenian, A. (1994) *Regional Advantage*. Cambridge, MA: Harvard University Press.

Schrader, S. (1991) 'Informal Technology Transfers between Firms: Cooperation through Information Trading' *Research Policy*, 20, pp. 889–910.

Singh, J. (2007) 'Asymmetry of Knowledge Spillovers between MNCs and Host Country Firms', *Journal of International Business Studies*, 38, pp. 764–86.

Song, J. and J. Shin (2008) 'The Paradox of Technological Capabilities: A Study of Knowledge Sourcing from Host Countries of Overseas R&D Operations', *Journal of International Business Studies*, 39, pp. 291–303.

Szulanski, G. (2003) *Sticky Knowledge: Barriers to Knowing in the Firm*, London: Sage Publications.

Tallman, S., M. Jenkins, N. Henry and S. Pinch (2004) 'Knowledge, Clusters and Competitive Advantage', *Academy of Management Review*, 29, pp. 258–71.

UNCTAD (2005) *Transnational Corporations and the Internationalization of R&D, World Investment Report*, New York and Geneva: United Nations.

Veugelers, R. and B. Cassiman (2004) 'Foreign Subsidiaries as a Channel of International Technology Diffusion: Some Direct Firm Level Evidence from Belgium', *European Economic Review*, 48, pp. 455–76.

Welch, D.E. and L.S. Welch (2008) 'The Importance of Language in International Knowledge Transfer', *Management International Review*, 48 (3), pp. 339–60.

Yang, Q., R. Mudambi, and K.E. Meyer (2008) 'Conventional and Reverse Knowledge Flows in Multinational Corporations', *Journal of Management*, 34 (5), pp. 882–902.

Zanfei, A. (2000) 'Transnational Firms and Changing Organisation of Innovative Activities', *Cambridge Journal of Economics*, 24, pp. 515–54.

Zedtwitz, M. von and O. Gassmann (2002) 'Managing Customer Oriented Research', *International Journal of Technology Management*, 24 (2/3), pp. 165–93.

9 Product development in multinational companies

The limits for the internationalization of R&D projects

Martin Heidenreich, Christoph Barmeyer and Knut Koschatzky

The organization of research and development (R&D) is becoming increasingly international. An indication of this is that from 2001–3, 16.7 per cent of all patents issued by the European Patent Office (EPO) belonged to foreign inventors. Correspondingly, 16 per cent of entrepreneurial R&D expenditure was invested in foreign countries (OECD 2007: 162 and 172). Multinational companies (MNCs) are the central protagonists of international innovation processes. They are responsible for most of the global R&D expenditures and are therefore an important channel for the transfer of technological knowledge across national as well as cultural and institutional borders (UNCTAD 2005: 119). The 'knowledge-based approach' of MNCs (Scaperlanda 1993, Zander 1998, Zander and Sölvell 2000) assumes that MNCs are the central arenas for the international transfer of knowledge: 'Firms are social communities that specialize in the creation and internal transfer of knowledge' (Kogut and Zander 1993: 625). MNCs are analysed as polycentric networks, in which internationally distributed subsidiaries produce innovations:

> MNCs ... have continuously extended their network of R&D locations and knowledge centers across the world. They are moving away from a single, self-contained, in-house center of knowledge towards a more distributed and open architecture of knowledge production and application.
>
> (Gerybadze 2004: 123)

Critics however have pointed out that worldwide distribution of innovation activities is limited by the fact that knowledge cannot flow freely and without constraints across national and organizational borders, Becker-Ritterspach (2006) and Patel and Pavitt (1991), for example, analyse the R&D activities of multinational businesses as 'an important case of "non-globalisation"' – a result that corresponds to the findings of Ruigrok and van Tulder (1995: 159), who observed that none of the Fortune 100 largest non-financial companies in the world 'is truly "global", "footloose" or "borderless" ... R&D remains firmly under domestic control'.

Even if now companies rely to a considerable extent on foreign R&D capacities – above all as a result of international mergers and acquisitions – Patel and

Vega (1999) and Le Bas and Sierra (2002) convincingly demonstrate that the technological competences in the homeland of multinational companies still play an important role for the internationalization strategies of MNCs. Only 10.5 per cent or 17 per cent, respectively, of the MNCs considered in these two studies were able to appropriate new technological competences abroad if they did not have such competences at their disposal in the homeland. Phene and Almeida (2008: 913) show that the technological abilities of subsidiaries depend above all on the national context and hardly at all on company-wide knowledge flows.

On the one hand, R&D efforts thus are increasingly internationalized. On the other hand, however, the headquarters of the companies and those competences concentrated in the country of origin still play a crucial role for the location of R&D. *The question therefore is how the observed internationalization of R&D can be reconciled with the crucial role of domestic locations and competences.*

This question can hardly be answered on the basis of existing empirical studies, because these rarely analyse concrete trans-border innovation processes (Zander and Sölvell 2000: 45). Therefore, in the following, we examine the possibilities and limitations of international R&D projects on the basis of two selected innovation case studies in order to analyse how organizations deal with the above-mentioned contradiction. We will analyse how international innovation projects are organized, how they can transcend national borders, which R&D activities are internationalized and which advantages and disadvantages international collaborations have for the projects we observed. Unlike other studies, we will not concentrate on the mere existence of foreign R&D centres (Zedtwitz and Gassmann 2002) or on the country of residence of an inventor that registers a patent (Patel and Vega 1999, Le Bas and Sierra 2002).

In order to analyse the possibilities and limitations of trans-border innovation projects, we propose in the following a conceptual framework which takes into account the cognitive, strategic and normative (or contextual) dimensions of trans-border innovation projects. These projects can be analysed as learning processes, as bargaining and power relationships and as institutionally and culturally embedded activities. In these three dimensions, innovation projects have to balance the advantages of spatially and organizationally concentrated projects and the advantages of distributed innovation projects (Section 1). On the basis of two international innovation projects,[1] we analyse how the companies deal with these contradictory requirements (Section 2). Faced with the necessity for cooperation in complex innovation projects, company-wide as well as external collaborations are selectively used for non-strategic tasks which can be easily decoupled from the core of the innovation project (Fritsch and Franke 2004). In the core of the observed innovation processes, MNCs choose spatially concentrated, intra-organizational forms of cooperation at one site or in the same region. Foreign locations and external companies are only involved for strategically less crucial tasks. Preferably, only the results of largely finished and already patented innovation projects are transferred to other locations or companies. Despite the possible advantages of international and collaborative innovation projects which could exploit company-wide and external knowledge resources, the exploitation of direct, informal communication and cooperation and the protection and development of the organizational core competences are often more important (Section 3). In one case however, we were able to observe a trans-border innovation project where the competences were spatially distributed – also the result of a carefully designed international management structure.

1 The dilemmas of distributed innovation processes

MNCs are important channels for the international transfer of technological knowledge across national, cultural and institutional borders. The assumed transformation from a 'space of places' to a 'space of flows' (Castells 2010) might be explained by the acceleration of innovation cycles and the necessity to adjust quickly to different markets world-wide and to use local competences. It is, however, an open question whether and to what extent MNCs succeed in trans-border learning processes because the advantages of spatial and social proximity and the interest of strengthening the domestic locations and controlling new technological competences are an important barrier for the internationalization of R&D processes. Therefore, two opposed theses can be formulated concerning the internationalization of R&D processes: on the one hand, it can be hypothesized that the company-wide distribution of R&D processes is essentially advantageous in an innovation-centred global knowledge economy (*distributed innovation thesis*; cf. Rammert 2006); on the other hand, the advantages of organizational control, spatial proximity and socially embedded mutual learning processes might favour the territorial and organizational concentration of R&D activities (*concentration and embeddedness theses*).

An adequate conceptual framework for internationalization processes requires that MNCs are not only analysed as knowledge-based networks but also as trans-border arenas for power, control and exchange relationships and as sociocultural and institutionally-embedded organizations. The knowledge-based approach refers to trans-national processes of learning and knowledge exchange and analyses the internalization of knowledge flows in MNCs (Granstrand *et al.* 1992, Scaperlanda 1993, Zander 1998, Monteiro *et al.* 2008); the micropolitical perspective examines the power and exchange relationships between headquarters and their foreign subsidiaries (Bouquet and Birkinshaw 2008, Andersson *et al.* 2007: 815); the cultural and institutionalist approaches emphasize the role of national institutions, business systems and organizational and management cultures (Barmeyer 2000, d'Iribarne 2001, Hofstede 2001, Whitley 1999).

In these three dimensions of headquarters–subsidiary relationships, R&D processes in MNCs face different dilemmas due to the complementary advantages and disadvantages of organizationally and spatially distributed and of concentrated and spatially embedded innovation processes. In the cognitive dimension, MNCs face the dilemma of trans-border learning and innovation opportunities and organizationally and territorially embedded learning. In the strategic dimension, they have to deal with the conflicting interests of protecting proprietary knowledge through the concentration and internalization of R&D

processes at one site and the necessity of including other subsidiaries and external partners (cf. also Narula and Santangelo 2009: 400 for the tension between knowledge sharing and knowledge expropriation). In the contextual dimension, R&D processes in MNCs face the tensions between the advantages of a socioculturally homogeneous development team and the strengths of diversity.

In the *cognitive dimension*, spatial and organizational proximity can facilitate mutual learning processes (Cooke *et al.* 2004). Spatial and organizational proximity reduces not only the cost of information, communication and other transactions, but also eases mutual understanding. Bathelt *et al.* (2004: 31) term local forms of the knowledge transfer as 'buzz': 'learning processes taking place among actors embedded in a community by just being there'. However, implicit, context-specific knowledge can also be transferred internationally within MNCs. For such global forms of the knowledge transfer international communication channels are especially important, which Bathelt *et al.* (2004: 31) term 'pipelines': 'knowledge attained by investing in building channels of communication ... to selected providers located outside the local milieu'. The decision regarding the distribution or the concentration of innovation processes thus is on the choice between the advantages of 'buzz' and 'pipelines'.

In the *strategic dimension*, the interest in trans-border, company-wide cooperation is based on the interests in developing foreign markets and enhancing the technological capabilities of the group by the access to specific competences of foreign companies and locations. This tension between organizational control and the development of new markets and competences shapes the decisions between internal development and external collaborations. On the one hand MNCs try to control strategic product and process technologies to avoid the loss of control connected with external collaborations, but on the other hand they try to exploit the competences and cost advantages of external partners. On the one hand, MNCs exploit the competences of foreign subsidiaries; on the other hand, even within a MNC, the headquarters or subsidiaries may have an interest in securing their own influence in the group through the concentration of R&D activities in the homeland:

> The headquarters of the federative MNC is always confronted with a dilemma whereby externally embedded subsidiaries give potential access to a variety of competences, but where increasing embeddedness may therefore lead to a reduced interest in contributing to the MNC's overall performance: that is, the balance shifts towards rent-seeking rather than profit-seeking.
>
> (Andersson *et al.* 2007: 816)

Not only subsidiaries, but also the headquarters themselves may try to maximize their influence through the concentration and control of strategic R&D activities.

In the *contextual dimension*, an MNC can also cooperate with other businesses in the regional and national territory in addition or alternatively to the use of company-wide, international competences with innovation processes. In this case, a common institutional and sociocultural framework may facilitate cooperation. Calculability and perhaps even trust-based relations may be an important asset for R&D networks within the same national or regional context. However, the flipside of this may be the renouncement of foreign competences, technologies and market chances as well as path-dependencies and locked-in-effects.

In conclusion: The organization of innovation processes in MNCs is faced basically with the choices between internal and external R&D ('make or buy') and between spatially distributed and concentrated innovation processes. On the one hand, R&D activities can be concentrated at a domestic site or even within the same organization. This embeddedness in a common sociocultural and institutional environment allows the exploitation of the advantages of territorial and organizational proximity. On the other hand, international R&D networks allow the utilization of distributed, company-wide and external competences. In the *cognitive dimension*, the territorial and organizational concentration of R&D processes is a prerequisite for reciprocal learning processes, interaction and the transmission of implicit knowledge. Learning and the transmission of implicit knowledge is, however, also possible over great distances especially within the same company – for instance through communication technologies and spatial mobility. Therefore MNCs face the challenge of choosing between those competence-widening strategies that are territorially and organizationally concentrated and those that are distributed – without being able either to renounce the advantages of organizationally and/or spatially embedded processes or the advantages of organizationally and/or spatially disembedded processes. In the *strategic dimension*, power and exchange relationships within the company are important for the organization of innovation processes. On the one hand, the control of strategic organizational competences and sometimes also rent-seeking interests of the headquarters favour the centralization of technological competences at one site; on the other hand, interests in quick technological advances favour distributed and externalized innovation processes and a higher autonomy of individual subsidiaries. In the *contextual dimension*, cooperation within a common institutional and sociocultural context may reduce the uncertainties of innovation processes, but can also lead to innovation barriers and lock-in effects.

It is has not yet been determined how multinational companies deal with these three dilemmas within the framework of international development projects. This point will be examined empirically in the following, taking the examples of two MNCs.

2 Limits and organization of distributed innovation projects: two case studies

On the basis of two (technologically and economically very diverse) international innovation projects in the automobile and IT industries (cf. Table 9.1), we will discuss how MNCs dealt with the above-mentioned dilemmas of distributed and concentrated innovation processes. In the first case, the central R&D activities were concentrated in one location; foreign R&D sites or external cooperation partners were involved only for peripheral tasks, which could be decoupled from the core of the innovation project. The second, a US company with a subsidiary in Germany, used also foreign R&D sites for strategic functions. The first company dealt with the above-mentioned dilemma by choosing an intensive cooperation with foreign partners on one location at the expense of more intensive international collaborations, while the American MNC took the route of intensive, trans-border R&D cooperation without external partners.

2.1 The regional embeddedness of a global development project

The German automotive company Auto-D spends approximately 3.4 per cent of its turnover (2005) on R&D and employs thousands of researchers and

Table 9.1 The organization of R&D processes in two MNCs

	Auto-D	*IT-USA*
The innovation project	Development of a fundamentally new drive technology	Development of a program for the integration of two different software systems
Organization of the research phase (actors, spatial concentration, team)	Introduction of external experts for the explorative approach to the subject at a very high level	'Creative' initial phase: Interdepartmental core team in a R&D laboratory in Germany (at first 4 persons), subordinated into two clearly defined competence areas, concentrated at one location
Cooperation with other locations within the company	New competence base required (not available within the company)	Support of high-rank US professional necessary for the approval of the project
Cooperation with external partners	Low (normal scientific contacts)	No
Organization of the product-development phase	Functionally structured project organization (for industrialization of new technology)	Local software development, global tests
Cooperation with other locations within the company	Relevant functions are concentrated on one site (400 employees), close cooperation with proximate production plant and headquarter	Implementation of new module in a complex system 'owned' by a dozen US laboratories: Approval by global requirement and architecture boards. Tests in US, China
Cooperation with external partners	Joint venture with an US competitor and a Canadian component developer	No

developers in all parts of the world (Germany, India, China, Japan, Russia, and the United States). In this company we analysed the development of a new drive technology on which the company has been working since the beginning of the 1990s and in which it has already invested more than €1 billion.

Since the 1990s, three companies have cooperated directly and through various subsidiaries on the project: (*a*) the German automobile company, Auto-D; (*b*) an American competitor, Auto-USA, one of the biggest automakers in the world, and (*c*) a Canadian company, B-Can, which develops and produces the core component of the new drive system. In 2007, company B-Can sold its development activities in the automobile industry to Auto-D and Auto-USA and now concentrates on the production of the core component of the new drive technology only. A common subsidiary of Auto-D and Auto-USA, which employs approximately 215 staff and shall here be called System-D, develops the necessary periphery components, coordinates the supply of the core component and assembles the different components into the new drive system for deployment in different vehicle types.

The cooperation of Auto-D with a foreign competitor is crucial to the success of the innovation process since the development of standards for a fundamentally new drive technology requires allies. This is also important for later market access strategies since the same infrastructures can be used for the vehicles of different carmakers. The cooperation with a competitor is regarded as unproblematic since the technology is in a pre-commercial phase. Cooperation also means that the billions of development expenditures can be shared.

System-D is located in southern Germany close to the headquarters and the principal plants of Auto-D. All activities for the development of the new drive technology are concentrated at this site, and all partners (Auto-D, Auto-USA, B-Can) are represented at this location with their own staff. Including the staff of Auto-D (160 personnel) and System-D (215), approximately 400 persons are employed at this site for the development of the new drive technology.

Proximity is central for the industrialization of the new technology, because the whole project focuses on the industrial utilization of known technology in a mass series production. Here, the close cooperation with the company headquarters and the neighbouring plants is crucial. Only by close cooperation with the production can the costs of the new drive technology be lowered so far that large-scale production becomes competitive. The deviation from previous standard procedures must be limited to a calculable extent since an automobile is a highly complex product of tens of thousands of individual parts that must combined in a reliable, durable and economic way. Innovation therefore is possible only in close coordination with the production – and this is eased considerably by spatial proximity.

Spatial proximity is however essential only for the strategic core of the project. Non-strategic technologies and components are developed and produced with external, often regional suppliers, often in order to reduce the price. For these tasks, the company relies especially on long-time cooperation partners. With innovative components, however, the price is less important than the possibility and necessity to secure and develop the core competences of the company as the basis of its strategic advantage. The detailed knowledge of externally-produced components is part of the core competence of the focal company (Auto-D). For the company it is vital to master the relevant knowledge and to monitor the performance of the suppliers. This implies that the company often prefers to develop the crucial components within the company.

The protection of strategic competences and competitive advantages thus has a clear priority over economic criteria, as well. In 2005, System-D for example was taken over completely by Auto-D and Auto-USA. The role of the Canadian cooperation partner, B-Can, which previously had equal standing, became subordinate in order to prevent the drain of automobile-specific knowledge into this company. Furthermore, System-D concluded an exclusive supplier agreement with its two parent companies, although it has already found other customers world-wide. System-D's independent market activities seemed to endanger competitive advantage of its parent companies.

In conclusion: The innovation project under review was initially dispersed on a national and international scale then became concentrated in a single location. On the one hand, this concentration highlights the crucial role of spatial proximity at the core of this global development project. A company founded by Canadian, German and American partners was in the centre of the project. Through the pre-market cooperation of these companies, industrial standards shall be set, the immense development costs shared and later a market and the required infrastructure created. Suppliers are also included in the development – up to now however only for non-strategic components (with the exception of the new drive technology itself), in order to use economies of scale and experience. On the other hand, the companies involved are trying to control the core competences of the new drive technology within the company or to develop at least the competence to judge, coordinate and monitor external activities. The development process analysed is therefore characterized by a double challenge: On the one hand it requires the coordination of development activities and the exploitation of heterogeneous knowledge from various locations, companies, professions and places. On the other hand, it requires the protection of strategic competences. The company faces these challenges by the spatial concentration of the innovation project and by the internalization of crucial competences and processes.

2.2 *A local IT-project in a global development context*

IT-USA is a huge US company in the field of information technology with a large development centre employing over 1,000 persons in Germany. The IT project we were able to observe aimed at the integration of two complex software products of IT-USA. The mother product is administered by an American laboratory, and as such cooperation between the two teams was extremely important. The cross-functional task explains some peculiarities of the project: Developers from two completely different 'software worlds' – the 'process world' and the 'database world' – had to cooperate. This was the central challenge of the project under examination; otherwise, its innovative content is considered to be rather low.

The development of the new software took place in a big software laboratory in a city in south-western Germany, in which the idea for this project also originated. This laboratory is part of a world-wide network of more than two dozen R&D locations. The German core project team consisted of four software engineers. In 2003 the project began with the design phase. At the beginning of the project, the direct interaction between the employees involved in the project was extraordinarily important for the development of ideas and the persuasion of superiors. One of the involved software engineers emphasized that from the beginning personal contacts with chief developers, architects or other technical executives were important in order to gain supporters in the company. In this design phase, which was the most creative phase, the involved software engineers worked in an explorative way, i.e. without specific guidelines. Communication in this phase was very intensive and informal and completely without external partners.

As the project advanced, it became more structured. The developers had to document their achievements more exactly. Clearer task definitions, goals and responsibilities were defined and the different phases were better documented. The project-specific knowledge basis became more transparent and new staff and teams could be integrated more easily into the project, which provided the basis for the expansion and internationalization of the project. Eventually, another team from the southwest German software laboratory was incorporated, and additional teams from the United States, Canada and China were responsible for testing and the interfaces with the mother product. This considerably increased the planning and coordination efforts, and for this reason a project manager was appointed. However, spatial proximity and the opportunities for direct, informal interaction still played a crucial role in the phase in which the prototype was integrated into the main product. Personal contacts helped the actors to agree on a common vision of the project and to understand what was important for the different partners.

In the project only one component of a much more complex information system was produced. Therefore, from the very beginning, the project had to be designed for integration into a complex global software environment. This integration began with the decision to provide funding for the continuation of the development project. Here, the merits of the project had to be defended and explained to the technical management of the MNC. The company has developed a global network of professionals, whose support is crucial for the decisions on new projects. In the company, not only the support of the hierarchy counts, but also the support of the technical level, as (mostly US) technical experts have to act as 'sponsors' of the planned project. The decision to advance the project thus would not have been possible without professional and personal contacts with the US decision-makers. In addition, the integration of the new product into the global software environment also had to be decided by special company-wide requirement boards. An 'architecture board' checked whether the module had the appropriate interfaces and whether it fit into the infrastructure.

In conclusion: The IT development project was on the one hand characterized by the spatial proximity of the development team, especially in the creative phase at the beginning of the project. On the other hand, the developed component was part of a global software product, for which a dozen other locations develop other components. Therefore, a careful, highly formalized cooperation with these laboratories was necessary in order to develop a component for such a complex software product. Also the political support of a new project required direct contact with the US management and experts. The MNC has systematized these intra-organizational and international contacts through a global managerial and professional hierarchy. At the suggestion of high-ranking professionals of the German laboratory, leading technical experts in the United States had committed themselves as 'sponsors' of the project under examination. Thereby the IT company was able to coordinate its technical activities through a global, though inner-organizational, hierarchy of technical experts. The integration of the software project in the company-wide development policies therefore was assured not only by managerial hierarchies, but also by a company-wide professional epistemic community. In the project analysed external partners played barely any role at all (only as pilot users).

The two companies under review (Auto-D, IT-USA) have R&D centres in different parts of the world. In a broad sense, the two innovation projects examined were internationally distributed. In spite of these common traits and in spite of the technical, economic and organizational differences between the two case studies, they raise some doubts on the vision of deterritorialized, disembedded and externalized innovation processes. In the first case, foreign subsidiaries and external partners were used above all for simple, peripheral activities – for example testing prototype vehicles in other parts of the world. The core activities – the development and industrialization of a new drive technology – were concentrated at a German location close to the headquarters of the German MNC. Also the US partner supported or accepted the strategy of facilitating the recombination of heterogeneous knowledge by the spatial and organizational concentration of technical competences. The spatial concentration of competences was initially also very important for the IT project observed. But the IT company succeeded in integrating the IT-project observed in global efforts for the advancement of a complex IT-product. A trans-border, company-wide hierarchy of technical experts and cross-border boards were essential in this respect pointing to the necessity of specific organizational prerequisites for distributed innovation processes. If external collaborations were used at all, they were limited in both cases to less critical functions, for example external testing functions and the supply of standardized components. External partners did not essentially contribute to the success of the innovation project. Only when an external partner was crucial to the success of the whole project (for example by facilitating access to the American market or by enforcing industry-wide technical standards, as in the case of Auto-US), was it involved on equal terms.

3 Conclusions

In the introduction, the thesis of a distributed innovation project was confronted with an 'embedding thesis', which stressed the advantages of spatial and organizational proximity. On the one hand, MNCs can be analysed as inner-organizational channels for the international transfer of technological knowledge, as arenas for trans-border power and exchange relationships and as disembedded, 'footloose' companies exploiting on a global scale the most promising technological competences and market opportunities. On the other hand, MNCs can exploit the advantages of institutionally, socioculturally and territorially embedded innovation processes in one region or even within one location in order to concentrate its core competences in its headquarters and a few selected domestic sites and to avoid the risks of collaborative and distributed innovation processes.

On the basis of two case studies in the automotive and IT-industries, the explanatory power and limitations of these two opposed theses were discussed in terms of cognitive, strategic and contextual dimensions. Given the considerable costs, risks and uncertainties of basic innovations, the companies are, on the one hand, dependent on trans-border cooperation within the company and with external partners. The carefully designed organizational infrastructure for the trans-border cooperation in US-IT and the common innovation project of two otherwise-competing automotive companies support the thesis of distributed innovation processes. On the other hand, however, both companies tried to exploit the advantages of spatially and organizationally concentrated competences as predicted by the embedding thesis (*cognitive dimension*). A crucial requirement for territorially concentrated learning processes at least in case of Auto-D was a common institutional environment. The company tried to achieve technological leadership in a new drive technology through cooperation with technologically advanced partners, the acquisition of companies and partners and the incremental enhancement of its own competences by a close cooperation with external companies and regional R&D institutes (*normative dimension*). In addition, both companies tried to limit the losses of control, power and competitive advantages in relation to the interest of appropriating and controlling crucial competences (*strategic dimension*). In this tension between distributed and embedded innovation processes, the car company relied predominantly on organizationally, spatially and socially 'encapsulated knowledge' and tried to restrict inter-organizational and trans-border cooperation to tasks at the periphery of the innovation process. In this case, an internationalization of innovation processes was therefore limited to strategically less crucial functions and tasks, which can be decoupled from the core of the new project. This indicates that even globally active companies rely on the spatial, organizational and social proximity of the researchers and developers involved – at least in the core of the innovation projects analysed.

However, the example of the IT-company under review shows that the problems interconnected with spatially distributed innovation processes can be overcome through the extensive use of new communication technologies, by frequent business trips, through international networks of scientific and technical experts and through adequate organizational devices which support global transparency, accountability and decision-making in trans-border development projects. Global competition for attractive R&D tasks, formalized decisions on new projects and on project advances and a formalized reputation hierarchy of technical experts proved to be especially important in this case, because they facilitated the organization of distributed innovation processes. Therefore, distributed and at the same time socially and organizationally embedded development projects are possible – but they seem to be very demanding and not at all self-evident, even in globally active MNCs with an international network of R&D centres.

Note

1 The case studies on which we draw upon in this chapter were performed in the framework of the project 'Regional Learning in Multinational Companies' in 2005–6 (financially supported by the Volkswagen Foundation). The selected projects were the 'most international ones' of the innovation projects analysed in ten German and French MNCs; most of the other projects took place only within one country. The interviews were carried out and synthesized by Jannika Mattes. We thank our interview partners in the two companies under review for their support.

References

Andersson, U., Forsgren, M. and Holm, U. (2007) 'Balancing Subsidiary Influence in the Federative MNC: A Business Network View', *Journal of International Business Studies*, 38, pp. 802–18.

Barmeyer, C.I. (2000) *Interkulturelles Management und Lernstile*, Frankfurt and New York: Campus.

Bartlett, C.A. and Ghoshal, S. (1989) *Managing across Borders: The Transnational Solution*, Boston, MA: Harvard Business School Press.

Bas, C. le and Sierra, C. (2002) 'Location versus Home Country Advantages in R&D Activities',. *Research Policy*, 31, pp. 589–609.

Bathelt, H., Malmberg, A. and Maskell, P. (2004) 'Clusters and Knowledge: Local Buzz, Global Pipelines and the Process of Knowledge Creation', *Progress in Human Geography*, 28, pp. 31–56.

Becker-Ritterspach, F. (2006) 'Wissenstransfer und -integration im Transnationalen Konzern', in U. Mense-Petermann and G. Wagner (eds) *Transnationale Konzerne: Ein neuer Organisationstyp?*, Wiesbaden: VS Verlag, pp. 153–87.

Birkinshaw, J. (2001) 'Strategy and Management in MNE Subsidiaries', in A. Rugman and T. Brewer (eds) *The Oxford Handbook of International Business*, New York: Oxford University Press, pp. 380–401.

Birkinshaw, J. and Lingblad, M. (2005) 'Intrafirm Competition and Charter Evolution in the Multibusiness Firm', *Organization Science*, 16 (6), pp. 674–86.

Bouquet, C. and Birkinshaw, J. (2008) 'Managing Power in the Multinational Corporation: How Low-Power Actors Gain Influence', *Journal of Management*, 34 (3), pp. 477–508.

Castells, M. (2010) *The Rise of the Network Society* (2nd ed.), Chichester: Wiley-Blackwell.

Cooke, P., Heidenreich, M. and Braczyk, H.-J. (eds) (2004) *Regional Innovation Systems: The Role of Governances in a Globalized World*, 2nd edition, London: Routledge.

Dunning, J.H. (1988) *Multinationals, Technology and Competitiveness*, London: Unwin Hyman.

Fritsch, M. and Franke, G. (2004): 'Innovation, Regional Knowledge Spillovers and R&D Cooperation', *Research Policy*, 33, pp. 245–55.

Gerybadze, A. (2004) 'Knowledge Management, Cognitive Coherence, and Equivocality in Distributed Innovation Processes in MNCs', *Management International Review*, 44 (3), pp. 103–28.

Granstrand, O. and Sjörlander, S. (1992) 'Internationalization and Diversification of Multi-technology Corporations', in O. Granstrand, L. Håkanson and S. Sjörlander (eds) *Technology Management and International Business: Internationalization of R&D and Technology*, Chichester: John Wiley & Sons, pp. 181–207.

Hofstede, G. (2001) *Culture's Consequences: Comparing Values, Behaviors, Institutions and Organizations across Nations*, 2nd edition, Thousand Oaks: Sage.

D'Iribarne, P. (2001) *Ehre, Vertrag, Konsens: Unternehmensmanagement und Nationalkulturen*, Frankfurt: New York: Campus.

Kogut, B. and Zander, U. (1993) 'Knowledge of the Firm and the Evolutionary Theory of the Multinational Corporation', *Journal of International Business Studies*, 24, pp. 625–45.

Monteiro, L.F., Arvidsson, N. and Birkinshaw, J. (2008) 'Knowledge Flows within Multinational Corporations Explaining Subsidiary Isolation and Its Performance Implications', *Organization Science*, 19, pp. 90–107.

Narula, R. and Santangelo, G.D. (2009) 'Location, Collocation and R&D Alliances in the European ICT Industry', *Research Policy*, 38(2), pp. 393–403.

OECD (2007) *Science, Technology and Industry Scoreboard*, Paris: OECD.

Patel, P. and Pavitt, K. (1991) 'Large Firms in the Production of the World's Technology: An Important Case of "Non-Globalisation"', *Journal of International Business Studies*, 22 (1), pp. 1–21.

Patel, P. and Vega, M. (1999) 'Patterns of Internationalisation of Corporate Technology: Location versus Home Country Advantages', *Research Policy*, 28, pp. 145–55.

Phene, A. and Almeida, P. (2008) 'Innovation in Multinational Subsidiaries: The Role of Knowledge Assimilation and Subsidiary Capabilities', *Journal of International Business Studies*, 39, pp. 901–19.

Rammert, W. (2006) 'Two Styles of Knowing and Knowledge Regimes', In J. Hage and M. Meeus (eds) *Innovation, Science, and Institutional Change: A Research Handbook*, Oxford: Oxford University Press, pp. 256–93.

Ruigrok, W. and van Tulder, R. (1995) *The Logic of International Restructuring*, London, New York: Routledge.

Scaperlanda, A. (1993) 'Multinational Enterprises and the Global Market', *Journal of Economic Issues*, 27 (2), pp. 605–16.

UNCTAD (2005) *World Investment Report 2005: Transnational Corporations and the Internationalization of R&D*, New York and Geneva: United Nations.

Whitley, R. (1999) *Divergent Capitalisms: The Social Structuring and Change of Business Systems*, Oxford: Oxford Univ. Press.

Zander, I. (1998) 'The Evolution of Technological Capabilities in the Multinational Corporation', *Research Policy*, 27, pp. 17–35.

Zander, I. and Sölvell, Ö. (2000) 'Cross-Border Innovation in the Multinational Corporation: A Research Agenda', *International Studies of Management and Organization*, 30 (2), p. 44.

Zedtwitz, M. von and Gassmann, O. (2002) 'Market versus Technology Drive in R&D Internationalization: Four Different Patterns of Managing Research and Development', *Research Policy*, 31 (4), pp. 569–88.

10 Innovation, generative relationships and scaffolding structures

Implications of a complexity perspective to innovation for public and private interventions

*Federica Rossi, Margherita Russo,
Stefania Sardo and Josh Whitford*

Introduction

An increased emphasis on the role of innovation as a primary driver of economic growth in contemporary knowledge-based economies has put the politics of innovation processes on the front burner. But just what exactly one thinks should be done depends crucially on the theory of innovation that is adopted. In this contribution, we explore how a view of innovation inspired by complexity theory can help us to understand whether we need coordinated interventions to support innovation and, if so, to understand how these can be designed.

Complexity theory is a developing area of research characterized by a wide – and increasing – range of interdisciplinary applications. As a result, the meaning and implications of this approach even within the relatively narrow field of innovation studies are still being negotiated, and different, sometimes conflicting, positions coexist. Therefore, in the next section, we describe what we mean by a complexity perspective to innovation, contrasting our approach and its policy implications both with the traditional 'linear' model of innovation and with more recent and broader 'systemic' approaches. Then, having broadly outlined the theoretical framework on which the analysis is based, we explore its implications for coordinated interventions in support of innovation, with reference to two case studies. Finally, we draw some concluding remarks for policy design.

A complexity perspective to innovation

Economic and organization theories have progressively moved beyond the traditional linear view[1] of innovation – which conceptualizes innovation as a sequence of well defined, temporally and conceptually distinct, stages – in favour of systemic approaches that interpret innovation as a complex process. In this latter approach, the analyst must pay attention to a multiplicity of actors, to the relationships between those actors, and to the social and economic context in which they are embedded. The influential literature on national systems of innovation, which emerged at the beginning of the 1990s (Freeman 1988, Lundvall 1992, Nelson 1993), has highlighted the interplay of a wide range of factors, organizations and policies influencing the capabilities of a nation's firms to innovate. At the same time, the focus on the cognitive aspects of innovation has fostered interest in interactions among agents as sources of new knowledge. Direct interactions among people are considered the main modes of transmission and creation of tacit knowledge (Hagerstrand 1970, Polanyi 1969), which is thought to be a key source of innovation. Researchers have begun to study various forms of cooperation between firms directed at developing innovations, including user-producer interactions (von Hippel 1978, Lundvall 1985, Rosenberg 1963, Russo 1985). The role of proximity – cognitive, technological, social or geographic – in fostering innovation processes has also been explored theoretically and empirically (Audretsch and Feldman 1996, Jaffe 1986, Lundvall 1992, Nooteboom 1999).

Paralleling the evolution of the academic discourse, policymakers' theoretical understanding of innovation processes has also evolved, particularly in Europe (Mytelka and Smith 2002). In line with a systemic approach to innovation, it has been acknowledged that innovation policies must be implemented through interventions that involve not only the activities of basic scientific research, development and commercialization of research outcomes, but also the productive activities of firms and the social and institutional contexts in which they operate. Interest in social interactions as a locus for innovation has led policymakers to assign particular importance to supporting the activities of 'clusters', intended as aggregations of organizations, as well as networks of cooperation among heterogeneous actors (Audretsch 2002).

However, despite the widespread attention dedicated to these issues, designing interventions that are consistent with a systemic approach to innovation often proves a challenge (Russo and Rossi 2009a). Indeed, the European Commission (2003) has explicitly admitted that many interventions claimed to be consistent with a systemic approach to innovation in fact owe much to the linear model. We argue that the solution lies in a conceptualization of innovation as a complex process. This entails, however, recognizing also that it is not possible to devise context-independent ways to support it. Two of us have argued elsewhere (Russo and Rossi 2009) that innovation theories should not be used to derive general 'policy recipes' but rather they should support policymakers in formulating and addressing questions that are appropriate to their particular socioeconomic and institutional contexts. Taking this step, however, requires an improved theoretical and empirical understanding of innovation processes, of the economic actors that drive them and of the channels through which communication processes take place and lead to the development and consolidation of innovations.

To help fill the gap between theoretical understanding and policy implementation, we elucidate the policy implications of a complexity theory understanding of innovation processes, drawing in particular on the dynamic interactionist perspective outlined by Lane and Maxfield (1997, 2005, 2009).

According to this perspective, processes of innovation are guided by (formal or informal) scaffolding structures that shape the rules guiding the operation of the market systems in which such innovations will be embedded, and that create the competence networks that sustain and reproduce necessary systemic functionalities. Scaffolding structures include organizations such as trade or professional associations, but also regular events such as exhibitions and trade fairs, as well as various kinds of publicly funded interventions. Such structures are essential if agents are to effectively manage uncertainty by jointly shaping the direction in which market systems develop (for example, by agreeing on technological standards). They often provide interaction spaces where agents can develop generative relationships that give rise to further innovations. Relationships have high potential to generate innovations when the agents share a common focus on the same artifact or process (aligned directedness) but differ in terms of expertise, attributions or access to particular agents or artifacts (heterogeneity). They also have high potential when agents have the chance to work together on a common activity (opportunities for joint action), as well as when they are able to carry out discursive interactions outside conventional exchanges confined to requests, orders, declarations, and such (right permissions). Agents must also seek to develop recurrent patterns of interactions from which a relationship can emerge (mutual directedness).

In the next section, we show what it means, in practice, to construct scaffolding structures as a means to support diffuse innovation processes. We present two cases, using micro-data on inter-organizational interactions studied through social network analysis. Although not itself explanatory, such analysis can help highlight certain features of inter-organizational interactions whose meaning and purpose can then be interpreted through the lenses of our theory of innovation. The analysis has been complemented by qualitative interviews.

Empirical analysis

The case studies discussed here concern two very different coordinated interventions in support of innovation, both of which have been implemented in Italian regions whose economic structure is characterized by the presence of clusters of firms organized in industrial districts. These are presented to illustrate what it means to devise interventions – both public and private – that take into account the complex nature of innovation processes.

A public policy intervention supporting heterogeneous innovation networks

The 'Innovazione Tecnologica in Toscana' programme, funded within the ERDF Innovative Actions framework (henceforth, RPIA-ITT), was implemented by Tuscany's regional administration in the period 2001–4; the programme was conceived as a pilot test for the use of further structural funds in the region.

RPIA-ITT intended to promote development in the regional economy through the creation of networks of organizations tasked with carrying out innovative projects. Project proposals were solicited within four action lines. The programme required heterogeneous networks (the call for tender requested each cooperation network to comprise at least four firms, one university or public research centre, and one public, private or mixed company having among its statutory aims the provision of services to firms) and encouraged participation by SMEs, which in fact constituted a large share of the actors taking part in the programme. Table 10.1 summarizes the main data on the programme.

For our present purposes, the relevant question is whether this intervention in fact fostered the creation of innovation networks that produced good quality project proposals and exploited them in ways that could give rise to further cascades of innovations. A few organizations played key roles in the policy programme. We set out to investigate these roles by studying the relationships between organizations involved in different projects. To do so, we constructed the two-mode network describing the participation of the 409 organizations involved in the programme, in the 36 (funded and non-funded) project proposals. From this network we extracted the one-mode network of relationships between the 36 projects (participation of the same organization to more than one project indicated a connection between these projects) as well as the one-mode network of relationships between the 409 organizations (participation of the same two organizations to the same proposal indicated a connection between these organizations).

Here we present a brief summary of our findings.[2] Apart from two isolated projects whose participants were not present in other networks (and which failed to secure funding), most projects were connected through one or more organizations in common. We focused in particular on the 58 organizations that were present in more than one project: these featured 177 times as project partners, out of a total of 528 participations (33.5 per cent).

We first noticed that many of them had already collaborated, before and outside the RPIA-ITT programme, on other projects funded by the European Commission, by the regional administration, and by national government agencies. Furthermore, many had also been involved in set of talks set up by the regional administration before the launch of the RPIA-ITT programme. This suggests that the projects were activated by organizations that were already accustomed to working with each other and with the regional administration.

The analysis performed on the one-mode network of 36 projects showed that there are several separate 'k-cores',[3] indicating groups of projects that have relatively dense connections with each other and sparse connections with projects outside the core.

Two of these k-cores were composed of projects mainly submitted to action lines 1 and 2; the funded projects in this group were assigned 45 per cent of the programme's total budget. The organizations connecting these projects, both located in Pisa, are the most central in the one-mode network of organizations[4] described above: Scuola Superiore S. Anna (an influential postgraduate research institution) and CPR (a research consortium whose partners include several local administrations and the main provincial academic institutions, including S.S.S. Anna itself).

Table 10.1 A synthetic overview of the RPIA-ITT programme

	Applications	Funded projects
Number of projects	36	14
Number of partners	528	264
Number of different organizations involved	409	203
Number of SMEs featuring as partners	295	129
Number of different SMEs involved	262	118
Organizations involved in more than one project	58	22
Budget (in euro)	15,504,764*	6,494,298**
Action lines	*% available resources*	
1 Promoting technology and difference of innovation in Western Tuscany	29	
2 Promoting technology and difference of innovation in the fashion industry: textiles, clothing, shoes	27	
3 Promoting technological development and industrial applications of optoelectronic technologies	21	
4 Promoting technological development and industrial agricultural, environmental applications of biotechnologies	23	

Notes
* of these, 11,661,951 euro were to be financed by the region
** of these, 4,703,029 euro were financed by the region

The third k-core was composed of seven projects that were promoted by a network of research centres that are specialized in optoelectronics, a field characterized by technological convergence in a vast range of applications. The interviews confirmed the presence, in the region, of an established network of prestigious public research institutions in this field (CEO, INOA, CNR-IFAC) and of a company, El.En., worldwide leader in laser technology. This is complemented by a tight fabric of SMEs involved in the production of high-technology optic instruments and of related software applications. In order to set up a large number of projects, these organizations were able to rely on their previous experience of successfully bidding for public funds, since optoelectronics had already been a focus of regional policy during the 1990s.

Therefore, the network analysis highlighted the important role played by some research centres and large firms (already accustomed to collaborating with each other and with the regional administration, and to monitoring funding opportunities) in the coordination of most project proposals.

The analysis of individual projects' networks and the qualitative interviews showed that the requirement of heterogeneous competences within each project enabled many organizations to interact with partners with whom they might not have worked otherwise. However, the recruitment of certain organizations, specifically small companies and university departments, proved difficult since both, for different reasons, were unaccustomed to collaborative innovation and were often ill-equipped to deal with the complicated administration of EU-funded programmes. In these cases, their involvement had to be mediated by a set of service providers. Despite having different structural characteristics, different behaviours and different objectives, these service providers engaged in activities (training, certification, research and technology transfer) that allowed them to weave close relationships with both manufacturing firms and other local actors (trade associations, local administrations, universities). These organizations can be defined as multivocal: they understand several languages – from academic research to the specific production technology – and they can interpret the needs of actors that might not even be able to express them. As such, they were essential in order to recruit actors with specific competences, and in many instances, they were also able to develop good quality project proposals and to effectively disseminate the projects' results.

A private technology broker sponsored by a group of large firms

Our second case study involves an organization – named Centro di Ricerca Innovazione Tecnologica (CRIT) – that acts as a 'technology broker' primarily but not exclusively for many leader firms in the Modenese and Emilian mechanical industry. A cross between an association and a firm, CRIT was an indirect consequence of a 1999 law that offered funding and incentives for universities to connect with other research centres in the region of Emilia Romagna. One proposal involved linking a network of university research centres to a 'science technology park' (Sardo 2009) that would be placed in Spilamberto, a town in an area densely packed with mechanical firms on the border between the provinces of Modena and Bologna. The project had the support of local governments (who saw a chance to rehabilitate a large swath of industrial land long in disuse), of the university, and of some of the larger mechanical firms in the region, 14 of which established CRIT in 2000. They each committed to paying what was for such leader firms a limited amount – €25,000 annually – to sustain the organization. The idea was that CRIT would have a small technical and administrative staff that could draw upon the expertise of its member-owners to analyse the demand for innovation in the region. Using that knowledge, it would then broker the demands for technology of the mechanical industry – especially of member firms – and sources of supply. The latter would mostly be located in the proposed technology park, which would host regional research centres and universities.

However, while efforts to establish the technology park have foundered amid political infighting, CRIT has not only survived, it has added 11 new members to its original 14 founders. We argue that it has done so because it has been able to remake itself as an organization that aims more generally to stimulate collaborative innovation, working primarily but not exclusively with member firms that are not direct competitors, but that often share some overlapping technologies and perhaps suppliers.

The most innovative feature of CRIT is the combination of activities in which it engages. CRIT combines services to firms of two basic sorts that we conceptualize as either 'switches' or 'spaces'. Switching is classic brokerage, in which CRIT is approached with a demand for a service or for information, uses data in internal databases or conducts an external search, and either provides the service using internal engineers or connects the client to an organization that can provide the desired service. Switching includes R&D projects, technology scouting, analyses of competitors' patenting patterns. CRIT serves instead as a space of potential interactions when it creates opportunities for open dialogue. CRIT does this by hosting events such as thematic working tables, seminars, technology tours, group training events, and meetings of technical directors. These events are sometimes proposed by CRIT, but are often born of initiatives proposed by member firms. The key is that they take place in a setting in which participants can openly share ideas, but are structured enough that the conversation will be limited to particular topics of technological relevance.

In the period 2000–8, there were 187 space events, against 295 switches. Some 169 organizations participated in just spaces events, 94 in just switches, while 60 took part in both sorts.

In order to understand the nature and dynamics of the interaction space enabled by CRIT – without which such interactions would have not occurred – we analysed the pattern of co-participation of organizations to the events. We created a two-mode network involving all CRIT events and all participants in the period 2000–8. From this we extracted several sub-networks on the basis of temporal intervals (different years) and/or of types of events (switch or space, or particular types of switch or space events). These sub-networks have also been transformed in one-mode networks. Here we present a brief summary of our findings.[5]

First, we observe that the network generated by serviced offered by CRIT grew around a nucleus of more active and central actors. Mechanical firms have the highest centrality[6] in space events; among these, member firms are even more central. The most central group is a nucleus of seven that are especially active: GD, IMA, Tetrakpak, Gruppo Fabbri, Selcom, System, and CMS. These are slightly more central than another also quite central group that includes Sacmi, Italtractor, Rossi Motoriduttori, CNH, and Datalogic. These are also, notably, the same firms that generally have a high centrality in switch events. But for switch events, we see high centrality also for non-members, including especially research centres and universities: the fact that they have very particular competences explains their occasional involvement in a seminar, or in a particular technical meeting.

Second, the analysis of line islands[7] within one-mode networks of participants over time shows that, even among central actors, there is a nucleus that is even more central and that tends to interact a great deal (and that has become even more stable since 2005, the year that CRIT became fully independent of the technology park). It is a nucleus whose activities are highly varied (by type of event, and therefore by the potentiality of interactions with other participants).

Third, there have been changes in the services requested over time. Initially, many firms asked for R&D projects and for technological scouting. Over time, the importance of space events has increased considerably, as if member firms learned how to best use CRIT over time. CRIT too learned from experience, by introducing new services some of which, if not important quantitatively, show that CRIT experiments in response to needs signalled by firms.

Conclusion: supporting collaborative innovation in a complexity perspective

Both case studies concern coordinated interventions that have been successful in promoting innovation in their specific contexts. As such, their interpretation in light of some concepts of complexity theory can help us derive some indications for policy design.

First of all, both interventions were inspired by fairly conventional views of innovation, but they ended up unfolding along unconventional lines.

In the case of RPIA-ITT, the setup of heterogeneous innovation networks was underpinned by a fairly rigid view of what would be an appropriate 'division of innovation labour' within the networks: according to the policymakers, the projects should have practiced technology transfer from universities and research centres – which would have developed relevant innovations – to firms that would implement them in particular applications; small firms would generally act as mere testers of innovations developed elsewhere.

But, the small firms' involvement went beyond the testing of new technological applications. Thanks to the mediation of service providers, the programme became a learning experience and firms became more likely to participate to collaborative innovation in the future. In addition, the university departments acquired a better knowledge of SMEs' competences and needs. The projects (even the planning of those that were not funded), provided a temporary space which allowed unusual interactions. The requirement of heterogeneity, which in the eyes of the policymakers should have simply allowed knowledge transmission from universities and research centres to firms, in fact served also to provide opportunities for further innovation.

In the case of CRIT, the main function of the technology broker according to its founders should have been to favour the match between their demand for technology and information and the supply of that knowledge available elsewhere. However, CRIT and its founders learned over time that the classic brokering function was not the most interesting way to use the organization. Rather, CRIT could allow members the right permissions and opportunities to talk to other organizations, creating a kind of public space which according to Lester and Piore (2004) favours innovation since it provides 'a venue in which new ideas and insights can emerge, without the risk that private appropriation will undermine or truncate the discussion'.

Therefore, both interventions were conceived as conventional technology transfer exercises, but much of their value added came from the creation of spaces for open-ended discussion, where the interpretative ambiguity (Fonseca 2002, Lane and Maxfield 1997) necessary for innovation could emerge.

This leads us to the second point: the importance of structuring interactions. In both cases, the space for interactions was designed (sometimes involuntarily) to provide the conditions that enhance generative potential. In the RPIA-ITT, the involvement of service providers allowed the recruitment of small firms and university departments that were relatively unaccustomed to dealing with each other, thus helping to achieve some degree of heterogeneity. In CRIT, heterogeneity is monitored by the members, which are careful to involve organizations that are not direct competitors. In both cases, opportunities for joint action and the right permissions for interaction were also present.

Within heterogeneous networks, an important role is played by mediating organizations capable of engaging in multivocality, as opposed to traditional brokering activities. Such mediators do not merely transmit information between agents that do not know each other, bridging a structural hole in the network (Burt 1992); they also facilitate direct exchanges among these agents. Service providers and the staff of CRIT are the agents able to perform this role in each case.

Third, both the RPIA-ITT programme and CRIT can be seen as scaffolding structures providing continuity in support of innovation processes that unfold in many cases over long temporal scales. Interventions supporting collaborative innovation generally need to last over a long period of time – the development of new technologies and the understanding of how to exploit them commercially are lengthy processes, after all. Especially in the case of radically new technologies that open up new market systems, scaffolding structures are important in order to foster the creation of the competences necessary to ensure that the technologies can be maintained and diffused. In the case of RPIA-ITT the short duration of the programme was perceived as a limiting factor, but not a critical impediment to innovation, since the main actors involved in the programme were able to exploit a wide range of policy instruments and managed to effectively use the regional policy infrastructure as a scaffold for their innovation activities. In the case of CRIT, it took its members several years to learn how to use the organization productively. This was made possible because firms had made a continuous commitment to participate in at least some sponsored activities.

Fourth, the comparison between these two cases highlights that there is no one-size-fits-all approach to sustaining innovation through collaborative processes. The two interventions considered were inspired by a fairly conventional view of innovation, but they worked because their implementation was tailored to the actual features of the local innovation systems. For example, the RPIA-ITT explicitly involved service providers, which are key actors in Tuscany's regional innovation system. The creation of CRIT probably would not have occurred without a critical mass of large local firms that are active in the same sector but are not in direct competition with each other. Despite these differences, one can still generalize to a conclusion: any successful coordinated intervention in support of innovation requires an effort to identify, *ex ante*, the key actors that are best able to construct networks of relationships that can support innovation processes by creating conditions that enhance the generative potential of key relationships.

Finally, improving the tools available for the analysis of collaboration networks can enhance our ability to monitor and support innovation processes. The analysis of dynamic temporal networks and of multi-level networks involving both organizations and individuals should help in this sense, as should the development of agent-based models to construct scenarios relevant to innovation policies. Better integration of these quantitative techniques with ethnographic research should also enrich our set of tools for policy design and analysis.

Notes

1 For an overview of the historical development of the linear model, see Godin 2006.
2 A more detailed analysis is presented in Russo and Rossi (2009).
3 For the notion of k-core (groups of connected vertices which have at least k links with each other) see Moody and White (2003).
4 Different measures of centrality (degree, betweenness and closeness centrality indexes: see Degenne and Forsé 1999, for definitions) relative to the same network led to similar results.
5 A more detailed analysis is presented in Russo and Whitford (2009).
6 Most analyses were performed using betweenness centrality indexes, but consistent results were found when degree centrality was used instead.
7 The computation of the line islands was done with Pajek (min. = 3, max. = 32).

References

Audretsch, D. (2002) *Entrepreneurship: A Survey of the Literature*, prepared for the European Commission, Enterprise Directorate General.

Audretsch, D.B. and Feldman, M.P. (1996) 'Spillovers and the Geography of Innovation and Production', *American Economic Review*, 86 (3), pp. 630–42.

Burt, R.S. (1992) *Structural Holes: The Social Structure of Competition*, Cambridge, MA: Harvard University Press.

Castells, M. (2010) *The Rise of the Network Society, Volume I: The Information Age*, Chichester: Wiley-Blackwell.

Degenne, A. and Forsé, M. (1999) *Introducing Social Networks*, London: Sage.

European Commission (2003) *Investing in Research: An Action Plan for Europe*, COM(2003)226.

Fonseca, J. (2002) *Complexity and Innovation in Organizations*, London: Routledge.

Freeman, C. (1988) 'Japan: A New National System of Innovation?', in G. Dosi, C. Freeman, R. Nelson and L. Soete (eds) *Technical Change and Economic Theory*, London: Pinter Publishers, pp. 38–66.

Godin, B. (2006) 'The Linear Model of Innovation: The Historical Construction of an Analytical Framework', *Science Technology Human Values*, 31 (6), pp. 639–67.

Hagerstrand, T. (1970) 'What about People in Regional Science?', *Papers of the Regional Science Association*, 24, pp. 7–21.

Hippel, E. von (1978) 'Users as Innovators', *Technology Review*, 80 (3), pp. 1131–9.

Jaffe, A.B. (1986) 'Technological Opportunity and Spillover of R&D: Evidence from Firms' Patents, Profits and Market Value', *American Economic Review*, 76 (5), pp. 985–1001.

Lane, D.A. and Maxfield, R. (1997) 'Foresight, Complexity and Strategy', in B. Arthur, S. Durlauf and D.A. Lane (eds) *The Economy as a Complex Evolving System II*, Reading, MA: Addison-Wesley.

Lane, D.A. and Maxfield, R. (2005) 'Ontological Uncertainty and Innovation', *Journal of Evolutionary Economics*, 15 (1), pp. 3–50.

Lane, D.A. and Maxfield, R. (2009) 'Building a New Market System: Effective Action, Redirection and Generative Relationships', in D. Lane, D. Pumain, S.E. van der Leeuw and G. West (eds) *Complexity Perspectives in Innovation and Social Change*, Berlin: Springer, Method Series 7, pp. 263–88.

Lester, R. and Piore, M. (2004) *Innovation: The Missing Dimension*, Cambridge, MA: Harvard University Press.

Lundvall, B.A. (1985) *Product Innovation and User–Producer Interaction*, Aalborg: Aalborg University Press.

Lundvall, B.A. (1992) *National Systems of Innovation: Towards a Theory of Innovation and Interactive Learning*, London: Pinter Publishers.

Moody, J. and White, D.R. (2003) 'Structural Cohesion and Embeddedness: A Hierarchical Concept of Social Groups', *American Sociological Review*, 68 (1), pp. 103–27.

Mytelka, L. and Smith, K. (2002) 'Policy Learning and Innovation Theory: An Interactive and Co-evolving Process', *Research Policy*, 31 (8/9), pp. 1467–79.

Nelson, R.R. (1993) *National Innovation Systems: A Comparative Analysis*, New York: Oxford University Press.

Nooteboom, B. (1999) 'Innovation, Learning and Industrial Organization', *Cambridge Journal of Economics*, 23, pp. 127–50.

Polanyi, M. (1969) in M. Grene (ed.) *Knowing and Being*, London, UK: Routledge and Kegan Paul.

Rosenberg, N. (1963) 'Technological Change in the Machine Tool Industry: 1840–910', *Journal of Economic History*, pp. 414–43.

Rossi, F. and Russo, M. (2009) 'Innovation Policy: Levels and Levers', in D. Lane, D.

Pumain, S.E. van der Leeuw and G. West (eds) *Complexity Perspectives in Innovation and Social Change*, Berlin: Springer, Method Series 7, pp. 311–27.

Russo, M. (1985) 'Technical Change and the Industrial District', *Research Policy*, December, 14 (6), pp. 329–43.

Russo, M. and Rossi, F. (2009) 'Cooperation Partnerships and Innovation. A Complex System Perspective to the Design, Management and Evaluation of an EU Regional Innovation Policy Programme', *Evaluation*, 15 (1), pp. 75–100.

Russo, M. and Whitford, J. (2009) 'Industrial Districts in a Globalizing World: A Model to Change, or a Model of Change'? *Materiali discussione DEP*, 615, Unimore. http://merlino.unimo.it/web_dep/materiali_discussione/0615.pdf.

Sardo, S. (2009) 'Brokeraggio tecnologico nel settore metalmeccanico in Emilia-Romagna: dal Parco Scientifico Tecnologico ex-SIPE a CRIT srl', *Materiali discussione DEP*, 614, Unimore.

11 The road to recovery

Investing in innovation for knowledge-based growth

Henry Etzkowitz and Marina Ranga

Introduction

This chapter discusses a strategy, centred on the entrepreneurial university, to address the current economic crisis. This crisis has deeper origins than the collapse of a financial bubble and is ultimately caused by gaps in the transition from an industrial to a knowledge-based society. An industrial mode of production has run out of steam in many countries, hastened by globalization challenges and increased competition. The spectre of a global incapacity to manage change haunts the innovation systems of societies irrespective of their national differences, developmental stage or level of success. Economic crises impelled by the downturn of significant industries from the 1970s onwards and persisting blockages to industrialization in many developing countries have brought transition to a knowledge-based society to the forefront as a universal aspiration.

Universities and other higher education institutions that are major producers of knowledge are thus seen in a new light as potential direct contributors to economic advance in addition to their traditional supporting roles. University institutional characteristics such as human capital flow-through of student generations and ability to integrate multiple missions such as education, research and technology transfer generate a significant innovation potential that often remains untapped. Involving new actors that do not traditionally participate directly in innovation such as the university, restructuring others to perform new roles and creating new networks and relationships may enhance the knowledge base and promote its utilization. Despite widely different university systems, a common direction of academic development towards an entrepreneurial university in a Triple Helix regime of co-equal institutional spheres (university–industry–government) may be discerned in response to these challenges and opportunities.

The Triple Helix model places the university in a leading role in knowledge-based societies, equivalent to government and industry that have held preeminent status from the eighteenth century (Etzkowitz 2008). A triple helix typically begins when an existing innovation regime, whether a single helix, based on industry, or a double helix of government-industry falls into a crisis that cannot be resolved within the existing framework. It is a guided social evolution, often driven by research-based policy interventions, in contrast to the accidental mutations of biological evolution. Innovation policies directed toward economic recovery may be based on constructing Triple Helix regimes, with a key role for the entrepreneurial university at the regional and sectoral levels. Entrepreneurship is thus expanded from a firm-based to an academic concept and from an individual to a group activity. The potential role of the entrepreneurial university in innovation policy and practice calls for a rethinking of the traditional drivers of innovation.

The new drivers of innovation

Economic and social challenges of recent decades have removed ideological blinders that obscured the true nature of innovation in both laissez-faire and statist societies and made explicit new drivers of innovation such as the polyvalent nature of knowledge, organizational evolution and collective entrepreneurship.

The polyvalent nature of knowledge

Knowledge is polyvalent, imbued with multiple simultaneous characteristics (Viale and Etzkowitz 2005). It is at one and the same time theoretical and practical, publishable and patentable. Hybridization among institutional spheres is also driven by the diminishing gap between 'blue sky' theoretical knowledge and practical 'knowledge for use'. The so-called 'Pasteur's Quadrant' exemplifies knowledge that is simultaneously theoretical and practical (Stokes 1998). Bohr's 'pure basic research' quadrant actually contains significant elements of Edison's 'pure applied research' quadrant, as the economic potential of basic research and the scientization of invention are realized.

These two individuals themselves are actually good examples of the increasingly conjoint nature of knowledge. Bohr was not an isolated theorist but was also directly involved in the effort to control the deleterious consequences of technology. He personally lobbied US Secretary of State James Byrne not to use the atomic bomb against Japan and hold a demonstration of its power instead. He was also one of the founders of the scientists' movement to control atomic weapons. Edison, the self-styled 'cut and try' inventor, employed scientists on his staff and was the discoverer of the 'Edison effect', an eponymous physical phenomenon. The polyvalent nature of academic knowledge became temporarily hidden in an 'ivory tower' academic model that emerged in the late nineteenth century but is now exploding in a variety of formats (Shapin 2008).

Organizational evolution

Innovation is no longer the province of any single society or organization, or confined to a particular level of development. New concepts for innovative organizations are regularly invented and reinvented to fit local circumstances through circulation and interaction among Triple Helix actors. For example, the

Brazil vision of incubators transcends simple support for firm formation and grew into an educational model of training groups of individuals to work as an effective organization. Thus, incubation became part of the teaching as well as the economic development mission of the university. Brazilian science policy specialists made the US model of incubating high-tech firms from academic research relevant to broader constituencies including industry, government (e.g. municipalities and industrial associations seeking to expand low and medium-tech clusters), and not-for-profit organizations seeking to address problems of poverty by encouraging the formation of cooperatives in the service sector.

Renewal of innovation systems proceeded from the technological implications of knowledge and the incapacity of existing organizational formats to effectively manage these consequences. Awareness of the technological potential of academic research grew as a consequence of Second World War R&D projects heavily involving academic scientists and their continuation in the postwar with ongoing government funding. The beginnings of professional technology transfer activities from university to industry created a confluence of interest from two sources in clarifying the ownership of academic generated intellectual property rights in the United States. On the one hand, technology transfer professionals within the university and, on the other, legislators interested in utilizing academic knowledge to enhance industrial productivity and innovation in response to competitiveness challenges came together to craft an amendment to the US patent law, the so-called Bayh–Dole Act of 1980.

Management of intellectual property became an explicit organizational task. The law turned over to the universities the intellectual property rights generated from federally funded research, with two basic conditions: (*a*) universities should make an effort to put that knowledge to use, and (*b*) the faculty and student inventors should receive a significant share of monies earned (Stevens 2004). In subsequent decades, measures similar to the Bayh–Dole Act have been instituted in various countries, including Japan, Germany, Denmark and Brazil, or are in the process of adoption, e.g. in Eastern European countries. Their common intent is to encourage the university to play a more direct role in industrial innovation, through technology transfer and creation of spin-off firms.

Collective entrepreneurship

Entrepreneurship is a broad and fundamental phenomenon that is inherent in the creation of any organizational format, irrespective of whether or not it is directly tied to a profit motive. Schumpeter (1951), for example, identified the US Department of Agriculture as an entrepreneur in recognition of its role in incentivizing and renewing American agriculture. Schumpeter implicitly recognized that the entrepreneurial phenomenon was broader than an individual act. The collective nature of entrepreneurship is particularly visible in high-tech entrepreneurship, which is virtually always a collective phenomenon, consisting of an 'entrepreneurial circle' of complementary individuals. Technical and business expertise backed by previous entrepreneurial experience constitutes the 'collective entrepreneur', as only rarely does a single individual embody all of these required elements.

A new high-tech firm typically takes off after collaboration is secured between persons with business and technical expertise, backed by an experienced entrepreneur, especially if the initial collaborators are relatively inexperienced. However, in the United States a strong ideology of individual entrepreneurship usually suppresses the contributions of collaborators and pushes a single individual to the forefront. For example, in the creation of the Apple origin myth, Steve Jobs moved to the foreground and Steve Wozniak, the technical collaborator, and Mark Merkula, the experienced semiconductor executive, who gave the original duo credibility with suppliers and financers, were elided (Freiberger and Swaine 2000). In Sweden, by contrast, collective entrepreneurship is openly accepted, as individuals are culturally inhibited from attempting an entrepreneurial act unless backed up by a group.

The academic entrepreneurial transition

The relatively new third mission of the university for economic and social development originated from the transition to the research university and the significant inventions that ensued. For example, the University of Toronto realized that it had to take responsibility for the discovery of insulin by its researchers, just as the University of Wisconsin did for the Babcock test for milk purity in order to ensure ethical manufacture. These universities were impelled to patent in order to be able to guarantee that these inventions were used for the public good. A sense of public responsibility and realization that the name of the university could be tainted by faulty substitutes led these universities to seek intellectual property protection.

The next step was the realization that university-originated inventions could become a source of income for the university as well as for the academic inventors. At MIT this realization came early in the twentieth century in response to business people looking around the campus for technology and poaching the discoveries of its staff. The University of Wisconsin was one of the earliest schools to establish an organization, a wholly controlled external entity, the Wisconsin Alumni Research Foundation, to systematically conduct technology transfer based on intellectual property protection (Apple 1989).

The origins of entrepreneurial science may be found in the university's teaching as well as in its research mission. For example, as more explicit methods of investigation were developed in organic chemistry, it became possible to train new researchers on a larger scale. Justus von Liebig invented the teaching laboratory at the University of Giessen in the mid-nineteenth century. Renovating an unused military barracks at the outskirts of the university, the teaching laboratory was organized with its rows of workbenches. Students were supervised by assistants under the general supervision of the professor in a format much as we are familiar with to this day. Students were given substances to investigate whose properties had not been previously established. As they conducted a repertoire of tests, new knowledge was created.

Some of the substances investigated were also found to possess useful properties. Liebig organized a firm to market one of these discoveries known as 'Liebig's extract', an edible substance with healthful properties that today would be called a 'functional food'. Liebig organized a second firm based on his theorizing about metallic-based fertilizers but farmers found the product useless. The firm failed and the theory was disconfirmed (Rossiter 1974). Nevertheless, an essential feature of the entrepreneurial university as a source of commercializable research and spin-off firms was adumbrated.

A first step for the university taking the role of entrepreneur is the ability to set its own strategic direction. The second step is a commitment to seeing that the knowledge developed within the university is put to use, especially in its local region. The entrepreneurial university takes in inputs and problems from the local environment and translates the outputs of academic knowledge into economic activity. This can take a variety of forms, from meeting local human capital and research needs to playing a collaborative role in establishing a strategy for knowledge-based regional economic development and participation in initiatives to implement that strategy. In Silicon Valley, after generations of firms were spun off from the original university start-ups, links with Stanford weakened and academic roles reverted to the traditional ones of supplying human capital and knowledge. In contrast, in Linköping, Sweden, the university took the lead in creating an organization that keeps high tech firms that it helped create in regular contact with the university.

However, when an economic crisis appears, especially in a high-tech region, the university is typically called into play to address the downturn by supplying new sources of technology as the basis of firm formation. It is at this point that Triple Helix coalitions are very useful in re-energizing the links between the knowledge and economic spheres. For example, during the downturn of the mid-1990s 'Joint Venture: Silicon Valley' was established, bringing together Silicon Valley's firms, governments and universities to brainstorm the potential of new fields for economic development. Out of these open public meetings a strategy was generated to promote the development of 'networking' firms.

The concept of the entrepreneurial university is often misinterpreted to imply the subordination of the university to business on the assumption that industry is inevitably the stronger partner. A counter hypothesis is that the entrepreneurial university increases its independence through its own income-generating capacities. The question of *who influences whom* in university–industry–government interactions is always an empirical one, with the answer weighted towards the actor with the most highly valued good under varying societal conditions. The dominance of industry over university in the industrial society is superseded in knowledge-based societies, where knowledge embedded in intellectual property gives its holder significant bargaining power in setting the terms of its utilization. The enhanced role of the university in a knowledge society calls for a reconsideration of previous theoretical frameworks in which political or economic institutions are dominant, whether individually or in collaboration.

The Triple Helix and regional development

The Triple Helix can be used as an analytical framework for the renewal of declining industrial regions and the growth of new knowledge-based regions. The case of New England in the early twentieth century provides a good example in this sense. A declining industrial region with significant financial institutions built upon the region's previous economic success, New England was also a knowledge region with a high concentration of universities. The New England Council, founded by the Governors of New England in the 1920s, representing university–industry–government, provided the impetus. The analysis of regional strengths and weaknesses and brainstorming among the Triple Helix actors in this venue, to take advantage of the former and fill gaps in the latter, led to the creation of a high-tech firm formation strategy from academic research and the invention of the venture capital firm to facilitate renewal (Etzkowitz 2002).

A comparison of contemporary north-east England to early twentieth century New England may be instructive to understanding the issues involved in the renewal of an older industrial region. The following two elements are relevant to this comparison:

Industrial policy choices between reviving older clusters, building new ones or both

By the 1930s the staff of the New England Council determined that the existing industrial base of New England had declined too far to be revived and that the focus of revival strategy should be on creating a new industrial base. In north-east England retention of the heavy engineering base and support of mid-tech industries, in chemical and automotive sectors, has been at least as important a focus as starting new industries. New England failed in its effort to attract automotive branch plants and the like and had little alternative other than utilizing its academic base to spin off new enterprises. North-east England was able to attract a Nissan auto plant and many of its engineering firms found a new lease on life working for the energy industry. Starting with a much smaller academic base than New England, the Northeast's dual strategy made sense (Etzkowitz *et al.* 2007)

Knowledge base and critical mass of science and technology resources

In contrast to New England that had a pre-existing critical mass of academic research with commercial potential, north-east England is struggling to create viable research clusters. North-east England starts with a smaller academic base and therefore might usefully think of aligning itself with neighbours to create critical mass. Newcastle University has attempted to create research centres with a critical mass of activity by partnering with Durham University and by working with innovative research institutes like the Centre for Life, supported by the European Union and the regional development agency as a mixed use of science museum, research centre, incubator facility and science park.

However, the efforts to develop research capacity are competing with other universities' and regions' attempts to build similar capabilities. For example, two key stem cell research groups left Newcastle for attractive offers elsewhere: one to Spain and the other to France. One investigator felt he was not offered an appropriate professorship; the other that adequate research facilities were not provided in a timely fashion. Since the research cluster was based on a few leading investigators, this loss was devastating to Newcastle's position in the field and likely the explanation for its non-appearance on the current list of UK stem cell research sites that includes Cambridge, Edinburgh, Liverpool, London and Manchester. However, Newcastle University Medical School is making up for the loss by recruiting stem cell researchers in related areas of investigation to human embryo stem cell research (Whitaker 2009).

The question remains whether there is any single niche area of high tech in which Newcastle, or even the entire north-east of England, can attain and hold world-class status? If not, is there an alternative strategy for success in this highly competitive arena of high-tech development? The ability of a city or region to hold research groups and firms in a high-tech field has been called 'stickiness'. A recent consultant's report (*Journal* 2009) labelled Newcastle a 'Science Super City' focused on a single area of expertise, nanotechnology, although just a few years ago Newcastle was recognized in the *New York Times* (2004) as one of a very few world centres of stem cell research.

Despite significant research strengths in energy, nanotechnology and stem cells, but without being able to capture a major national laboratory to date, it is unlikely that there will be any single area of technology and science, in which north-east England can achieve word class critical mass on its own. Given difficulties in achieving critical mass locally, Newcastle's nascent technology clusters might link with partners elsewhere to create 'virtual technology regions' as Copenhagen and southern Sweden have done with their Oresund project and its subsidiaries, such as the Medicon Valley Academy that encourages collaboration in a high-tech field across a broader region (Hospers 2006).

A potential UK example would be to link the highly profitable gaming expertise in Dundee with Teesside University's training programmes in gaming and Newcastle's start-ups as a first step towards creating viable concentration of expertise that may be called 'Hadrian's Valley' as it comes to be achieved, just as Silicon Valley was labelled after the initial rise of the semiconductor industry. This concentration of resources should be supported as a building block of a larger project with the potential to replicate the scale of Silicon Valley, from different technological platforms (Etzkowitz and Dzisah 2008a). Adapting the Canadian Centres of Excellence model, across geographically dispersed universities, could create sufficient academic critical mass to engender self-sustaining high-tech growth. To conserve green space and promote interaction, Hadrian's Valley should be linked by upgraded rail service, like the 'mag-lev' line linking Shanghai and its airport that is currently being extended to a broader region.

A critical mass of academic and government research in emerging areas of science and technology has been the basis of successful science cities.

Competitors in California, for example, are spending three billion dollars to gain leadership in stem cell research and commercialization through a public bond issue, borrowing funds in the expectation, based on a venture capital model, that even a single spectacular success of a start-up firm could return the investment (Etzkowitz and Rickne 2009).

UK Science City strategy

Gordon Brown launched the UK Science City project in 2004 when he was Chancellor of the Exchequer. It was based upon a successful effort to develop a science park in York that had made up for significant employment losses in existing industries by developing new firms in software and biotechnology. The label of Science City was given to five other reindustrializing cities in the United Kingdom in an attempt to encourage local initiative (Birmingham, Bristol, Manchester, Nottingham and Newcastle). However, no significant new block of funds was allocated to support this effort. The various science city projects have struggled, individually and collectively, to find their way. Newcastle Science City, led by Newcastle University, in partnership with the Newcastle City Council and the One Northeast regional development agency, focused on a property redevelopment scheme to attract science-based firms to the region, in parallel with enhancing the university's research capabilities in four science city theme areas: ageing and health, energy and environment, molecular engineering, and stem cells and regenerative medicine.

Much of the strategy for Newcastle Science City is based on building new facilities for firms in the expectation that the opportunity to interact with academics in related fields will be a sufficient attractor. A few years ago, the former operational leadership of the Centre for Life attended industry conferences in an effort to determine what factors would lead firms to locate R&D facilities in Newcastle. They felt frustrated in their efforts to make a 'build it and they will come' strategy of constructing a first class facility to attract a significant number of science-based firms work and were soon replaced with new leadership. The centre has perforce changed its strategy from a joint firm/academic lab locale to a primarily academic research facility, taking an indigenous development approach.

The abandoned headquarters site of Newcastle Brown Breweries was cleared to provide space for new construction. Tearing down Newcastle Brown Breweries instead of giving it other more profitable uses such as renovating it into a tourist attraction, as Dublin did with the old Guinness brewery, is a missed opportunity. Newcastle could have taken advantage of the international reputation of the Newcastle Brown brand to attract visitors. The original building might be reconstructed as was done for bombed out historical sites in many European cities following the Second World War. Rebuilding the brewery as a tourist destination would likely provide a better return on investment than the projected £23 million 'Science Gateway' office block, a remnant of a larger-scale construction vision expected to be largely private sector funded that came to naught in the recent downturn. Instead, public funds are to be utilized in an effort to achieve a piece of the original plan.

A human resources development strategy should be pursued before, or at least in parallel, with facility building. Science Gateway puts 'the cart before the horse' creating an impressive structure to house science-based companies rather than expanding the process to create such firms. But, where is the horse to pull the cart? Successful science cities, like Silicon Valley and Boston, were built on creating the conditions for firms to spin-off from academic and government research. Unsuccessful science cities built impressive science parks first. Some were able to attract multi-national firms to locate facilities, at least temporarily, but eventually, like Stockholm's Kista Science Park, have adopted a strategy of founding a new information technology-focused university as the basis for producing spin-offs. Others, like One Northeast, the regional development agency in north-east England, supports the 'Professors of Practice' experiment, linking Newcastle University and firms (Etzkowitz and Dzisah 2008b).

To jump-start the Science City, the Professors of Practice experiment was started in 2006 by Newcastle University and One Northeast at the suggestion of the Triple Helix Research Group – a 'science city' think-tank located in the Newcastle University Business School. The traditional Professor of Practice model, based on bringing distinguished practitioners into the university as teachers, was adapted from teaching to research by attracting scientific entrepreneurs with high academic credentials and research experience who had started successful firms and gained significant management experience. In the Science City theme areas mentioned above, four Professors of Practice (PoPs) were appointed. Working half time in each venue, the PoPs serve as role models for faculty while maintaining their industrial inspiration. They have initiated various projects, such as drawing together the university's drug discovery experts to undertake larger projects and attract higher levels of funding, a new doctoral programme integrating business, engineering and medical disciplines to train future academic and industrial leaders in the medical devices field (NESTA 2009).

The next step in developing the PoPs model is to extend it down the academic ladder by appointing post doctoral fellows and lecturers as Researchers of Practice (RoPs), who will work half time in an academic unit and half time in the business development side of the university e.g. technology transfer office, incubator facility or science park. The RoPs will involve their students in analysing feasibility of technology transfer projects and in developing business plans with firms in the university's incubator facilities, along with traditional academic tasks.

Birmingham Science City is organizing an innovative human capital initiative called the Science City Interdisciplinary Research Alliance (SCIRA), to link Birmingham and Warwick universities, establish critical mass and encourage a new interdisciplinary ethos within and among the theme areas of the science city (advanced materials, energy futures, translational medicine and IT). The £80 million to be provided by the West Midlands Regional Development Agency – Advantage West Midlands – will be used for new research equipment and infrastructure. To encourage cross-institutional links through dual appointments and interdisciplinary development, eight SCIRA fellows have already been appointed (out of an expected 15–20). The SCIRA fellows participate in a 'college' across the whole science city and they are also expected to promote significant industrial interactions and public engagement in the West Midlands region and beyond.

A knowledge-based strategy to address the economic crisis

In response to the 1930s depression, governments followed John Maynard Keynes advice and built public works such as schools, dams and post offices to employ people and revive consumer demand (Rose 1993). Later on, at the very close of the Second World War, W.H. Beveridge (1945) presented a report to His Majesty's Government 'Full Employment in A Free Society'. He duly noted that an incipient depression prior to the First World War and the Great Depression of the 1930s were only cured by full wartime mobilization. Beveridge posited that addressing deep unmet needs in British society for education, housing, health and other social goods could create demand sufficient to achieve the objective.

In the current crisis, building post-offices is no longer an adequate strategy for renewal in the Internet era. In addition to renewing physical infrastructure, government should invest in knowledge and innovation as the most effective substitute for war to cure the deep economic downturn. Government has taken a larger role in society, as industry by itself has proven a sub-optimal economic actor. However, just like a single helix government strategy was inadequate to transcend the 1930s depression, a double helix of government–industry may be insufficient to bridge the transition between the industrial and the knowledge-based modes of production. A third helix – the entrepreneurial university as the core of knowledge-based regional development efforts, needs to be brought into the picture. Science cities need to move from labels to realities through sustained funding from both public and private sources.

The beginnings of a knowledge-based strategy for renewal need to be expanded upon. For example, the US government already allocated about 2.5 per cent of GDP ($800 billion) for the economic stimulus package, while the European Union mobilized 1.5 per cent of the GDP (€200 billion) for this purpose, estimating that the real spending is about 3.3 per cent when other expenditures are taken into account. The money is intended to support existing and new industries, particularly in the area of green energy and low-carbon consumption, preserve jobs and boost professional skills. Although criticism regarding the insufficiency of the funding has been often expressed, the longer-term effects of the stimulus packages still remain to be seen. Knowledge investment returns are highly promising, but also highly unpredictable.

The forces of 'creative destruction' identified by Joseph Schumpeter (1942) work at an ever more furious pace during a downturn. The 'Valley of Death' between invention and innovation that deepens in downturn is the tip of an iceberg of an underlying innovation gap. Investments in knowledge also need to recognize the need to move to a new technological base. Some measures have already been adopted in this respect, such as the United Kingdom's plan to extend broadband access nation-wide.

Paradoxically, the downturn is a propitious moment for high-tech firm formation. People who are laid off, especially from high-tech firms are available to explore starting new ventures. The blockages induced by the slow-down or even disappearance of private venture capital to support firm formation can be lessened by public venture capital, which can equalize the flow of venture capital among regions, and also make funds available when private sources are frozen.

The reconfiguration of various interface entities associated with the entrepreneurial university, such as the science parks, research centres and technology transfer offices can enhance innovation. For example, technology transfer offices at some less research-intensive universities started to broaden their remit by encouraging the early phases of developing research with commercial potential, as well as the later phases focused on harvesting commercially-ready findings. It may also be productive to greatly expand technology transfer capabilities at highly successful research universities to provide outreach service for the majority of faculty researchers, who may have only a moderate interest in technology transfer. This would supplement the usual focus on serial entrepreneurs of a relatively small technology transfer office on a research-intensive campus.

Spontaneous generation of knowledge-based high-tech regions is a myth, just as self-regulating capitalism is a chimera. Behind every successful initiative is an often-lost history of Triple Helix interactions, public–private collaboration and investment. Today, in the face of escalating economic crisis, there is a return to Marx and Keynes' ideas. Although Marx explained capitalism's unsustainable bubbles, he did not delineate a clear path to a new mode of production (Perez 2003). Keynes justified an activist role for government in the downturn, but did not take on board Schumpeter's thesis of 'creative destruction' and the need to replace old industries with new ones. A Triple Helix strategy of university–industry–government interactions can be a driver of innovation at the cusp of transition between industrial and knowledge society.

References

Apple, R. (1989) 'Patenting University Research: Harry Steenback and the Wisconsin Alumni Research Foundation', *ISIS*, 80 (3), pp. 374–94.

Benneworth, P. and G. Hospers (2007) 'The New Economic Geography of Old Industrial Regions: Universities as Global–Local Pipelines', *Environment and Planning*, 25 (6), pp. 779–802.

Beveridge, W.H. (1945) *Full Employment in a Free Society*, New York: W.W. Norton.

Etzkowitz, H. (2002) *MIT and the Rise of Entrepreneurial Science*, London: Routledge.

Etzkowitz, H. (2008) *The Triple Helix: University–Industry–Government Innovation in Action*, London: Routledge.

Etzkowitz, H. and J. Dzisah. (2008a) 'Unity and Diversity in High-tech Growth and Renewal: Learning from Boston and Silicon Valley', *European Planning Studies*, 16 (8), pp. 1009–24.

Etzkowitz, H. and J. Dzisah (2008b) 'Professors of Practice and the Entrepreneurial University', *International Higher Education*, 49 (fall).

Etzkowitz, H. and A. Rickne (2009) 'Science Policy and Direct Democracy: Proposition 71, California's Experiment in Stem Cell Innovation', paper presented at the Triple Helix 7 Conference, Glasgow, 17–19 June.

Etzkowitz, H, L.M. Ranga, J. Dzisah, Y. Lu and C. Zhou (2007) 'Evaluation of International Knowledge-based Entrepreneurship Programmes and Policy Recommendations for the North East of England', Synthesis Report prepared for the Newcastle City Council, November 2007.

Freiberger, P. and M. Swaine (2000) *Fire in the Valley: The Making of the Personal Computer*, New York: McGraw Hill.

Hospers, G. (2006) 'Borders, Bridges and Branding: the Transformation of the Oresund Region into an Imagined Space', *European Planning Studies*, 14 (8), pp. 1015–33.

NESTA (2009) *The Connected University: Driving Recovery and Growth in the UK Economy*. Online, available at: www.nesta.org.uk/the-connected-university.

'Our Tiny Technology Needs Critical Mass' (2009) *Journal*, 29 May, p. 5. Online, available at: www.journallive.co.uk.

Perez, C. (2003) *Technological Revolutions and Financial Capital: The Dynamics of Bubbles and Golden Ages*, Cheltenham: Edward Elgar.

Romer, C. (1991) *What Ended the Great Depression?*, Cambridge, MA: NBER Working Paper No. W3829.

Rose, N. (1993) *Put to Work: Relief Programs in the Great Depression*, New York: Monthly Review Press.

Rossiter, M. (1974) *The Emergence of Agricultural Science: Justus Liebig and the Americans*, New Haven: Yale University Press.

Schumpeter, J. (1942) *Capitalism, Socialism and Democracy*, New York: Harper.

Schumpeter J. (1951) *Essays on Economic Topics*, Port Washington: Kennikat Press.

Shapin, S. (2008) *The Scientific Life: A Moral History of a Late Modern Vocation*, Chicago: University of Chicago Press.

Stevens, A. (2004) 'The Enactment of Bayh-Dole', *Journal of Technology Transfer*, 29 (1), pp. 93–9.

Stokes, D. (1998) *Pasteur's Quadrant: Basic Science and Technological Innovation*, Washington, DC: Brookings Institution Press.

'Therapeutic Cloning may be Permitted at Newcastle University, UK' (2004) *New York Times*, 18 June. Online, available at: www/medicalnewstoday.com\articles\9606.php.

Viale, R. and H. Etzkowitz (2005) 'The Third Academic Revolution: Polyvalent Knowledge: the DNA of the Triple Helix', theme paper for the fifth Triple Helix Conference. Online, available at: www.triplehelix5.com.

Whitaker, M. (2009) Interview with Henry Etzkowitz, Director of Business Development, Newcastle University Business School.

12 Institutions of higher education as multi-product firms

An empirical analysis

T. Austin Lacy

Universities have long been regarded as multi-product firms in that they seek to produce multiple outputs simultaneously. The theoretical and empirical literature on institutions as multi-product firms has focused almost exclusively on the joint production of a familiar trio of outputs: teaching, research, and service, with no attention paid to the entrepreneurial role of universities. Additionally, significant research exists on the presence of scale economies in institutions. Using data for a sample of US universities from the National Center for Education's (NCES) Integrated Postsecondary Education Data System (IPEDS) – which contains extensive data for all US research institutions – the National Science Foundation's Integrated Science and Engineering Resources Data System (NSF Web-CASPAR), and the US Licensing Activity Survey of the Association of University Technology Managers (AUTM), I attempt to test the presence of scale economies in both traditional university outputs and the more recent, entrepreneurial outcomes. The analysis used OLS and negative binomial regression, with fixed-effects for institutions and years, to examine the association between various institutional characteristics and the production of three of the four outputs. My empirical results suggest that while graduate enrollments provide constant returns to entrepreneurial outcomes, they will provide diminishing returns to educational outputs. Conversely, the results indicate that large undergraduate enrollments will negatively impact educational production, while both large and small institutions are effective at achieving the entrepreneurial goals.

A number of studies have advanced the conception of institutions of higher education as multi-product firms. Multi-product firms are those that seek to simultaneously produce multiple outputs from multiple inputs. In the case of postsecondary institutions, observers usually highlight a familiar trio of outputs: teaching, research, and service.[1] While these are the traditional goals of higher education, over the past two decades research universities began to engage in activities of a more entrepreneurial nature. With the introduction of this agenda, these institutions began to produce outputs that serve the goals of economic development through fostering university and industry connections. In institutions' entrepreneurial pursuits, outputs often took the form of patents, licenses, and start-up companies, utilizing the technology transfer office (TTO) for the vehicle to bring scientific inventions to the market. In both theoretical and methodological terms, empirical investigation into multi-product firms poses more challenges than a single output perspective. However, if one ignores institutions' other outputs, resulting models can lead to a misunderstanding of a single outcome of interest. That is, studies using higher education institutions as the units of analysis should attempt to account for the multi-dimensional nature of these organizations.

Concurrent with the shift to entrepreneurialism was an increasing demand for institutions to become more accountable. This often manifested itself in the call for colleges and universities to be more 'efficient'. While the notion of efficiency for policymakers and the public is likely different from that of economists, efficiency is a typical measurement in studies that employ a firm based approach. However, as Worthington (2001) notes, in the case of education, ignoring the potential for inefficiency within these organizational types may lead to misunderstandings of the nature of their economies.

The goal of this study is to empirically assess the presence of scale economies in research universities, while paying mind to multiple goals and outputs. Using panel data on 103 US institutions, this analysis tests the influence of various measures of university size and capacity on three outputs: undergraduate education, research, and entrepreneurialism. Because this study draws on several strands of literature, the following review is presented in two parts. The first section of the literature review focuses on institutions as multi-product firms and studies that focus on specific outcomes associated with technology transfer. The second section discusses the literature on scale economies in higher education.

Literature review

Early studies of efficiency in institutions grew out of cost studies in multi-product firms, often treating colleges and universities as single-output organizations as measured by undergraduate enrollment (e.g. Bowen 1980). While scant, some later research acknowledged these shortcomings and attempted to account for institutions' multiple outputs, varying in their methodology and findings.

To date, much of the quantitative research explicitly studying institutions as multi-product firms focused on European institutions. These studies primarily used nonparametric techniques which preclude the types of hypothesis testing found in many social-science analyses in the United States. Using full-time equivalent (FTE) enrollment as an input and degrees awarded an output, Anthanassapoulos and Shale (1997) found 45 efficient institutions in their British sample. Consistent among the inefficient institutions was a pursuit of research outputs beyond a point consistent with their established missions, implying that inefficiency may occur when institutions pursue outputs dissonant with their purpose. Continuing on the work of Anthanassapoulos and Shale, Johnes (2005) used Data Envelopment Analysis (DEA) to analyze 109 British institutions, indicating six universities as having 'best operating practices'.

In analyzing the relationship of state revenue shares on US 4-year institutions, Robst (2001) found inefficiency to decrease as the undergraduate population grows, and to increase as the graduate population grows. There was no difference in efficiency across Carnegie classifications and between institutions that receive smaller or larger shares of state revenue.

The economic development literature in the United States is ripe with studies whose focus was the efficiency and production of universities' TTOs. TTOs are the offices that many US universities established to identify potentially profitable inventions, secure patents, and facilitate the transfer of licenses and intellectual property to the marketplace. For data, researchers consistently turned to the annual survey from AUTM. The survey asks member institutions to report information on various outcomes associated with technology transfer (i.e. patents, licenses, and invention disclosures). In a synthesis of existing studies, Siegel and Phan (2005) reviewed the existing literature on TTOs, showing that researchers use a variety of quantitative and qualitative techniques to assess their outputs.[2] Of these studies, several are particularly notable.

A 2002 study by Thursby and Kemp used DEA to compute efficiency scores and then ran a logistic regression to determine whether or not a TTO was efficient, finding faculty quality, number of TTO staff, private control, and lack of a medical school positively impacted their binary measure of efficiency. Incorporating two years into the analysis indicated that in 1991, 48 of 112 TTOs were inefficient and by 1996 the number decreased to 28, suggesting that over time TTOs became more efficient. In another study that year, Thursby and Thursby supplement AUTM data with their own survey, finding academic entrepreneurialism, as measured by a professor's willingness to patent, led to growth in licensing and patenting outcomes.

Siegel *et al.* (2003) found that invention disclosures and TTO staff size positively affect the number of licensing agreements an institution produces. Negatively influencing this outcome is the amount of annual legal expenditures. However, on the output of licensing revenue, the number of invention disclosures has a greater impact, TTO size drops from significance, and legal expenditure has a positive impact. From their field interviews, they found that the importance of patents was not consistent for all industries. Thus it may be that invention disclosures are the universities' primary inputs for technology transfer (Siegel *et al.* 2003).[3] While one cannot underestimate the contributions that the AUTM survey has made to the understanding of university TTOs, it is not without its limitations. All institutions that are technology transfer 'success stories' are AUTM members. This self-selection may exclude institutions that engage in technology transfer less effectively and at the margins. To my knowledge, the true effects of the selection bias are unknown; however, one should interpret the existing evidence with caution when applying it to the larger university population. Also hampering these analyses is the economic development literature's consistent omission of university characteristics and outputs more familiar to higher education researchers. By ignoring the diversity, complexity, and importance of the contexts in which university entrepreneurialism occurs, researchers are unlikely to approach true understanding of institutional operations and outputs.

Economies of scale

Since the 1920s, research on colleges and universities has looked into the presence of economies of scales in higher education institutions. In a synthesis of 60 years of research, Brinkman and Leslie (1986) summarize the findings on scale economies in higher education, concluding that enrollments are but one factor influencing costs. In reviewing research universities, they found substantial differences across studies addressing this sector of higher education. Despite these discrepancies, Leslie and Brinkman concluded that '[economies of scale] typically will be experienced by a representative group of private universities of varying sizes, but the same cannot be said for public research universities' (1986: 18). In the 20 years since Leslie and Brinkman's synthesis, researchers continued to investigate the presence of scale economies in postsecondary institutions. Cohn *et al.* (1989) found ray economies of scale in both public and private institutional sectors at mean levels of output and input. Below average sized institutions were more costly than average sized institutions. When measuring the single output of research in private institutions, scale economies were not found. Recognizing the multi-product nature of colleges and universities, they warn that an increase in undergraduate enrollment alone may not be cost-effective; expansion only produces greater efficiency when growth occurs in more than one output.

Laband and Lentz (2003) found economies of scale throughout all products in both public and private US institutions. In a study of 196 PhD granting institutions, Koshal and Koshal (1995) concluded that economies of scale are present in these institutions. Their study also suggested that institutional quality, as measured by average SAT, explained a large proportion of the variance of total cost. In another study, the same authors found liberal arts colleges to benefit from economies of scale, concluding that the optimal enrollment at these institutions is 2,343 undergraduate FTE and 88 graduate FTE (Koshal and Koshal 2000).

Sav (2004) tested the presence of scale economies for undergraduate, graduate, and professional education across institutional type for both public and private US institutions. He found that undergraduate economies exist in private, non-research institutions while graduate economies are only present in research institutions.

Little evidence exists on the presence of scale economies at the departmental level. Using data from the American Association of University Data Exchange (AAUDE), Dundar and Lewis (1995) analyzed the scale and scope economies of departments at 18 research universities finding that for most fields, product specific economies of scale exist for the outputs of teaching and research.

Research design

The primary research question is: how does size influence the various teaching, research, and entrepreneurial outputs of US postsecondary institutions? Recognizing the multiple outputs of institutions, the focus of this study is to investigate how various factors related to size influence the different outcomes. While the aforementioned output of service is certainly an output for many, if not all postsecondary institutions, the difficulty in quantifying this output at the institutional level led to its omission from the analysis.

In order to examine the efficiency of US institutions, the following factors will be examined:

Undergraduate enrollment: undergraduate enrollment at US research institutions is quite varied, consisting of both small institutions, such as Rice University and the Catholic University of America, and large, complex institutions like Pennsylvania State University and the University of Texas at Austin. As the number of undergraduates increases, to what extent are these institutions able to capture economies of scale? While it is plausible that a more focused institution may lead to a greater production of outputs, it is also conceivable that with size and complexity comes the ability to better address the competing outputs.

Graduate enrollment: while undergraduate enrollment is one indicator of size and complexity, the size of graduate enrollments also ranges across institutions. Due to the heterogeneity across research universities, graduate enrollments are not necessarily correlated to undergraduate enrollments. I hypothesize that graduate enrollments will have a positive effect on teaching, research, and entrepreneurial outcomes. Graduate students are often employed as teaching assistants, which may contribute to the output of undergraduate credit hours. Likewise, they are frequently employed as research assistants, a valuable input for increases in research and commercializable discoveries.

Student to faculty ratio: in addition to these measures of sheer size, the study also accounts for institutional capacity. For some outputs, a lower student to faculty ratio may lead to greater outputs. For example, the fewer students an institution has in respect to faculty may give the latter increased opportunity to concentrate on research and entrepreneurial outputs. On the other hand, a lower ratio may lead to smaller classes and a decrease in the educational output.

In addition to these size variables, controls for Scholastic Aptitude Test (SAT) and average professor's salary were included in the model. While this study's interest lies in the effects of size, these time-varying covariates inclusion appears necessary to account for both institutions investment in faculty and quality.

Data: data for this study were drawn from three sources: IPEDS, NSF Web-CASPAR, and AUTM's Annual US Licensing Activity Survey. The years of analysis were limited to 2003–6 due to lack of consistent data from NCES. Because of the interest in universities' shift to entrepreneurial goals, universities in the study are restricted to US institutions that are AUTM members, resulting in 379 institution-years. Some AUTM members were removed from the analysis because of their level of reporting (at the higher-education system level rather than the institutional level), their absence of undergraduate education, and appearance in only one report. In 2005, the survey only reported the indicator for invention disclosures over a three-year period. For this data, previous and later surveys, which included annual estimates, aided in deriving the 2005 estimates. Table 12.1 provides descriptive statistics for all independent and dependent variables.

Table 12.1 Descriptive statistics

Variable	Description	Source	Mean (std. dev.)	
Undergrad	Undergraduate credit hours divided by undergraduate FTE (logged)	NCES (derived)	3.492	(0.180)
Research	Federal R&D expenditures divided by total faculty (logged)	NSF/NCES (derived)	11.045	(0.801)
Invention disclosures	Number of invention disclosures submitted to the university TTO	AUTM	9.069	(94.224)
Student to faculty ratio	Undergraduate students divided by total faculty	NCES (derived)	7.987	(4.800)
Undergraduate FTE	Undergraduate full-time-equivalent (logged)	NCES	9.592	(0.658)
Graduate FTE	Graduate full-time-equivalent (logged)	NCES	8.100	(0.648)
Average professor's salary	The average salary of all faculty (logged)	NCES	11.284	(0.164)
SAT composite	SAT 75th percentile	NCES	1,274.378	(108.530)

For the sake of comparison of multiple outputs, models were run using the three individual university output variables as dependent variables. The variable for undergraduate education was derived by dividing the NCES indicator '12-month instructional activity credit hours: undergraduates' by an estimated FTE undergraduate enrollment. NCES calculates the numerator by multiplying the credit hour value of individual courses by the number of students enrolled in the course and summing across the institution. The construction of the denominator of undergraduate enrollments required the creation of a variable of full-time-equivalent enrollment. To account for part-time enrollments, the number of part-time students was divided by three and then added to the number of full-time students.

Research is derived from the NSF indicator 'federal R&D expenditures at universities and colleges' divided by the NCES indicator 'full-time faculty, grand total'. Though universities produce research that is not captured by this indicator, due to the competitive nature of federal research grants, they serve as an adequate proxy for the level of quality research at an institution. Dividing this value by total number of faculty serves to not privilege larger institutions.

Last, from AUTM, the variable 'invention disclosures' is an indicator of institutional entrepreneurialism. Much of the technology transfer pipeline occurs outside of the university in areas where the institution can have little, if no effect (Matkin 1990). Thus the number of invention disclosures is the economic-development factor over which institutions have the most control. This was not divided by faculty members since research suggests that these outcomes are typically the product of a few 'star scientists'.[4]

All independent variables were gathered and derived from NCES' IPEDS. Student to faculty ratio is the ratio of total full-time faculty divided per 100 undergraduate student FTE. The undergraduate and graduate FTE variables were calculated as previously described. The variable for SAT is the institutions combined SAT seventy-fifth percentile score. Institutions whose applicants predominantly take the ACT test were converted to SAT scores using tables from the College Board's website. The variable for 'average faculty salary' was converted to constant 2006 dollars using the Consumer Price Index.

Methods: to test hypotheses about the influence of university characteristics, ordinary least squares (OLS) and negative binomial regression with fixed effects for both institution and year. The fixed effects were modeled by creating dummy variables for all institutions and years were used. Fixed effects accounts for unobserved institutional characteristics and potential temporal influences. Because of the presence of heteroskedasticity, institutions' errors were clustered across time, producing robust standard errors, which, when coupled with the clustering addresses this issue.[5]

Models 1 and 2 are expressed with the following equation:

$$y_{it} = \beta_1 x_{it1} + \ldots + \beta_k x_{itk} + u_{it}, \, t = 1, 2, \ldots, T$$

Where y is the output of interest, x the vector of explanatory variables, t the unit of time in the model, i the cross-sectional unit identifier, and u_{it} the error term.

As the data is panel in nature, the unit of analysis is institution-years, each observation being a single institution at a specific point in time.

A Hausman test was conducted to determine whether fixed or random effects should be used. In the Hausman test, the null hypothesis is random effects. In all cases, the null was rejected hypothesis, preferring the fixed effects models.

Because invention disclosures are count data by nature, the negative binomial estimator was used for this outcome. Prior to using the negative binomial model, a Poisson model was constructed and tested for overdispersion, the assumption that the mean is equal to the variance. I failed to reject the null hypothesis of no overdispersion, which required the use of the negative binomial model. The negative binomial model introduces an individual, unobserved effect into the conditional mean to account for the overdispersion. For this analysis, the Hausman, Hall and Griliches (HHG) (1984) model for a fixed effect negative binomial model was used.[6] The mass function for the HHG negative binomial model is expressed as:

$$f(y_{it}|\lambda_{it}, \theta_i) = \frac{\Gamma(\lambda_{it} + y_{it})}{\Gamma(\lambda_{it})\Gamma(y_{it} + 1)} \left(\frac{\theta_i}{1 + \theta_i}\right)^{y_{it}} \left(\frac{1}{1 + \theta_i}\right)^{\lambda_{it}}$$

Where the covariates enter in the link function:

$$ln\lambda_{it} = \beta x_{it}$$

While the negative binomial estimator uses maximum likelihood estimation, its results are presented alongside the linear regressions.

For all models, specification tests were run to determine the need for polynomial terms. Quadratic terms were included for graduate enrollment in the outcome of undergraduate credit hour production and for undergraduate enrollment in the outcome of invention disclosures.

Results

The results from this study indicate that several institutional size characteristics have statistically significant effects on the production of institutions' teaching, research, and entrepreneurial outcomes. Table 12.2 presents the results of the regression models.

Returning to the hypotheses, in the production of undergraduate credit hours, undergraduate FTE enrollment has a negative effect on per capita credit hour output. That is, as institutions become larger, they produce relatively fewer credit hours. In contrast, graduate FTE enrollment has a positive effect, up to a certain point. An F-test of the joint significance of graduate FTE enrollment and graduate FTE enrollment squared produce an F-statistic of 4.16 with a p-value of 0.044. This finding indicates that as graduate enrollments increase, the production of undergraduate credit hours increases and then begins to decrease. The inflection point, or the point at which the combined enrollment effect is zero, is an enrollment of 7.611 graduate students (logged values). Figure 12.1 displays the quadratic relationship with all other values held at their mean.

Table 12.2 OLS and negative binomial results including fixed effects for both institution and year (standard errors in brackets)

	Model 1	Model 2	Model 3
	Undergrad	Federal R&D	Invention disclosures
Student to faculty ratio	0.000 (0.000)	−0.047** (0.010)	−0.003 (0.002)
Undergraduate FTE enrollment (logged)	−0.580** (0.178)	−0.514 (0.287)	−10.435** (3.781)
Undergraduate FTE enrollment (logged and squared)			0.536* (0.200)
Graduate FTE enrollment (logged)	1.560* (0.764)	−0.268 (0.149)	0.705** (0.216)
Graduate FTE enrollment (logged and squared)	−0.102* (0.050)		
SAT	0.000 (0.000)	0.000 (0.000)	0.000 (0.000)
Average professor's salary (logged)	0.004 (0.072)	0.042 (0.266)	0.137 (0.296)
Constant	3.307 (2.657)	18.004 (3.660)	46.623* (18.263)
Observations	379	379	379
Number of universities	103	103	103

Notes
* significant at 5 percent; ** significant at 1 percent.

While the initial hypothesis was that there would be increasing returns to graduate enrollment, the quadratic nature of the variable is particularly striking. I offer two possible explanations for the diminishing returns of graduate enrollment to undergraduate education. On one hand, expansion of graduate enrollments may be part of an institutional shift away from undergraduate education. On the other, schools with higher graduate enrollments may have a greater proportion of professional graduate students who do not contribute to the production of undergraduate credit hour production.

Model 2, estimating the influence of the size characteristics on Federal R&D per capita finds a negative influence in the student to faculty ratio. It is plausible that as the number of students relative to the number of faculty increases, faculty must focus on the educational demands of their institution, at the expense of engaging in competitive research. A notable non-finding is the absence of the effect of graduate enrollment. While a larger sample size may be able to untangle this effect, it appears that graduate enrollment does not provide the input to competitive, federal research. It may be that this research is driven by post-doctoral students, an understudied group in higher education research. Unfortunately, little data and information are collected about this class of students.

Last, turning to the number of invention disclosures the results indicate a positive influence of graduate enrollment. This finding suggests that institutions with high levels of graduate enrollments may be positioned to utilize them as an input towards the production of entrepreneurial outcomes. The effect of undergraduate enrollment is again, somewhat mixed. The number of invention disclosures decreases until undergraduate enrollment reaches 9.729 logged values.[7] Figure 12.2 presents the quadratic effect.

This finding suggests that both small, focused institutions and large, complex universities are at an advantage in pursuing commercialization.

None of the models indicated an influence of the control variables for faculty salary and SAT score. While these control variables had no hypotheses attached to

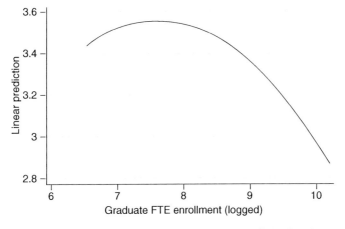

Figure 12.1 Undergraduate credit hours: quadratic effect of graduate enrollment.

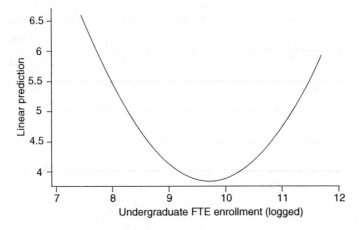

Figure 12.2 Invention disclosures: quadratic effect of graduate enrollment.

them, it may be that within the subsample of AUTM, the data are unable to capture the variance that exists in these variables in US institutions. That is, across the elite institutions that comprise much of AUTM, there is likely the substantial difference in SAT and faculty pay that one would expect among all institutions.

Conclusions

There is need for continued studies into universities' efficiency in producing multiple outputs. Insufficient and inconsistent data are likely to persist as the most vexing issue researchers will continue to encounter. While the use of fixed-effects accounts for much of the unobserved heterogeneity in institutions, future studies may wish to incorporate other institutional characteristics beyond size. Models that include additional institution-years would provide a greater sample size and likely produce more robust estimates. Further investigation into consist-ent variables within the IPEDS database, as well as the creation of algorithms and cross-walks between variables over different survey iterations, will be needed as researchers pursue these questions.

Importantly, continued restriction samples to AUTM members poses serious analytic constraints for studies of TTOs' outputs. AUTM is the major organiza-tion that surveys indicators relating to the entrepreneurial outputs of institutions, but this sample restriction likely creates significant selection biases. Future work should attempt to explicitly account for and model this bias.

Despite these shortcomings, the analysis provides some valuable information to policy makers. The economy of scale discovered for invention disclosures suggests that both small and large institutions are able to most effectively engage in entre-preneurial activity. While the positive effect of large institutions is unsurprising, as by sheer size they may produce more disclosures, the effect of small institutions is quite striking. It may be that curricula at these small institutions are focused and conducive to university entrepreneurialism. Further studies should attempt to account for the influence of curricular differences on university outputs.

The effect of graduate enrollment on invention disclosures seems to confirm that these students may provide a valuable labor input towards the new university purpose of economic development.

While these findings are revealing, they must be taken in respect to the effect the variables have on the other outputs, notably undergraduate education. While these findings predict that there would be constant returns to graduate education for invention disclosures, at a certain point this may come at the expense of undergraduate education. Likewise, expanding the size and scale of one's university through the expansion of undergraduate enrollment may yield greater entrepreneurial outcomes, but the findings suggest that such gains might come at the expense of undergraduate educational outcomes.

The initial success of licensing and patenting at a few universities led to a rapid expansion of state policies attempting to leverage research universities for the purposes of economic development (Plosila 2004). Policymakers seeking to leverage their postsecondary institutions for economic development purposes should be mindful of which research institutions are best to target for state level programs, considering how these goals may negatively impact the educational purposes of the universities. This analysis suggests that research institutions focusing on graduate education and have low levels of undergraduate education may be best positioned to achieve these two goals.

Notes

1 Consistent with other studies of multi-product firms, this chapter uses the terms outputs and products interchangeably.
2 One cannot overstate the contribution of Siegel and Phan in bringing together this literature.
3 Donald Siegel reiterates this finding in Siegel and Phan (2005).
4 See Zucker and Darby (1996, 2006) and Zucker et al. (2002, 2003).
5 Due to the nature of the estimator, this correction is unavailable for the fixed effect negative binomial model.
6 For a thorough discussion of the fixed-effects negative binomial model and its limitations see Allison and Waterman (2002).
7 The F-statistic for these terms is 7.64 with a p-value of 0.006.

References

Allison, P. and Waterman, R. (2002) 'Fixed-effects Negative Binomial Regression Models', *Sociological Methodology*, 32, pp. 247–65.
Anthanassapoulos, A.D. and Shale, E. (1997) 'Assessing the Comparative Efficiency of Higher Education Institutions in the UK by Means of Data Envelopment Analysis', *Education Economics*, 5 (2), pp. 117–34.
Bowen, H.R. (1980) *The Costs of Higher Education*, San Francisco: Jossey-Bass Publishers.
Brinkman, P.T. and Leslie, L.L. (1986) 'Economies of Scale in Higher Education: Sixty Years of Research', *Review of Higher Education*, 10 (1), pp. 1–28.

Cohn, E., Rhine, S.L.W., and Santos, M.C. (1989) 'Institutions of Higher Education as Multiproduct Firms: Economies of Scale and Scope', *Review of Economics and Statistics*, 71 (2), pp. 284–90.

Dundar, H. and Lewis, D. (1995) 'Departmental Productivity in American Universities: Economies of Scale and Scope', *Economics of Education Review*, 14 (2), pp. 119–44.

Hausman, J., Hall, B.H., and Griliches, Z. (1984) 'Econometric Models for Count Data with an Application to Patents–R&D Relationship', *Econometrica*, 52 (4), pp. 909–38.

Johnes, J. (2005) 'Data Envelopment Analysis and its Application to the Measurement of Efficiency in Higher Education', *Economics of Education Review*, 25 (3), pp. 273–88.

Koshal, R.K. and Koshal, M. (1995) 'Quality and Economies of Scale in Higher Education', *Applied Economics*, 27 (8), pp. 773–8.

Koshal, R.K. and Koshal, M. (2000) 'Do Liberal Arts Colleges Exhibit Economies of Scale and Scope?', *Education Economics*, 8 (3), pp. 209–20.

Laband, D.N. and Lentz, B.F. (2003) 'New Estimates of Economies of Scale and Scope in Higher Education', *Southern Economic Journal*, 70 (1), pp. 172–83.

Matkin, G.M. (1990) *Technology Transfer and the University*, New York: Macmillan Publishing Company.

Plosila, W.H. (2004) 'State Science-and Technology-based Economic Development Policy: History, Trends, and Development, and Future Decisions', *Economic Development Quarterly*, 18 (2), pp. 113–26.

Robst, J. (2001) 'Cost Efficiency in Public Higher Education Institutions', *Journal of Higher Education*, 72 (6), pp. 730–50.

Sav, G.T. (2004) 'Higher Education Costs and Scale and Scope Economies', *Applied Economics*, 36 (6), pp. 607–14.

Siegel, D.S. and Phan, P.H. (2005) 'Analyzing the Effectiveness of University Technology Transfer: Implications for Entrepreneurship Education', in: G. Libecap (ed.) *University Entrepreneurship and Technology Transfer: Process, Design, and Intellectual Property*, San Diego: Elsevier, pp. 1–38.

Siegel, D.S. Waldman, D., and Link, A. (2003) 'Assessing the Impact of Organizational Practices on the Relative Productivity of University Technology Transfer Offices: An Exploratory Study', *Research Policy*, 32 (1), pp. 27–48.

Thursby, J.G. and Kemp, S. (2002) 'Growth and Productive Efficiency of University Intellectual Property Licensing', *Research Policy*, 31 (1), pp. 109–24.

Thursby, J.G. and Thursby, M.C. (2002) 'Who is Selling the Ivory Tower? Sources of Growth in University Licensing', *Management Science*, 48 (1), pp. 90–104.

Worthington, A.C. (2001) 'An Empirical Survey of Frontier Efficiency Measurement Techniques in Education', *Education Economics*, 9 (3), pp. 245–68.

Zucker, L.G. and Darby, M.R. (1996) 'Star Scientists and Institutional Transformation: Patterns of Invention and Innovation in the Formation of the Biotechnology Industry', *Proceedings of the National Academy of Sciences*, 93 (23), pp. 12709–16.

Zucker, L.G. and Darby, M.R. (2006) 'Movement of Star Scientists and Engineers and High-Tech Firm Entry', NBER Working Paper.

Zucker, L.G. Darby, M.R. and Armstrong, J. (2002) 'Commercializing Knowledge: University Science, Knowledge Capture, and Firm Performance in Biotechnology', *Management Science*, 48 (1), pp. 138–53.

Zucker, L.G., Darby, M.R. and Brewer, M.B. (2003) 'Intellectual Human Capital and the Birth of US Biotechnology Enterprises', *American Economic Review*, 88 (1), pp. 290–306.

13 A complexity theory perspective on scientific entrepreneurship engineering and empirical investigation in German-speaking Europe

Philipp Magin and Harald F.O. von Kortzfleisch

Supporting scientific entrepreneurship from a complexity theoretical point of view

A complexity perspective on scientific entrepreneurship

The nature of entrepreneurship is complex and nonlinear per se (Alvarez and Busenitz 2001: 768). No matter which school of entrepreneurship is chosen for reference, the fundamental definition of entrepreneurship as the exploitation of opportunities to accomplish original and value-generating innovations implies tremendous complexity.

For instance, taking a resource-oriented perspective on entrepreneurship (Volery 2005 based on Penrose 1959) means understanding the entrepreneurial process as an allocation of input factors (resources) in an original way (Alvarez and Busenitz 2001). Therefore, the entrepreneur needs to *recognize* an opportunity (Shane and Venkataraman 2000 see also Murphy *et al.* 2005), get access to required resources (Brush *et al.* 2001) and organize these resources within a firm (Alvarez and Busenitz 2001) to create original and valuable new combinations. According to the resource-based theory, the entrepreneur has to address several dimensions to cover competitive advantages with his resource-combination on a long-term basis (Alvarez and Busenitz 2001 based on Peteraf 1993). For instance, to prevent his once reached competitive advantage the entrepreneur needs to develop inimitable resources (Itami 1987) like social networks (Barney 1995) or distinctive knowledge (Barney 1991, Grant and Baden-Fuller 1995) to ensure sustainable heterogeneity. This multi-dimensional solution space for strategic decisions and development directions is a highly complex environment with nonlinear causalities which makes it difficult or even impossible to predict certain outcomes, or behave according to standardized patterns.

Not only the nature of entrepreneurship in general, but also the specific characteristics of scientific entrepreneurship make this field even more complex. We understand scientific entrepreneurship as the recognition of business or social opportunities and their exploitation from within an academic, knowledge-intensive environment to carry out highly original and valuable combinations and put them through in terms of innovations.

This definition reveals three main causes for complexity, which are (*a*) the characteristics of the academic environment; (*b*) the kind of actors/promoters; and (*c*) the characteristics of the innovations themselves.

First, the academic environment is complex per se. Complexity refers to the traditional mission of universities as sources of knowledge creation and dissemination (Bok 2003, Geisler 1993). The creation of new knowledge implies non-linear causalities, which are a driver for complexity (Higgings 2006: 191ff.).

Second, the actors within the academic community are extensively different in nature, and also with regard to their different roles: researchers, educators, learners, co-researchers, etc. For scientific entrepreneurship, this implies that there is a very heterogeneous group of addressees with different expectations, who not only have to be served with appropriate measures for each subgroup (Magin and von Kortzfleisch 2008: 3), but also should be matched with one another to create entrepreneurial teams with multiple experiences and capabilities.

Third, scientific entrepreneurship often leads to innovative knowledge-intensive and technological innovations (Kulicke 2006) which are likely distinctive due to their originality, and whose exploitation is even more complex because of their uniqueness (Kohn and Spengler 2007: 9). Taking all the case studies of entrepreneurial ventures with roots in academic institutions (see e.g. O'Shea *et al.* 2007 for the MIT Case), there is a tremendous portion of innovations carried out by those ventures, for which e.g. a market or even a demand did not exist before.

Supporting scientific entrepreneurship in the field of controversy between chaos and order

Interest in supporting scientific entrepreneurship

The increasing interest in entrepreneurship as a research discipline of its own initially referred to the traditional mission of universities as sources of knowledge creation and dissemination (Bok 2003, Geisler 1993). A major step to enforce and support entrepreneurial activities themselves within academia, was taken by some US universities during the 1970s which started not only researching entrepreneurship, but developed a so-far missing practical perspective on this discipline (Anderseck 2004, Vesper and Gartner 1997). Their main objective was to enable and support the commercialization of academic knowledge by awaking the target groups' attention, arousing interest in starting-up own companies, and by teaching specific entrepreneurship related skills (Klandt and Volkmann 2008). Since the middle of the 1990s, European universities also began to adopt such approaches (Anderseck 2004) and concentrated on the development of appropriate initiatives to integrate the practice of entrepreneurship into the academic community.

With regard to those initiatives, the complexity of scientific entrepreneurial activities not only challenges the scientific entrepreneurs themselves, but also those who aim to support such activities in politics or within the academic community. Scientific entrepreneurship supporting initiatives often take place in an environment with nonlinear causalities and influences are (hopefully) bidirectional, not only from supporters to entrepreneurs, but also from entrepreneurs to those who manage and operate supporting initiatives. For instance, the duration of such initiatives is considerably longer than an average pre-seed phase of a start-up. That means entrepreneurs themselves evolve not only with respect to their own entrepreneurial process, but also change their role in the overall supporting initiative. Probably they switch sides and become supporters themselves, as mentors for other entrepreneurs or founders.

Taking these bidirectional influences and nonlinear causalities into consideration, complexity theory provides us an effective framework to describe interdependencies and options within such a complex environment.

Methods from complexity sciences within the field of entrepreneurship

While some researchers have already begun to apply complexity theory to some areas within the broader field of entrepreneurship research, scientific entrepreneurship in particular has not been discussed from a complexity science point of view so far. Nevertheless, related contributions to other facets of entrepreneurship research provide some adoptable thoughts on how to link complexity theory to the field of entrepreneurship. One example here is social entrepreneurship.

Goldstein *et al.* (2008) discuss the area of social entrepreneurship through a complexity lens. Also Massetti (2008) uses ideas from complexity sciences to develop a generic framework for social entrepreneurship.

A key insight developed within such studies, is that it is not only any heroic, single leader who solves social problems by carrying out a social innovation in a somehow mythic manner, but that it rather is a system of nonlinear interdependencies framing the social entrepreneurial process and creating social value (Goldstein *et al.* 2008: 10 see also Hazy *et al.* 2007). Taking their rationale, the social entrepreneur does not need to be a lonesome actor at all, but his core ability might be to create a broad and lively social network in which he is able to leverage several resources, triggering resonance and synchronization among stakeholders within the network, and in the end foster the emergence of social innovation (Goldstein *et al.* 2008: 12).

The importance of building and leveraging a complex system of stakeholders, particularly seems to be true for the scientific entrepreneur. The scientific environment with its very heterogeneous actors means of course a great potential for distinctive, break-through innovations, but this also implies tremendous complexity for the scientific entrepreneurs and for leaders within academia that aims to foster entrepreneurial activities. Matching the inventors of high-potential business ideas with the right people who have the abilities to market these often complex products or services and get them through as break-through innovations, is just one example for complexity within supporting measures.

Massetti (2008) makes use of methods from complexity sciences to define what social entrepreneurship is and what makes it distinctive from a sort of rather profit-oriented opportunity exploitations. She argues that researchers failed to grasp the (full) nature of social entrepreneurship appropriately (2008: 1f.) when trying to refer the distinctiveness of social entrepreneurship to specific traits (see e.g. Drayton 2002, Roberts and Woods 2005) or specific characteristics of the (social) business model (see e.g. Dart 2004, Harding 2004). In contrast to approaches that concentrate on making distinctiveness measurable, Massetti applies a continuum perspective from complexity sciences to the field of entrepreneurship research and proposes a 'social entrepreneurship matrix', a two-dimensional space, in which social entrepreneurs can be positioned more flexibly and with respect to their specific mission and their business model (2008: 4). Each dimension spans a field of controversy between two extremes: A fully socially-driven mission versus a fully market-driven mission and a fully non-profit requiring business model versus a fully profit-requiring business model. Entrepreneurial ventures can be positioned within this field. The better a venture balances social and market drive as well as ensuring independent financing, the better it fits to what researchers name social entrepreneurship.

Against this background, understanding fuzziness and addressing it appropriately has also significant importance for supporting scientific entrepreneurship. In order to foster entrepreneurial spirit and entrepreneurial activities within academia it is essential to first understand where entrepreneurial ventures need support and second to introduce measures to provide support within these fields, effectively and efficiently. One key question in this context is, how to balance between a deeply structured supporting plan for ventures from the very first stage throughout the entrepreneurial process up into the growth stage and, on the other hand providing enough space for creativity, fortuity and luck. Taking into consideration that entrepreneurship is somehow both creativity and rationality (with respect to the decision-making process for instance), supporting initiatives need to find a balanced position in the field of controversy between chaos and structure.

An engineering approach to excel at complexity of supporting scientific entrepreneurship

Engineering as a problem-solving principle

In spite of the emergence of inter-institutional programs and networks like the European Foundation for Entrepreneurship Research (EFER), the Foerderkreis Gruendungsforschung (FGF) in Germany or the German EXIST program which is a governmental program to foster scientific entrepreneurship, in Germany the rise of this discipline has primarily taken place locally and without using evaluated, commonly accepted approaches – in the sense of best practices. The FGF asserts in its annual report that there is an intense deficit of a systematic, goal-oriented and lasting approach to anchoring scientific entrepreneurship in German universities (2007: 6). Intentionally similar to this assertion, Twaalfhoven (2004) also figures out these deficits with regard to scientific entrepreneurship throughout the European system of universitary education. Furthermore, the practice of entrepreneurial activities in the academic community is characterized by a high degree of closeness and therewith a lack of transparency (Grichnik *et al.* 2009: 6).

With respect to the discussed challenges to which supporting initiatives are faced to and the intense complexity which requires a more holistic and systematic understanding of where ventures from within academia need support and how support can be provided, we propose to look at scientific entrepreneurship as an engineer would. Applying the engineering paradigm to scientific entrepreneurship might be a way to excel at complexity of supporting entrepreneurial ventures from within the academic environment.

The essence of the engineering paradigm is to operate complex tasks by using principles, methods and instruments (sometimes named tools) and evaluate their effectiveness and efficiency to profit from learning effects.

The adoption of the engineering paradigm to naturally non-engineering disciplines can be identified in different sciences, like computer sciences with a special focus on the process of software development and in economic sciences, for example for the process of service (re-)development. In software development, there was a necessity to systematize the development process in order to meet the continuously growing requirements. This systematization led to the evolution of the so called software engineering discipline (Balzert 1982, Dumke 2003). In service (re-)development, the situation was similar. The exceeding complexity of product-service bundles, the increasing convergence of markets, and the increasing intensity of competition required distinctive concepts to differentiate the own company against competitors. The systematization of this process then led to the so called service engineering approach (Bullinger and Scheer 2002).

In both disciplines, shifting to an engineering concept has been an implication to counteract the former unsystematic and heuristic ways of operating on complex tasks and enable the establishment of common standards, best practices and lasting progress.

Scientific entrepreneurship engineering

The adoption of the engineering paradigm to scientific entrepreneurship should cope with two important requirements, which scholars have identified as core to enhance this discipline (Twaalfhoven 2004, Witt 2006). As we mentioned above, engineering disciplines make use of principles, methods and instruments. Principles in this trilogy can be understood as the basic assumptions or the starting points to which methods and instruments should refer to. The two core requirements for scientific entrepreneurship, indentified by evaluating the current state of play, at the same time represent the two major principles of scientific entrepreneurship engineering.

The integration principle

The first core requirement is a holistic comprehension of scientific entrepreneurship (Magin and von Kortzfleisch 2008: 9ff.). Current approaches on how to support scientific entrepreneurship try to identify the core dimensions in which entrepreneurial ventures request support from respectively one point of view. To enable improvements in conceptualizing initiatives, it is necessary to integrate these perspectives into a single framework. On this note, *integration* is the first principle of scientific entrepreneurship. Magin and von Kortzfleisch (2008) propose such an integrative framework, based on a qualitative analysis of the self-description of more than 120 initiatives in German-speaking Europe (see Figure 13.1). This framework includes 12 action-fields, which represent the core dimensions in which scientific entrepreneurs may request respective support. One additional action-field concentrates on the evaluation of the initiative.

The systematization principle

The second core requirement is to bring systematization into initiatives. *Systematization* in this context does not necessarily refer to bringing more structure into initiatives, but to approach the balance of structure and chaos systematically. Similar to the situation of computer sciences or the theme of service development before the adoption of engineering paradigms, the current state of scientific entrepreneurship is rather unsystematic, heuristic-driven and in some cases obviously inefficient. Therefore we define systematization as the second principle of scientific entrepreneurship engineering.

To systematize scientific entrepreneurship initiatives, we propose a multilevel model for each action-field. The 12 plus one action fields resulting through the integration of current approaches define the space of required measures to cover scientific entrepreneurship holistically. Each action-field implies specific and evaluable objectives (see Figure 13.2). In order to reach these objectives, specific methods are required. Methods in engineering disciplines can be understood as specific ways or strategies of operating on tasks. Instruments are some kind of tools which are applied according to the particular method. The way of identifying methods which support the particular action-field is mainly theoretical driven and makes use of findings in associated disciplines (Magin and von Kortzfleisch 2008: 24f.).

To examine scientific entrepreneurship as an engineering discipline opens up the opportunity for improvements in conceptualizing initiatives holistically as well as their evaluation. Furthermore, the increase in systematization implied by the adoption of the engineering paradigm eases the comparison between existing initiatives.

Comparative analysis of scientific entrepreneurship initiatives

Purpose

Based on the concept of scientific entrepreneurship engineering, the purpose of this comparative analysis is to identify characteristic patterns among scientific

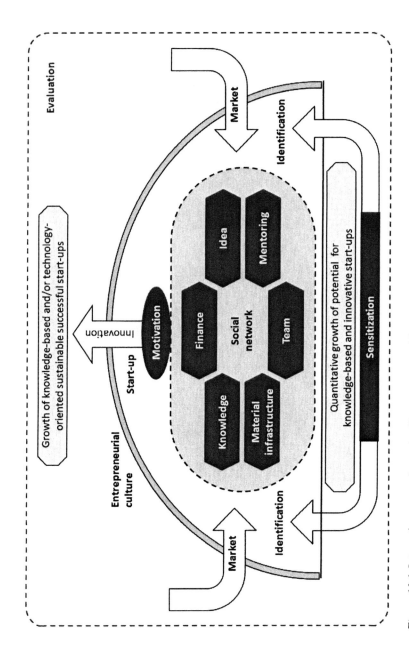

Figure 13.1 Integrative approach to scientific entrepreneurship.

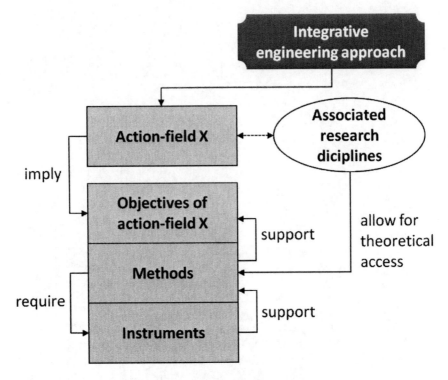

Figure 13.2 Engineering approach to scientific entrepreneurship.

entrepreneurship initiatives. The identification of such patterns will provide a distinctive understanding of how scientific entrepreneurship is practically operated in different countries and therewith in different institutional and governmental settings. This study should be understood as a first research contribution to approach complexity within the field of scientific entrepreneurship by identifying patterns and generating implications for further research.

Sample

The sample in this comparative study includes 183 institutions in Austria, Germany and Switzerland, which are engaged in any kind of activities to support scientific entrepreneurship. Among these institutions, there is a focus on universities and colleges of higher education (see Figure 13.3). These two types of institution are central in all three countries in the context of scientific entrepreneurship.

The initial points to compose this sample were the Internet presence of the countries' responsible ministries. From there on all institutions that were reachable via hyperlinks were included into the sample.

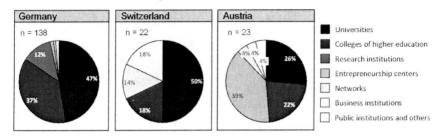

Figure 13.3 Constitution of sample.

Method

By definition, scientific entrepreneurship inherits the functional aspects of activities, which aim to support entrepreneurial intentions among academics. Due to the functional nature of scientific entrepreneurship, this study concentrates on those qualitative data, which document concrete activities of institutions engaging in the promotion of entrepreneurship in academia. The internet presence of each institution was observed qualitatively regarding specific activities which can be qualified as instruments according to the engineering paradigm. Each identified activity was allocated to exactly one (the most appropriate) action-field of the integrative framework for scientific entrepreneurship. For each combination of institution and action-field a binary result was assigned (0 if there weren't any activities which could be allocated to the respective action-field; 1 if there was at least one activity which could be allocated to the action-field). Then, the result matrix shows which institution operates at least one instrument for the respective action-field.

In order to compare the initiatives implemented and operated by different institutional types, frequencies of this result matrix were calculated per type of institution and per action-field. Based on these frequencies, the distribution of institutions that are engaged in a particular action-field was calculated for each action-field and country. Furthermore, the average number of action-fields in which a particular type of institution is engaged in was calculated per type of institution and country.

Results

In Austria, entrepreneurship centers play the major role in supporting scientific entrepreneurship (see Figure 13.4). Beside the action-fields of evaluation and team, entrepreneurship centers operate instruments for all action-fields. Especially in those fields that require highly individual support like mentoring, finance, identification or physical infrastructures, they provide in each case over 80 percent of all support activities in the respective field.

In Germany, universities and colleges of higher education are dominant in scientific entrepreneurship supporting initiatives. They both operate instruments

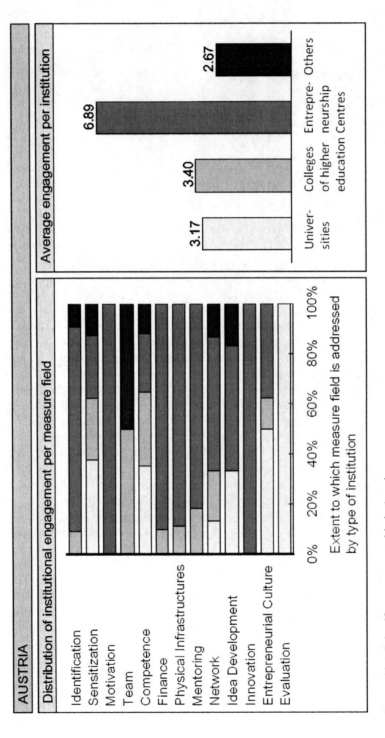

Figure 13.4 Scientific entrepreneurship in Austria.

in every action-field and provide over 70 percent of all activities in each action-field (see Figure 13.5). Research institutions and business institutions are less important than in Austria and Switzerland. Furthermore, it can be noticed that all types of institutions in Germany are more diversified in their engagement than in Austria and Switzerland. This is indicated by the constantly higher average engagement factor per institution (illustrated by the bar chart on the right of Figures 13.4, 13.5 and 13.6).

In Switzerland, the situation with regard to the average engagement is converse to the one in Germany. Swiss institutions are rather specialized. For each type of institution, the average engagement is more than 30 percent below the value of this indicator for the respective type of institution in Germany. The bar chart on the left of Figure 13.6 illustrates that those action-fields which require a highly individual support, are primarily operated by business institutions and networks whereas those which can be operated on a collective level, are under the responsibility of universities and colleges of higher education. The whole amount of scientific entrepreneurship supporting activities is divided and assigned to specific kinds of institution with respect to their specialization on only some action-fields.

Three empirical approaches of addressing complexity within the field of scientific entrepreneurship

This summary of the results highlights that there are different approaches to conceptualize, implement and operate such scientific entrepreneurship supporting initiatives in these three countries. In the following paragraphs, we interpret the findings of our study and give one possible characterization for each of the three identifiable approaches.

In Austria, the entrepreneurship centers were exposed to be the dominant type of institution. Austria has nine of these country-wide spread centers. Each center is an allocation of scientific entrepreneurship activities in the respective region. They all operate on a similar concept and provide a similar range of supporting activities. For this reason, they can profit from synergies in the development and evaluation of their activities. Despite a common underlying concept, they are independent from each other. With respect to the similarity of their instrumental portfolio and therewith the synergies from which they can profit, we label this approach the *Scientific Entrepreneurship Franchising Approach*.

For Germany, we have asserted that universities and colleges of higher education are central in scientific entrepreneurship initiatives. Other types of institutions, like business or research institutions complement the portfolio of the two dominant types of institutions with specific instruments according to their core abilities. Furthermore, it can be noticed that the initiatives in Germany are more and more organized as clusters, which means that closely located institutions of higher education (universities and colleges of higher education) combine their activities and operate as one another's complementing partners in a local or regional cluster. The institution, for itself, does not provide a specific portfolio

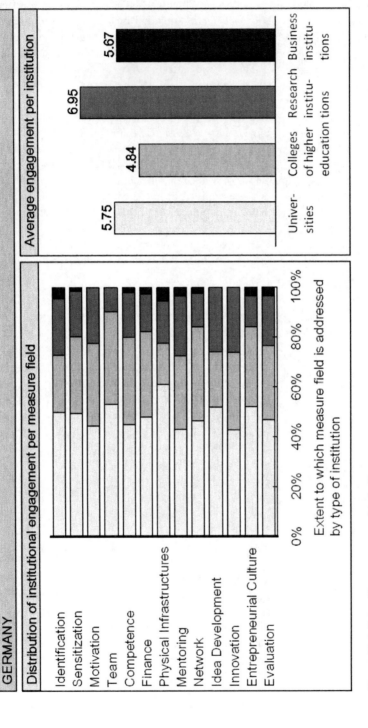

Figure 13.5 Scientific entrepreneurship in Germany.

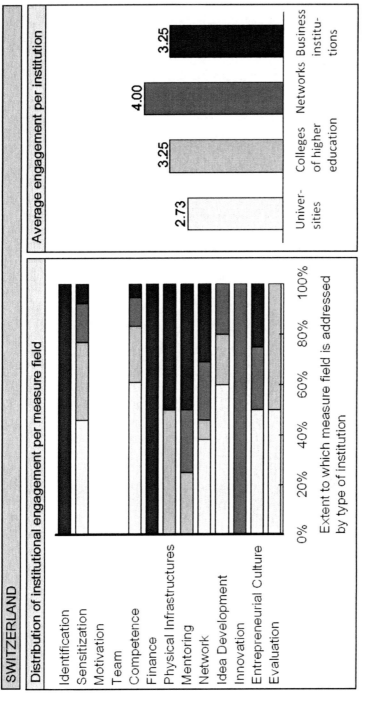

Figure 13.6 Scientific entrepreneurship in Switzerland.

of activities, but the cluster operates a set of instruments to support scientific entrepreneurship. Due to this characteristic way of the institutions' engagement and their mutual and complementing effect on the regional or local scientific entrepreneurship landscape, we name this as the *Scientific Entrepreneurship Clustering Approach*.

For Switzerland, we have pointed out a converse situation to the one in Germany. The Swiss institutions engaging in scientific entrepreneurship are not as diversified in their instrumental portfolio as those in Germany. In Switzerland, the original institutions of higher education (universities and colleges of higher education) focus on action-fields which can be targeted on an aggregate level. The business institutions and networks on the other hand are specialized on action-fields that require a rather individual targeting. There is no particular type of institution, which inherits a leading role in supporting initiatives. Different types of institutions collaborate in a regional network and complement one another with particular instruments according to their focus area. In contrast to Germany, where we also have noticed a characteristic way of regional or local collaboration of different institutions, the Swiss institutions are rather independently organized and operate their instruments on their own. In Germany, the interlinkage between the institutions' activities is considerably more intensive than the cross-institutional collaboration in Switzerland.[1] With respect to the rather loose and unconsolidated collaboration between institutions within a regional network, this approach is named as the *Scientific Entrepreneurship Regional Network Approach*.

Limitations and implications for further research

As one of the first studies that compare scientific entrepreneurship initiatives cross-nationally and applies methods from complexity science to the field of scientific entrepreneurship research, there are some limitations given.

The first refers to the method used to compose the different national samples. We started with the Internet presence of the countries responsible ministries and employed a non-famous sampling method by following hyperlinks and therewith spinning a network of interlinked institutions of higher education, which then represent the respective national samples. The underlying assumption of this procedure is that institutions engaging in scientific entrepreneurship are interlinked virtually on the Internet. The second limitation refers to the source of data. We base our qualitative analysis on the institutions' web presences. So the assumption is that all institutions describe their activities on their websites in comparable ways with regard to the precision, actuality and vocabulary of descriptions. The third limitation refers to the way of analyzing the qualitative data. We analyze the descriptions given on the web presences and assign binary numbers to each institution with respect to the described activities in each action-field. Qualitative research like this naturally inherits interpretation and subjectivity to some degree.

But despite the given limitations, our comparative study provides new and deep insights into the way institutions in German-speaking Europe conceptualize, implement and operate activities to support scientific entrepreneurship. The interpretation of the empirical-based findings led to three different, country-specific approaches, which we characterized and distinguished from one another.

This study is based on the idea of understanding scientific entrepreneurship as an engineering discipline. The findings are an important contribution to the development of a broader theory of scientific entrepreneurship engineering.

In order to get a deeper understanding of the critical success factors of scientific entrepreneurship supporting initiatives, further research should build on the findings provided within this chapter and focus on developing methods and instruments to identify and measure the core drivers for complexity within the field of scientific entrepreneurship.

Furthermore the qualitative approach used in this study should be extended in a way to determine how institutions position instruments within the field of controversy between chaos and structure in each action-field. Those findings might provide implications for finding out which action-fields require rather structured approaches and which should be addressed just by triggering creativity and even fostering chaos to some extent.

Besides extending the research depth as suggested, it would be really worthwhile to investigate the applicability of the three approaches to scientific entrepreneurship in other countries. This could enable researchers to make an empirically-based statement on the causes of the existence of these approaches. Further research should also concentrate on the consequences of different ways of supporting scientific entrepreneurship. If different approaches could be evaluated and benchmarked against one another, practical implications in terms of concrete advice regarding how to conceptualize, implement and operate scientific entrepreneurship supporting initiatives could be given on a very stable basis.

Note

1 One finding that supports this statement is that in Germany, the institutions of most clusters communicate their activities in concert and appear as one collective institution whereas in Switzerland, there are less indications of such a close collaboration between institutes.

References

Alvarez, S.A. and Busenitz, L.W. (2001) 'The Entrepreneurship of Resource-based Theory', *Journal of Management*, 27, pp. 755–75.

Anderseck, K. (2004) 'Institutional and Academic Entrepreneurship: Implications for University Governance and Management', *Higher Education in Europe*, 29 (2), pp. 193–200.

Balzert, H. (1982) *Die Entwicklung von Softwaresystemen: Prinzipien, Methoden, Sprachen, Werkzeuge*, Mannheim: BI-Wissenschaftsverlag.

Barney, J.B. (1991) 'Firm Resources and Sustained Competitive Advantage', *Journal of Management*, 17, pp. 99–120.

Barney, J.B. (1995) 'Looking Inside for Competitive Advantage', *Academy of Management Executive*, 9, pp. 29–61.

Bok, D. (2003) *Universities in the Marketplace: The Commercialisation of Higher Education*, Princeton, NJ: Princeton University Press.

Brush, C.G., Greene, P.G. and Hart, M.M. (2001) 'Creating Wealth in Organizations: The Role of Strategic Leadership', *Academy of Management Executive*, 15 (1), pp. 64–78.

Bullinger, H.-J. and Scheer, A.-W. (eds) (2002) *Service Engineering*, Berlin: Springer.

Dart, R. (2004) 'The Legitimacy of Social Enterprise', *Non-Profit Management and Leadership*, 14 (4), pp. 411–24.

Drayton, W. (2002) 'The Citizen Sector: Becoming as Entrepreneurial and Competitive as Business', *California Management Review*, 44 (3), pp. 120–33.

Dumke, R. (2003) *Software Engineering*, 4th edition, Wiesbaden: Friedr. Vieweg & Sohn Verlag.

European Foundation for Entrepreneurship Research (EFER) (2009) *Publishing on the Internet*, Hilversum. Online, available at: www.efer.eu/web/index.htm (accessed 22 June 2009).

Förderkreis Gründungsforschung e.V. (FGF) (2007) *FGF-Jahreschronik 2006*, Bonn.

Geisler, R.L. (1993) *Research and Relevant Knowledge: American Research Universities since World War II*, Oxford, UK: Oxford University Press.

Goldstein, J.A., Hazy, J.K. and Silberstang, J. (2008) 'Complexity and Social Entrepreneurship: A Fortuitous Meeting', *E:CO*, 10 (3), pp. 9–24.

Grant, R.M. and Baden-Fuller, C. (1995) 'A Knowledge-based Theory of Inter-firm Collaboration', *Academy of Management*, best paper proceedings, pp. 17–21.

Grichnik, D., von Kortzfleisch, H.F.O. and Magin, P. (2009) 'Open Scientific Entrepreneurship Engineering – Ein offener, ganzheitlicher und systematischer Ansatz zur Unterstützung von Existenzgründungen aus Hochschulen', in A. Walter and M. Auer (eds) *Academic Entrepreneurship: Unternehmertum in der Forschung*, Wiesbaden: Gabler Verlag, pp. 167–92.

Harding, R. (2004) 'Social Enterprise: The New Economic Engine?', *Business Strategy Review*, (winter), pp. 40–3.

Hazy, J.K., Goldstein, J. and Lichtenstein, B. (eds) (2007) *Complex Systems Leadership Theory*, Boston, MA: ISCE Publishing.

Higgins D. (2006) 'Theoretical Assumptions of Knowledge Creation', *Irish Journal of Management*, 27 (2), pp. 189–213.

Itami, H. (1987) *Mobilization Invisible Assets*, Cambridge, MA: Harvard University Press.

Klandt, H. and Volkmann, C. (2006) 'Development and Prospects of Academic Entrepreneurship Education in Germany', *Higher Education in Europe*, 31 (2), pp. 195–208.

Kohn, K. and Spengler, H. (2007) *KfW-Gründungsmonitor 2007*, Frankfurt am Main: KfW Bankengruppe.

Kulicke, M. (2006) *EXIT – Existenzgründungen aus Hochschulen, Forschungsbericht Nr. 555*, Berlin: Bundesministerium für Wirtschaft und Technologie (BMWI) der Bundesrepublik Deutschland.

Magin, P. and von Kortzfleisch, H.F.O. (2008) *Methoden und Instrumente des Scientific Entrepreneurship Engineering*, Lohmar: Eul Verlag.

Massetti, B.L. (2008) 'The Social Entrepreneurship Matrix as a 'Tipping Point' for Economic Change', *E:CO*, 10 (3), pp. 1–8.

Murphy, P.J., Liao, J. and Welsch, H. (2005) 'A Conceptual History of Entrepreneurial Thought', *Academy of Management Proceedings 2005*, A1–A6.

O'Shea, R., Allen, T.J., Morse, K.P., O'Gorman, C. and Roche, F. (2007) 'Delineating the Anatomy of an Entrepreneurial University: The Massachusetts Institute of Technology Experience' *R&D Management*, 37 (1), pp. 1–16.

Penrose, E.T. (1959) *The Theory of the Growth of the Firm*, New York: Wiley.

Peteraf, M. (1993) 'The Cornerstones of Competitive Advantage: A Resource-based View', *Strategic Management Journal*, 13, pp. 363–80.

Roberts, D. and Woods, C. (2005) 'Changing the World on a Shoestring: The Concept of Social Entrepreneurship', *University of Auckland Business Review*, 11, pp. 45–51.

Shane, S.A. and Venkataraman, S. (2000) 'The Promise of Entrepreneurship as a Field of Research', *Academy of Management Review*, 25, pp. 217–26.

Twaalfhoven, B.W.M. (2004) *Red Paper on Entrepreneurship*, European Foundation for Entrepreneurship Research (EFER).

Vesper, K.H. and Gartner, W.B. (1997) 'Measuring Progress in Entrepreneurship Education', *Journal of Business Venturing*, 12 (5), pp. 403–21.

Volery, T. (2005) 'Ressourcenorientierter Ansatz von Entrepreneurship: Ressourcen sind der Kern des Wettbewerbsvorteils', *KMU Magazin*, 9, pp. 12–4.

Witt, P. (2006) *Stand und offene Fragen der Gründungsforschung*, Vallendar: WHU – Otto Beisheim School of Management, Förderkreis Gründungsforschung e.V.

14 Promoting effective university commercialization

William Allen[1] and Rory O'Shea

Introduction

The study of university spin-off activity has become a significant part of economic and management studies as the path from research to commercialization and is recognized as increasingly important to a country's economic development. Entrepreneurial innovations in universities stimulate the economy through the development of new products, the creation of new industries and by contributing to employment and wealth creation (O'Shea *et al.* 2008). In areas like biotechnology, university research has led to the development of completely new industries that retain strong academic ties, highlighting how effective the relationship can be. In other areas, research and the transfer of technology within universities has been closely linked to the needs of local industry.

Previous literature devoted to studying university spin-offs has covered a number of different perspectives. According to O'Shea *et al.* (2007), such perspectives include studies that explain spin-off creation in terms of the individuals who are involved in such activities; those that emphasize the organizational characteristics and resource endowments of universities; those that argue that social norms and institutional behaviour are determinants of spin-off activity; and those that argue that the wider social and economic context facilitates the creation of spin-off ventures. Although such work has undoubtedly provided universities with an improved understanding and awareness of spin-off behaviour, academic institutions are still faced with the challenge of identifying and replicating a process that facilitates successful entrepreneurial activities (O'Shea *et al.* 2008).

The objective of this chapter is to assist in this challenge and explore the factors that contribute to the ability of academics to successfully commercialize opportunities for spin-off activity within universities. In doing this, we build on an existing conceptual framework outlined by O'Shea *et al.* (2008) and draw on additional existing literature to explore the factors in universities that can act as either a trigger for, or a barrier to, the successful implementation and commercialization of spin-off ventures. We then use these perspectives to inform a study of a single university, University College Dublin (UCD), which has been relatively successful at generating spin-off companies. Based on this study, we explore the factors that can either stimulate or inhibit an academic's ability to commercialize a spin-off venture within a university environment.

For this chapter we define, spin-off companies as 'technology-based firms whose intellectual capital originated in universities or other public research organizations' (OECD 2001). The term of 'academic entrepreneur' is used when referring to an individual that discovers a commercial opportunity while being full-time employed at university as a PhD student, postdoctoral researcher, lecturer or professor. Following on from this, we generate suggestions on how universities can improve their level of successful spin-off creation.

Theoretical development

O'Shea *et al.* (2008) organized the body of theory and research on university entrepreneurship into six streams as follows: (*a*) individual attributes as determinants of spin-off activity; (*b*) organizational determinants of university spin-off activity; (*c*) institutional determinants of spin-off activity; (*d*) external determinants of spin-off activity; (*e*) the development and performance of university spin-offs; (*f*) the economic impact of spin-offs. Integrating these perspectives into a conceptual university spin-off framework O'Shea *et al.* (2008) suggested that spin-off creation not only varies due to the characteristics of individual academics but also as a result of variation in environments and university contexts. This framework identifies four different factors influencing the rate of spin-off activity within universities: (*a*) engaging in entrepreneurial activity (individual characteristics studies); (*b*) the attributes of universities such as human capital, commercial resources and institutional activities (organizational-focused studies); (*c*) the broader social context of the university, including the 'barriers' or 'deterrents' to spin-offs (institutional and cultural studies); (*d*) the external characteristics such as regional infrastructure that impact on spin-off activity (O'Shea *et al.* 2008).

In the context of this chapter, our aim is to focus on the specific climate conditions that facilitate academic entrepreneurship.[2] Four factors have been identified:

1   *Appropriate use of rewards*: developing effective incentives for academics is an area which requires careful consideration for universities engaged in spin-off activity. Implemented correctly, the use of such incentives can contribute greatly to enhancing an academic's willingness to expose themselves to the risks associated with entrepreneurial activity. According to Scanlon (1981), the most obvious reward within an organization is monetary, but other types of reward are also important to people and can include: increased levels of participation; recognition; more individual freedom; increased responsibilities; and opportunities for learning and growth. In the case of university spin-offs, research has shown that academic institutions that give higher percentages of royalty payments to their faculty members positively impact on spin-off activities (Link and Siegel 2005). In contrast, Di Gregorio and Shane (2003) found evidence that technology transfer policies within certain universities which allocate a higher share of inventor's royalties actually

decrease spin-off activity because of the increased opportunity cost in engaging in firm formation. In terms of non-monetary rewards, academics engaged in spin-off activity may also find personal satisfaction as a result of their contributions to local economies. Research conducted by the Association of University Technology Managers shows that successful spin-off ventures from American academic institutions have contributed 280,000 jobs to the US economy between 1980 and 1999 (Shane 2004).

2   *Management support*: this factor relates to the willingness of universities to facilitate and promote the pursuit of spin-off activities. Such support can take many forms, and may include the provision of necessary resources or expertise, and institutionalizing entrepreneurial activity within the university's culture. According to Barney (1986), certain organizational cultures enable firms to do and be things for employees that could not be done as well without such a culture. As a result, a culture that is seen to support innovation is essential in order for entrepreneurial activity to be successful. When conducting a study of spin-off activities within Massachusetts Institute of Technology (MIT), O'Shea *et al.* (2007) suggest that the university's founding mission and institutional support of entrepreneurial activities played a major part in the development of academic entrepreneurship at MIT. This view is also supported by Djokovic and Souitaris (2009) who suggest that 'the changing role of universities towards commercialization activities combined with governmental and institutional support mechanisms is creating a fertile ground for the seeds of university spin-offs'.

Either acting as a tacit presence or as a more proactive force that builds and celebrates a climate of entrepreneurial endeavour on campus, academic leadership can also add a strong sense of purpose to a university's character and establish an entrepreneurial heritage. Creating such a common leadership vision was important for the National University of Singapore (Poh-Kam 2009), because it provided a clear sense of direction when it came to technology transfer. Ideas emerged from the laboratories, guided by the commitment of charismatic academics who had a strong belief in what they were doing and what could be achieved. This factor would support our suggestion that there is a definite need for a management supported entrepreneurial culture to exist within a university intent on improving its spin-off activity. In contrast to this, Ndonzuau *et al.* (2002) believe that some cultural factors such as the 'publish or perish' drive can act as a strong inhibitor to academic entrepreneurship. Indeed, Thursby and Kemp (2002) found that many faculty inventions with commercial opportunity are not pursued in a spin-off capacity due to the unwillingness of academics to delay publication while they wait for the patent and licensing process to be completed.

Academic vision is also a precursor to improving technology transfer activities, informing the university on the structures and polices it will need to move forward. For example, Yale saw itself as a 'contributing institutional citizen' in economic development as opposed to being a passive observer (Breznitz 2007). This vision was the catalyst for transformation and the emergence of academic entrepreneurial behaviour which now flourishes at the university.

3   *Resource availability*: it is essential that academics know there is an availability of resources within universities to support their entrepreneurial behaviour. Wernerfelt (1984) classified resources as both tangible and intangible. whilst Barney (1991) divided such firm resources into three categories: physical capital resources; human capital resources; and organizational capital resources. In relation to university specific resources, Clarysse *et al.* (2005) suggest the following categories: organization resources; human resources; technological resources; physical resources; and financial resources. O'Shea *et al.* (2005) also examined in detail the effect of university resources on spin-off performance and classify such resources into four types: organizational; financial; human capital; and commercial resources. According to Wright *et al.* (2006), universities face a number or resource constraints in creating successful spin-offs but they cite access to venture capital as the most important, with access to other forms of financing also figuring highly. The pecking order hypothesis assumes that internal funding is preferred over external sources, and where these are insufficient, debt is preferred over equity to avoid ownership dilution (Wright *et al.* 2006). In addition to this, Clarysse *et al.* (2005) refer to networking resources as having an important role to play in spin-off activity. These would consist of the social network available within the university which could facilitate interaction between various parties interested in the spin-off process. From these perspectives, it is apparent that a necessary condition of a university's ability to create university spin-offs is a function of its ability to attract a significant amount of financial, human capital and physical resources to support spin-off related activities.

4   *Organizational structure*: research universities cannot depend on vision alone and in many instances structural changes will be required. Support programmes and services need resources but it is also important that programmes and schemes are organized in an integrated and coordinated manner to ensure maximum effectiveness. Promoting a supportive environment for academic entrepreneurship is also about workplace conditions, commitment to training, and a general willingness among academic staff to 'pitch in' and contribute with all kinds of activity. Research on knowledge spill-over theory of entrepreneurship, by O'Gorman *et al.* (2008), suggests several crucial reasons why technology transfer officers (TTOs) play a critical role as an intermediary and gatekeeper of the technology transfer process. The first is that scientists with new knowledge might underinvest in commercialization activity as they do not see the benefits of commercialization; and second those with the new knowledge may not recognize the commercial potential of the knowledge or fail in their attempts to commercialize the new knowledge due to a lack of market understanding.

Given that academics might not be aware of the viable opportunities that are available, the role of the TTO professional is to guide them and help them realize their full potential. Run effectively, the TTO can define roles and responsibilities, structures and processes that support executive decision-making. Typically these conform to a three-pronged governance model based around leadership, organization and alignment of key processes. Without these pillars, investment initiatives will be at risk.

The challenge for universities is to create a TTO with the right skill set. There is a danger that staff can lack the business acumen to deal with the commercial world and enable technology transfer. From a policy perspective, it is a balancing act. The TTO should drive pro-active commercialization but not at the expense of a commitment to the highest academic standards and performance. Following on from this, many universities have now adopted approaches whereby TTOs work very closely with departments and academics to proactively identify opportunities that may have significant market applications (Wright *et al.* 2004). Given that academics typically know relatively little about the business of technology commercialization but usually have a high psychological ownership for their inventions, the support of a TTO has a key role in recognising the commercialization potential of a technology (Powers and McDougall 2005).

A number of recent studies suggest that universities need to improve their technology transfer structures if they are serious about promoting entrepreneurial development on campus (Litan *et al.* 2007). According to a study by Siegel *et al.* (2004), there is widespread belief that there are insufficient structures to promote technology transfer. The report finds that firms expressed great difficulty in dealing with university TTOs on IPR issues, citing the inexperience of the TTO staff, the lack of general business knowledge and their tendency to overstate the commercial value of the patent. In addition, his research revealed that TTOs usually do not actively recruit individuals with marketing skills, but more often they looked for expertise in patent law and licensing or technical expertise. Siegel concludes that university inflexibility has led many firms and scientists to completely avoid working with the TTO. They found that university TTOs appear to do a better job of serving the needs of large firms than small, entrepreneurial companies. According to his research, inefficiencies that exist in the TTOs need to be addressed so as not to prohibit academics from transitioning into entrepreneurial careers.

Research method

Study approach

In this chapter, we utilize a multiple case study approach examining four spin-off ventures at varying stages in the spin-off process based in University College Dublin (UCD), one of Ireland's most dynamic and modern universities (awarded second place in the *Sunday Times University Guide* 2008).[3] The university is organized into a number of colleges and schools across a range of disciplines with academics typically performing both teaching and research roles.

NovaUCD is the Innovation and Technology Transfer Centre at UCD and is responsible for the implementation of the UCD policies relating to intellectual property and for the provision of advice on the identification, protection and exploitation of intellectual property resulting from UCD Research. A key priority for NovaUCD is to work with UCD researchers in taking their innovative ideas from proof-of-principle to full commercial success (NovaUCD 2008). The NovaUCD Campus Company Development Programme (CCDP), which is now in its fourteenth year, is the main support programme run by NovaUCD for academic entrepreneurs who are spinning out campus companies.

As outlined by Eisenhardt (1989), the case study is a research strategy which focuses on understanding the dynamics present within single settings and typically combines data collection methods such as archives, interviews, questionnaires, and observations. In this study the authors selected four spin-off ventures based in University College Dublin in order to gain widespread understanding of the factors that affect commercialization within that university. The four ventures selected, at the time of writing, are positioned at varying development stages of the spin-off process. Two cases are successfully established spin-off companies, whilst the other two are still in the process of completing commercialization. The decision to focus on spin-offs of varying age was intentional in order to enhance the generalizability of our model. It is prudent to choose such cases in varying situations and polar types in which the process of interest is transparently observable (Eisenhardt 1989).

Each of the four cases studied originated from different faculties to one another. In order to maintain anonymity for each of the individuals interviewed, the cases shall be referred to as: A, B, C and D.

Data was collected using in-depth semi-structured interviews with a founding academic from each of the four cases selected. Triangulation (Jick 1979) was aided by collection of archival data including university-level information provided by the UCD website and archival documents.

The authors acknowledge that research may be subject to minor limitations and biases as result of the limited number of spin-offs included in the research sample. The findings are based primarily on the views of the relevant interviewees within each case and as such may be subjective in relation to individual opinions, interpretations and experience.

Data analysis – factors that stimulate/inhibit the ability of academics to commercialize spin-off ventures

We present our analysis of the four cases studied, focusing on the four factors as outlined in our theory development: (*a*) appropriate use of rewards; (*b*) management support; (*c*) resource availability; and (*d*) organizational structures within an academic institution. In each of these factors, we explore the elements which either simulate or inhibit spin-off activity.

Factor 1: appropriate use of rewards

All four cases immediately spoke of the potential monetary rewards that a successful spin-off venture could provide and indicated that this was a primary incentive for each case in the pursuit of commercialization. It is important to note that in most cases once an idea for spin-off is commercialized, the university acquires a 15 per cent equity stake in the company whilst the founding team retain the remaining 85 per cent. There can also be additional equity agreements in relation to patent royalties from the success of spin-off ventures.

Non-monetary rewards were also highlighted as important. The founder of Case A was 'an academic who doesn't believe that research should stay in the lab'. Citing the potential for evaluating technology in a real-world environment as a significant reward for spin-off activity, Case A enthused: 'The key thing is developing technology and evaluating that technology in the real world.' These sentiments were echoed by Case C who described themselves as being 'as much of a technologist as an academic' and listed one of their main research goals as the understanding of how their work can be applied outside of the university.

The possibility of making a contribution to economy through the creation of employment was another form of reward that was highlighted. Case A stated that the creation of employment through the success of their venture has presented them with a great deal of personal satisfaction: 'I've been fortunate enough to be involved in something that's created quite a significant number of jobs – that's rewarding enough in itself.'

According to all four cases studied, with the exception of the obvious financial rewards and perhaps the personal reward of seeing their own technology commercialized, there is little or no other level of reward offered by university in relation to the commercialization of spin-off ventures. This apparent lack of institutional sponsored rewards is a factor that all four cases agreed could potentially inhibit or at least nullify an academic's aspirations to pursue commercialization. Case B in particular cited the absence of a university recognized reward system as having a negative effect on spin-off activity: 'Some incentives for academics would certainly aid commercialization – perhaps in the form of a recognized award for the academic in question.'

Factor 2: management support

When questioned about the level of support for spin-off ventures within the university the four cases responded with somewhat mixed views in relation to the existence of an entrepreneurial culture. All four cases agreed that in general, there was quite a positive attitude within the university in relation to spin-off activity. Case D went so far as to suggest that the 'rising stars' in UCD at the moment are those involved in commercialization activity. They did however admit that such an enthusiastic culture may be faculty specific and as such could differ for other academics.

Approved allocation of time for spin-off related activity was seen as a significant support mechanism by all four cases. According to the founder of Case A, academics can usually spend 20 per cent of their time on activity that is reasonably related to their research interests, but in this particular example the academic was able to negotiate further time from their faculty to spend working on the commercialization of their own specific venture. Such support was instrumental in the initial activity in establishing Case A as a viable venture. Likewise, Case B confirmed that they were also allocated specific time to work on their project by their faculty in the lead up to the actual commercialization of the venture.

Although the university is undoubtedly aiming to create a culture that embraces the pursuit of spin-off activity, each of the cases studied allude to an underlying culture that actually makes commercialization rather difficult. This inhibiting factor relates to the conflict an academic faces when tasked with the choice of pursuing publication of their research as opposed to pursuing commercialization of the idea in question. Case B and Case C both raised this as major issue for any academic who may be considering a spin-off venture. According to them, if publications are viewed as being the mainstay of an academic's career, then it is obvious that many will feel the need to pursue publications over spin-off activity. As a result, cases B, C and D cited this as a concern which universities must address if the level of spin-off activity is to be increased.

Management expectation levels was also raised by cases A and D as a management support related issue. Despite being allocated a specific amount of time for spin-off activity, the academics were still expected to fulfil their own full time university responsibilities as well. Essentially both cases were expected to fit their full working week into the reduced time allocated for their academic tasks and also work on the commercialization of their spin-off project. In response, Case D suggested that more time needs to be spent by management within the university trying to match certain types of academic individuals with specific workloads as certain academics would be happy to carry out high levels of administrative work whilst others have a natural inclination towards entrepreneurial activity.

The academic involved with Case C also spoke of the university's attempts to promote an entrepreneurial culture in favour of spin-off activity but was slightly sceptical about how effective such attempts actually are: 'for every person that you meet that is enthusiastic about what you are doing, you will meet an awful lot more that are not.' In agreement, Case C specifically referred to such attitudes as 'barrier building'.

Factor 3: resource availability

Although there are a wide range of university resources that can potentially encourage and aid the commercialization of spin-off ventures, the authors will focus on the principal resources highlighted by the cases as being fundamental in their own bids for commercialization.

The founder of Case A spoke about the availability of human capital resources as being a key driver. With Case A, the academic was able to utilize the skills of their research student to the point where they are now regarded as a joint founder of the venture. Case A saw their role as an academic as being quite evangelical, meaning that there was also a strong need for someone to engage in a more day-to-day role, such as writing the code for the project. 'One of the critical things is to have someone who is going to be able to develop and run with the project for commercialization', said the academic. Without access to a research student to carry out such a role, it would have been a lot more difficult for the academic involved with Case A to pursue the venture. The academic involved with Case D also cites the availability of human capital resources as a key stimulus to commercialize. In the instance of Case D, the academic feels that they are able to draw on advice and expertise from different sources within the university and is currently collaborating with one of the founders of Case A in various aspects of their own venture. The academic involved with Case C, meanwhile, received assistance from two MBA students within UCD in relation to their initial thoughts on commercialization. This was arranged by NovaUCD and according to Case C proved to be of great benefit to them in their decision to pursue commercialization.

Opinions on the availability of other resources vary across the four cases. Of utmost importance is potential funding. Although Case A bemoaned the availability of funds when they were initially pursuing commercialization the other three cases have complimented the role of Enterprise Ireland in providing funding for potential spin-off ventures, which, depending on the type of spin-off opportunity, can provide various levels of financial support.[4] Cases B, C and D also referred to NovaUCD and the wide range of other resources that are provided for academics by providing facilities that allow them to network with others with similar interests and objectives.

It is of vital importance that the relevant resources needed for the commercialization of spin-off ventures are perceived to be available by academics if a university is to be successful in its attempts to improve its level of spin-off activity. Unfortunately, this study reveals that some of these resources are not always readily available, which could have an inhibiting effect on the spin-off process within UCD.

According to the founder of Case A, the availability of financial resources (or lack thereof) can present a large barrier to those seeking the commercialization of spin-off opportunities: 'the single thing that prevents nine out of ten projects from proceeding to any level of commercialization is the gap between the initial idea and funding being granted.' In outlining their concerns over the availability of such funding, Case A highlighted their own search for initial venture capital which took approximately 12 months. Although the academic in this case had a university salary to facilitate them, the research partner was totally reliant on the funding for the project. Its absence could have had a negative impact on the successful completion of the commercialization drive. This experience differs from the other cases in relation to financial resources and it must also be taken into consideration that Case A was founded before all of the other cases and faced different funding issues compared to more recent spin-off ventures.

None of the cases highlighted the availability (or lack thereof) of any other types of resources as inhibiting the abilities of academics to pursue commercialization.

Factor 4: organizational structures

The existence of a supportive organizational structure for entrepreneurial activity is essential for any university looking to increase the commercialization of spin-off ventures. Often an organizational structure like this is facilitated through the existence of a Technology Transfer Office, such as NovaUCD in the case of this particular study. All four cases have had varying levels of interaction with NovaUCD and three of the four cases – B, C and D – spoke highly about the facilities and services provided.

The focus point of NovaUCD for academics looking to commercialize their spin-off ventures would appear to be the Campus Company Development Programme (CCDP). Over the past 12 years, approximately 130 projects and 190 individuals have completed the programme, the aim of which is to assist academic entrepreneurs in the establishment and development of knowledge-intensive enterprises by providing the skills necessary to transform ideas into commercially feasible ventures (NovaUCD 2008).

As already outlined, three of the four cases were quite complimentary in their views on the organizational structure within the university that supports the commercialization of spin-off ideas. In particular, NovaUCD was commended on the role that it plays in such a process. The founder of Case A did however voice certain concerns over the facilities provided. While acknowledging that NovaUCD has evolved and improved with age, Case A believe that certain aspects related to the TTO still inhibit commercialization as opposed to providing a stimulus. According to Case A; spin-off ventures are very much 'left to their own devices' in relation to many of the tasks that need to be completed in relation to commercialization. Case A believed that many academics would find the tasks associated with spin-off activity rather daunting and as such they may not feel confident enough in their ability to pursue commercialization. It should however be noted that Case A initially dealt with NovaUCD a number of years earlier than the other cases so this could suggest that the organizational structure has improved since the commercialization of Case A.

Conclusions

In this chapter we utilized an existing literature-based framework put forward by O'Shea *et al.* (2008) and highlighted four key factors which can either stimulate or inhibit the ability of academics to commercialize a spin-off venture. The themes that emerge from this study can be moulded into a framework that will create an academic environment where spin-off activity is more likely to happen. It is clear that in order to support academics and encourage them to pursue commercialization and spin-offs, universities must endeavour to maximize the use of those resources that stimulate commercialization whilst also acknowledging and addressing the factors that may inhibit such commercialization.

Successful spin-off programmes demand an enduring commitment from all of the university to promote academic entrepreneurship. It is a commitment that needs nurturing and is built upon incentives, rewards, and the investment of resources in the education of faculty. By contrast, unsuccessful attempts are frequently marked by the unwillingness of the institution to consider the importance of commercial development as either necessary or desirable.

Only through the cultivation of strong leadership and supporting policies can a university hope to break down the walls between ivory towers of innovation and the commercial world. By setting out a vision and orchestrating its realization through the synchronization of operational resources, commercialization can be embedded in the culture of the university. Universities will be unable to develop successful spin-off programmes without changing the mindset of 'academics'. To do this effectively, universities must ensure that the governance structure is in place to support culture change. Academics will not understand the legitimacy of commercialization unless it is supported from the top of the institution, as well as by proper communication structures, leadership, and performance measures. Simply sponsoring activities is not enough. There has to be a tangible strategy, redirecting, establishing new priorities, and re-allocating resources to optimize the technology transfer process. And academic leaders must play their part in changing the status quo, helping overcome obstacles that inevitably arise in an institutionalized environment where there is resistance to change.

In order for policy makers to encourage academic entrepreneurship, an entrepreneurial approach to the identification, protection and commercialization of university intellectual property must be undertaken. The governance of the TTO is of central importance in achieving and sustaining technology transfer success and influencing stakeholders. Universities that achieve an entrepreneurial orientation will be more likely to succeed. A culture must be cultivated where start-up ventures are valued and peer support from fellow entrepreneurial academics is on hand, championing commercialization activity. Direct exposure to role models will heighten entrepreneurial activity. However, it must also be noted that the essential character of technology transfer lies not only in the formal structures and programmes which universities construct, but also in the agency of academics to engage in the entrepreneurial process. A sense of obligation and commitment from academic staff to the entrepreneurial process is what distinguishes top entrepreneurial universities from the rest. Therefore it is important that the absence of a rewards system in a university do not act as a barrier towards engaging in the entrepreneurial process.

We have seen in this volume the many positive elements within a university that can foster the entrepreneurial ambition that ultimately leads to spin-offs and commercialization. Unfortunately, there are still some elements within universities that hinder progress. A focus on traditional elements of academia, for example, may inadvertently inhibit spin-off activity. If universities genuinely want to increase the level of spin-off activity it may be beneficial for them to consider the following: (*a*) the introduction of meaningful rewards for academics who are successful in commercialization; the creation of an entrepreneurial culture with pro-active support systems in addition to career prospects for academics interested in spin-off activity; (*b*) the provision of readily available resources, financial, human capital or otherwise in order to encourage academics who may consider the route of spin-off creation; (*c*) the implementation of an organizational structure that supports, promotes and improves the commercialization of spin-out ideas within the university environment; and (*d*) the elimination of the risk associated with the enactment of engaging spin-off activity.

Notes

1 Corresponding author.
2 In a detailed review of entrepreneurial literature, Hornsby *et al.* (1999) and Hornsby *et al.* (2002), identified conditions that facilitate entrepreneurship which included; Appropriate Use of Rewards, Management Support, Resource Availability and Organizational Structure
3 *Sunday Times* University of the Year is an annual award given to a university or other higher education institution by the *Sunday Times*, based on league table positions in addition to the institution's contribution on a local, national and international level (*Sunday Times* 2008).
4 Enterprise Ireland is the government agency responsible for the development and promotion of the Irish indigenous business sector. One of Enterprise Ireland's fundamental goals is to help both companies and researchers based in third level institutions to engage in research. Their aim is to facilitate collaborative links between enterprise and the research community that lead to the practical application of research in business, yielding benefits to both groups (Enterprise Ireland 2008a).

References

Barney, J. (1986) 'Organizational Culture – Can It Be a Source of Competitive Advantage?', *Academy of Management Review*, 11 (3), pp. 656–65.

Barney, J. (1991) 'Firm Resources and Sustained Competitive Advantage', *Journal of Management*, 17 (1), pp. 99–120.

Breznitz, D. (2007) *Innovation and the State: Political Choice and Strategies for Growth, in: Israel, Taiwan and Ireland*, New Haven, CT: Yale University Press.

Clarysse, B., Wright, M., Lockett, A., Velde, E. van de and Vohora, A. (2005) 'Spinning out New Ventures: A Typology of Incubation Strategies from European Research Institutions', *Journal of Business Venturing*, 20, pp. 183–216.

Djokovic, D. and Souitaris, V. (2009) 'Spinouts from Academic Institutions. A Literature Review with Suggestions for Further Research', *Journal of Technology Transfer*.

Enterprise Ireland (2008a). Online, available at: www.enterprise-ireland.com/ResearchInnovate (accessed 17 August 2008).

Eisenhardt, K.M. (1989) 'Building Theories from Case Study Research', *Academy of Management Review*, 14 (4), pp. 532–50.

Gregoio, D. di and Shane, S. (2003) 'Why Some Universities Generate More TLO Startups than Others?', *Research Policy*, 32 (2), pp. 209–27.

Hornsby, J.S., Kuratko, D.F. and Montagno, R.V. (1999) 'Perception of Internal Factors for Corporate Entrepreneurship: A Comparison of Canadian and US Managers', *Entrepreneurship: Theory and Practice*, 24 (2), pp. 9–24.

Hornsby, J.S., Kuratko, D.F. and Zahara, S.A. (2002) 'Middle Managers' Perception of the Internal Environment for Corporate Entrepreneurship: Assessing a Measurement Scale', *Journal of Business Venturing*, 17, pp. 253–73.

Jick, T.D. (1979) 'Mixing Qualitative and Quantitative Methods: Triangulation in Action', *Administrative Science Quarterly*, 24, pp. 602–11.

Link, A.N. and Siegel, D.S. (2005) 'Generating Science-based Growth: An Econometric Analysis of the Impact of Organizational Incentives on University–Industry Technology Transfer', *European Journal of Finance*, 11 (3), pp. 169–81.

Litan, R., Mitchell, L, and Reedy, E.J. (2007) 'The University as Innovator: Bumps in the Road', *Issues in Science and Technology*.

Ndonzuau, F.N., Pirnay, F. and Surlemont, B. (2002) 'A Stage Model of Academic Spin-off Creation', *Technovation*, 22 (5), pp. 281–9.

NovaUCD Technology Transfer (2008). Online, available at: www.ucd.ie/nova/services/protection.htm.

O'Gorman, C., Byrne, O. and Pandya, D. (2008) 'How Scientists Commercialize New Knowledge via Entrepreneurship', *Journal of Technology Transfer*, 33 (1), pp. 23–43.

O'Shea, R.P., Allen, T.J., Chevalier, A. and Roche, F. (2005) 'Entrepreneurial Orientation, Technology Transfer and Spin-off Performance of US Universities', *Research Policy*, 34, pp. 994–1009.

O'Shea, R.P., Allen, T.J., Morse, K.P., O'Gorman, C. and Roche, F. (2007) 'Delineating the Anatomy of an Entrepreneurial University: The Massachusetts Institute of Technology Experience', *R&D Management*, 37 (1), pp. 1–16.

O'Shea, R.P., Chugh, H. and Allen, T.J. (2008) 'Determinants and Consequences of University Spin-off Activity: A Conceptual Framework', *Journal of Technology Transfer*, 33 (6), pp. 653–66.

Poh-Kam Wong (2007) 'Towards an Entrepreneurial University Model to Support Knowledge-based Economic Development: The Case of the National University of Singapore', *World Development*, 35 (6).

Powers, J. and McDougall, O. (2005) 'University Start-up Formation and Technology Licensing with Firms that Go Public: A Resource Based View of Academic Entrepreneurship', *Journal of Business Venturing*, 20 (3), pp. 291–311.

Scanlon, B.K. (1981) 'Creating a Climate for Achievement', *Business Horizons*, 24, March–April, pp. 5–9.

Shane, S. (2002). 'Executive Forum: University Technology Transfer to Entrepreneurial Companies', *Journal of Business Venturing*, 17, pp. 537–52.

Shane, S. (2004) *Academic Entrepreneurship: University Spin-offs and Wealth Creation*, Cheltenham, UK: Edward Elgar.

Siegel, D.S., Waldman, D. and Link, A. (2003) 'Assessing the Impact of Organizational Practices on the Relative Productivity of University Technology Transfer Offices: An Exploratory Study', *Research Policy*, 32, pp. 27–48.

Siegel, D.S., Waldman, D.A., Atwater, L. and Link, A.N. (2004) 'Toward a Model of the Effective Transfer of Scientific Knowledge from Academicians to Practitioners: Qualitative Evidence from the Commercialization of University Technologies', *Journal of Engineering and Technology Management*, 21 (1–2), pp. 115–42.

Sunday Times University Guide (2008). Online, available at: www.timesonline.co.uk/tol/life_and_style/education/sunday_times_university_guide/article4831936.ece.

Thursby, J. and Kemp, S. (2002) 'Growth and Productive Efficiency of University Intellectual Property Licensing', *Research Policy*, 31, pp. 109–24.

Wernerfelt, B. (1984) 'A Resource-based View of the Firm', *Strategic Management Journal*, 5, pp. 171–80.

Wright, M., Birley, S. and Mosey, S. (2004) 'Entrepreneurship and Technology Transfer', *Journal of Technology Transfer*, 29 (3–4), pp. 235–46.

Wright, M., Clarysse, B., Lockett, A. and Binks, M. (2006) 'University Spin-out Companies and Venture Capital', *Research Policy*, 35 (4), pp. 481–501.

15 Comparing university organizational units and scientific co-authorship communities

*Uwe Obermeier, Michael J. Barber,
Andreas Krueger and Hannes Brauckmann*

Introduction

A co-authorship network of scientists at a university is an archetypical example of a complex evolving network. Collaborative R&D networks are self-organized products of partner choice between scientists.

Various theoretical frameworks have been used to describe recent academic research. Though these frameworks differ in many respects, they all point to a more collaborative R&D 'network mode' of knowledge production. Concepts such as 'Mode 2' (Gibbons *et al.* 1994), 'Academic Capitalism' (Slaughter and Leslie 1997), 'Post-Academic Science' (Ziman 2000), or the 'Triple Helix' (Etzkowitz and Leydesdorff 2000) not only refer to external collaborations of universities with industry, government and other actors, but report changed practices inside academia.

In universities, the isolated researcher in the ivory tower has been widely replaced by interdisciplinary teams in collaborative research projects (Wuchty *et al.* 2007). Crossovers and co-operations between different scientific disciplines, different organizational units, and external actors are common and increasing phenomena of academic reality (Guimerà *et al.* 2005). Since collaborative research has become the dominant and most promising way to produce high-quality output (Jones *et al.* 2008), collaboration structures are also a target for research and management design (Bozeman and Lee 2005).

Collaborative research projects, co-authored publications, or multidisciplinary excellence networks in universities point to the peer network mode of today's knowledge production.

Our study measures these outputs by looking at co-publications between different organizational units of University College Dublin (UCD). With this, we try to reconstruct the interdisciplinary publication culture inside this university.

The study is set up as an academic research project but is intended to contribute to the self-monitoring mechanisms of UCD as well. We assume that demand from academic management for consultancy will grow due to rising legitimatory needs. Therefore it is important to embed network analysis in a methodological and ethical-regulatory context to evaluate its applicability.

Data, method, results

The combination of co-authorship and SNA is often used for mapping, analysing, and evaluating scientific research activity (Olmeda-Gómez *et al.* 2008, Barabasi *et al.* 2002, Yousefi-Nooraie *et al.* 2008).

Data of our study here reflect the publication output of UCD's permanent academic staff during the period 1998–2007, drawn from the ISI Thompson Citation Index (Web of Knowledge). We include publications from the Science Citation Index (SCI), Social Science Citation Index (SSCI), and the Arts and Humanities Index (A&HCI), which are classified as normal articles, reviews, letters and notes.

In UCD, all members of academic staff belong to schools or research units, which are organized in a college structure. For the purposes of our study, we measure interdisciplinarity as co-publication activity of UCD authors belonging to different colleges and schools.

Researchers with publications within and between colleges 1998–2007

Within Table 15.1 we have researchers with at least one publication covered by the Citation Index. We can observe within the College of Arts and Celtic Studies (1) there are 62 researchers publishing covered by the SCI (Science Citation Index), College of Business and Law (2) is represented with 44 researchers, College of Engineering, Mathematical and Physical Sciences (3) we have 139 researchers, College of Human Sciences (4) we find 97 researchers. The highest members of (SCI) publishing researchers comes from the College of Life Sciences (5) represented by 291 researchers.

In Figure 15.1 we only focus on researchers with co-authored publications within UCD. Every node represents a researcher with at least one co-authored publication, a link between two nodes represents at least one co-published paper. We can observe that within College of Business and Law nodes (2), 23 per cent of the researchers are connected by co-authorship. Within College of Human Sciences, nodes (4) we observe 31 per cent of researchers are connected by co-authorship. The highest level of co-authorship between researchers within UCD we can observe within College of Engineering, Mathematical and Physical Sciences (3) with 54 per cent and the nodes (5) College of Life Sciences with 76 per cent.

Table 15.1 Researchers with at least one publication within UCD

College	1	2	3	4	5	6
Count	62	44	139	97	291	9

Source: UCD Research.

Notes
(1) College of Arts and Celtic Studies, (2) College of Business and Law, (3) College of Engineering, Mathematical and Physical Sciences, (4) College of Human Sciences, (5) College of Life Sciences, (6) Others.

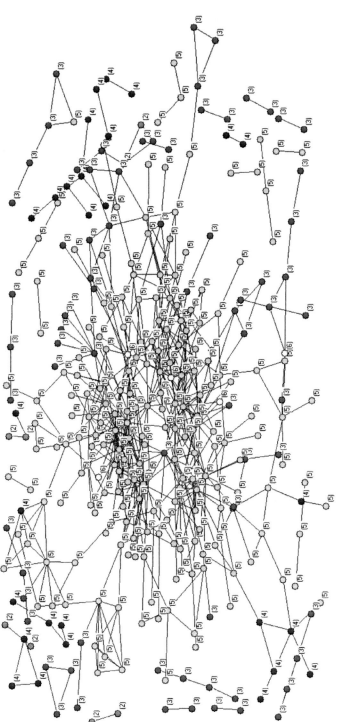

Figure 15.1 Co-authored publications within and between different colleges 1998–2007 (source UCD Research).

Notes
Nodes (3) College of Engineering, Mathematical and Physical Sciences, Nodes (2) College of Business and Law, Nodes (5) College of Human Sciences, Nodes (4) College of Life Sciences, Nodes (6) others.

Table 15.2 Researchers with co-authorship within UCD

College	2	3	4	5	6
Count		10 (23%)	74 (53%)	30 (31%)	222 (76%) 5

Notes
(2) College of Business and Law, (3) College of Engineering, Mathematical and Physical Sciences, (4) College of Human Sciences, (5) College of Life Sciences, (6) Others.

A high level of co-authorship from authors outside their own college we can observe within the College of Engineering, Mathematical and Physical Sciences with 34 per cent, and College of Life Sciences with 27 per cent of connected researchers at UCD.

In Table 15.4 and Figure 15.2 we focus on co-authorship between different schools to investigate interdiciplinarity. The highest level of co-authorship between different schools we can observe within the schools of the College of Life Sciences. The most central schools are Medicine and Medical Sciences, School of Biomolecular and Biomedical Science and the School of Agriculture, Food Science and Veterinary Medicine.

The high level of co-authorship is interesting between the School of Mathematical Sciences with the School of Agriculture, Food Science and Veterinary Medicine, as well with the School of Medicine and Medical Science. Another interesting high number of co-authored papers we can observe between the School of Computer Science and Informatics and the School of Business.

Using co-authorship as an indicator for mapping the interdisciplinary cooperation structure of the university shows that within and between single colleges the co-publication activity differs significantly. We observe for example:

- There are no co-authored publications with or within the College of Arts and Celtic Studies, at least not as covered by the Citation Index.
- The highest amount of intra-college co-authorship we can observe within the College of Life Sciences.
- The amount of co-publications between different schools of that college points to working interdisciplinary cooperation structures between thematically close disciplines (especially the schools of Medicine and Medical Sciences, Biomolecular and Biomedical Science, Agriculture, Food Science and Veterinary Medicine frequently co-publish with other schools of that college).

Table 15.3 Authors with co-authorship outside their own college

College	2	3	4	5	6
Count	4 (40%)	25 (34%)	5 (17%)	61 (27%)	5

Notes
(2) College of Business and Law, (3) College of Engineering, Mathematical and Physical Sciences, (4) College of Human Sciences, (5) College of Life Sciences, (6) Others.

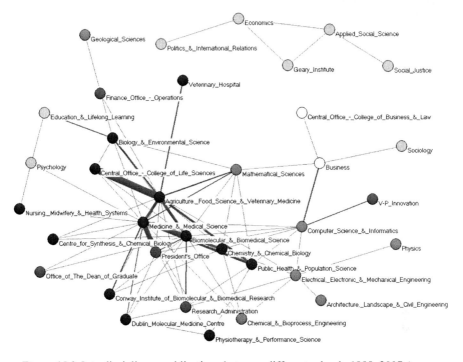

Figure 15.2 Interdisciplinary publications between different schools 1998–2007 (source UCD Research).

- The College of Life Sciences also shows the highest amount of interdisciplinary co-publications with other colleges, especially with the College of Engineering, Mathematical and Physical Sciences pointing to working interdisciplinary cooperation structures between fields which are not thematically close.
- This intra-college cooperation is concentrated in particular schools, i.e.

> the School of Agriculture, Food Science and Veterinary Medicine and the School of Medicine and Medical Science (both College of Life Sciences) frequently co-publish with the School of Mathematical Sciences (College of Engineering); or
> the School of Computer Science and Informatics (College of Engineering) and the School of Business (College of Business and Law) are frequent co-publishers.

These observations lead us to assume that there are specific profiles of schools and individuals with skills crucial to connect and contribute to other schools/individuals. This ranges from cooperation between thematically close areas to bridging huge disciplinary distances. We can easily identify 'champions' of interdisciplinary co-publication, brokers between disciplines, and areas where this is not working at all.

Table 15.4 Frequency of co-authorship between different schools

Co-auth	School 1 name	School 2 name
100	Central Office – College of Life Sciences	Agriculture, Food Science and Veterinary Medicine
71	Medicine and Medical Science	President's Office
64	Medicine and Medical Science	Biomolecular and Biomedical Science
40	Biomolecular and Biomedical Science	Agriculture, Food Science and Veterinary Medicine
39	Chemistry and Chemical Biology	Biomolecular and Biomedical Science
38	Medicine and Medical Science	Agriculture, Food Science and Veterinary Medicine
36	Public Health and Population Science	Agriculture, Food Science and Veterinary Medicine
34	Public Health and Population Science	Medicine and Medical Science
21	Biology and Environmental Science	Agriculture, Food Science and Veterinary Medicine
18	Veterinary Hospital	Agriculture, Food Science and Veterinary Medicine
16	Mathematical Sciences	Agriculture, Food Science and Veterinary Medicine
15	Conway Institute of Biomolecular and Biomedical	Medicine and Medical Science
15	Research Administration	Biomolecular and Biomedical Science
14	Computer Science and Informatics	Business
13	President's Office	Biomolecular and Biomedical Science
12	Dublin Molecular Medicine Centre	Medicine and Medical Science
10	Mathematical Sciences	Medicine and Medical Science
10	V-P Innovation	Computer Science and Informatics
9	Education and Lifelong Learning	Biology and Environmental Science
8	Biology and Environmental Science	Biomolecular and Biomedical Science
7	Central Office – College of Life Sciences	Biomolecular and Biomedical Science
7	Dublin Molecular Medicine Centre	Research Administration
7	Chemistry and Chemical Biology	Medicine and Medical Science
7	Chemical and Bioprocess Engineering	Biomolecular and Biomedical Science
6	Biomolecular and Biomedical Science	Centre for Synthesis and Chemical Biology
6	Dublin Molecular Medicine Centre	Biomolecular and Biomedical Science
6	Psychology	Medicine and Medical Science
6	Chemical and Bioprocess Engineering	Chemistry and Chemical Biology
5	Finance Office – Operations	Agriculture, Food Science and Veterinary Medicine
5	Chemistry and Chemical Biology	Agriculture, Food Science and Veterinary Medicine
5	Geological Sciences	Biology and Environmental Science
4	Physiotherapy and Performance Science	Medicine and Medical Science
4	Computer Science and Informatics	Public Health and Population Science
4	Chemistry and Chemical Biology	Computer Science and Informatics
4	Electrical, Electronic and Mechanical Engineering	Agriculture, Food Science and Veterinary Medicine

What is the position of the brokerage individuals?

From these first results we started to investigate the positions of the brokerage individuals, the ones connecting the different organizational units. Are they the central ones within their own organizational units, schools? Is there any relation to the thematic fields and what does it look like?

Heinze *et al.* (2009) distinguishes two main lines of thought. Network brokerage argues that people who are placed at the intersection of heterogeneous social groups have an increased likelihood of drawing upon multiple knowledge sources, leading to the generation of new ideas (Burt 2004). Whereas proponents of cohesive collaborative networks argue for the benefits of trust, shared risk taking and easy mobilization in facilitating information and knowledge transfer. According to these studies, individuals with cohesive ties are likely to be involved in innovations (Uzzi and Spiro 2005, Gloor 2006).

We analysed the schools strong in interorganizational co-authorship (see above). We especially looked at the schools from the College of Life Sciences and the schools from the College of Engineering, Mathematical and Physical Sciences; and from the College of Business and Law we looked at the School of Business.

We extracted the schools from the overall network and calculated the weighted centrality of the individuals. We compared this to the weighted centrality within the overall network removing the lines inside the schools. In order to use the centrality values and the ranks we calculated Pearson's Correlation Coefficient and Spearman's Rank Correlation with Pajek.

We have to keep in mind that the number of vertices is small and we don't have other universities to compare but the results show a clear tendency:

- within thematically close schools the brokerage individuals are also, within their own unit, more central;
- there is a negative correlation for individuals brokering between thematically distant schools.

To further investigate the role that the UCD colleges and schools play in determining co-authorship, we infer collaborative communities from the structure of the

Table 15.5 Comparison of degree centrality of individuals within selected schools to overall network

School	Number of vertices	Pearson Correlation Coefficient	Spearman Rank Correlation
Agriculture, Food Science and Veterinary Medicine	64	0.43	0.34
Computer Science and Informatics	16	−0.13	−0.15
Mathematical Sciences	13	−0.38	−0.34
Medicine and Medical Science	68	0.39	0.40
Business	9	−0.35	−0.64
Chemistry and Chemical Biology	9	0.59	0.27

network itself. We focus our attention on the bipartite network of authors and publications formed by linking each publication to its author. This network is not connected; we restrict investigation to the largest component, containing 234 authors.

The identification of communities from network structure is a topic of great recent interest. Formulation of the problem presents two main challenges. First, the notion of community is imprecise, requiring a definition to be provided for what constitutes a community. Second, community solutions must also be practically realizable for networks of interest. The interplay between these challenges allows a variety of community definitions and community identification algorithms suited to networks of different sizes, as measured by the number of vertices n or edges m in the network (for useful overviews, see Newman 2004b, Danon *et al.* 2005 and Fortunato and Castellano 2008).

A prominent formulation of the community-identification problem is based on the modularity Q introduced by Newman and Girvan (2004). The quality of communities given by a partition of the network vertices is assessed by comparing the number of edges between vertices in the same community c to the number expected from a null model network. Good quality communities then have more intra-community links than would be expected from the model, and fewer inter-community links than expected.

As we consider a bipartite network of authors and publications, we use a bipartite null model (Barber 2007). Formally, this is

$$Q = \frac{1}{m} \sum_c \sum_{i,j \in c} \left(A_{ij} - \frac{k_i k_j}{m} \right)$$

where the A_{ij} are components of the adjacency matrix for the network, the k_i are the degrees of the network vertices, and the sums over i and j are restricted to run over the authors and publications, respectively.

Community identification is then a search for high modularity partitions of the vertices into disjoint sets. An exhaustive search for the globally optimal solution is only feasible for the smallest networks, as the number of possible partitions of the vertices grows far too rapidly with network size. Several heuristics exist to find high-quality, if suboptimal, solutions in a reasonable length of time. Here, we use a two-stage search procedure:

- Agglomerative hierarchical clustering, where small communities are successively joined into larger ones such that the modularity increases. This stage is based on the so-called fast modularity (FM) algorithm (Clauset *et al.* 2004), adapted to work with bipartite networks.
- Greedy search, where vertices are moved amongst existing communities to ensure the resulting partition is at a local optimum of modularity. This stage uses the bipartite, recursively induced modules (BRIM) algorithm (Barber 2007).

The coarse structure is found with FM, with incremental improvements provided by BRIM.

Using the above search procedure, we identify communities of authors and publications for UCD. The community partition found has modularity $Q=0.93$ (compared to a maximum of $Q=1$), indicating a clear community structure in the network. There are 53 communities, in contrast to the 5 colleges and 26 schools corresponding to authors present in the network component. In Figure 15.3 we show the distribution of community sizes, as measured by the numbers of authors in the communities.

To further compare the structurally identified network communities to the organization in terms of colleges or schools, we use the normalized mutual information I_{norm} (Danon *et al.* 2005). The normalized mutual information allows us to measure the amount of information common to two different partitioning schemes. We explain it for the schools; the same explanation holds, *mutatis mutandis*, for the colleges. We take one of the partitions to be authors in the found communities, and the other to be a partition defined by the schools. The value of I_{norm} increases monotonically from zero to one as the correlation between the two partitions increases. At the extremes, we have $I_{norm}=1$ when the found communities match the school-based communities and $I_{norm}=0$ when they are independent.

Comparing the colleges to the structurally identified communities produces $I_{norm}=0.20$, indicating that the colleges reveal little about scientific collaborations. The comparison for schools gives $I_{norm}=0.56$, indicating that the schools partially, but by no means completely, describe the collaborative structures at UCD.

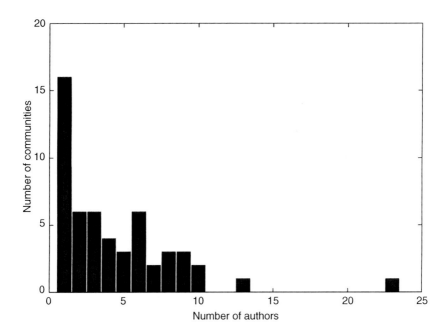

Figure 15.3 Frequency of community size, as measured by the number of authors in the communities (the 53 communities range from having just a single author to as many as 23 authors).

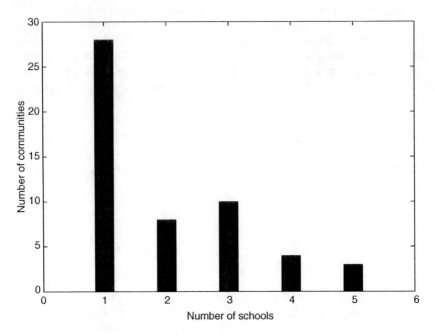

Figure 15.4 Diversity of the authors in communities can be assessed by examining the number of schools found in each community. Nearly half of all communities involve cross-school co-authorship.

Note
The 28 single-school communities include all 16 of the single-author communities.

To clarify the limitations of the explanative power for the schools, we consider the diversity of the authors in the communities. In Figure 15.4 we show how many communities have authors from different numbers of schools. Most common is that all authors in a community are from the same schools, but this includes the 16 communities which contain only one author. Excluding these, we determine that nearly 70 per cent of the multi-author communities involve inter-school collaboration, with several involving authors from five different schools. Partitioning the authors based on the schools alone cannot account for this, indicating that neither the colleges nor the schools provide a suitable understanding of the scientific collaborations at UCD.

Discussion

However, the kind of conclusions we can draw from these first results need to be put into context. It is not only the concern about anonymity of the individuals as discussed for example by Borgatti and Molina (2003, 2005), which needs consideration.

Scientific disciplines are differently represented in (co-)authorship databases such as the ISI Thompson Citation Index. This heavily biases our data and, accordingly, our results. To interpret these results appropriately, we need to relate them to the different publication cultures in the scientific disciplines.

These differences manifest themselves in various aspects, e.g. (co-)authorship is less frequent in theoretical than in experimental fields etc. In this context, we can just illustrate this point using some anecdotal evidence for different disciplinary publication cultures: van Raan (2005) estimates that in the medical sciences the share of publications covered by the Citation Index is about 80–95 per cent, whereas in the social sciences it is much lower, in psychology 50–70 per cent coverage, in the engineering fields it is about 50 per cent.

From this we know that our results are heavily biased. There is much (co-) publication activity, which is not recorded in the database, and, accordingly, is not appearing in our results. Furthermore, Laudel (2002) also shows that co-publication as an indicator does not cover important aspects of research collaboration: most collaboration within a university is never rewarded by a co-authored publication. This means, even if the data would be weighted against scientific publication cultures, co-publication would need to be complemented by other indicators of interdisciplinary cooperation dynamics. Laudel and Gläser (2006) have presented in detail how and where scientific practice of research collaborations slips through the nets of current evaluation instruments. This especially refers to using publications as the only measure for assessing research collaboration and performance within an over-simplified and wrongly standardized evaluation approach.

To avoid these pitfalls, we need to discuss the first results of our study in a methodological and ethical-regulatory context, especially in communicating them to university management. The methodological context concerns the discipline- and organization-dependent usability of bibliometric approaches like (co-) publications. The ethical-regulatory context refers to requirements of protecting personal data and ensuring transparent management procedures. Only while taking into account the methodological context of SNA-based higher education research, can university management practice start to translate these findings into strategic options for university development.

Acknowledgement

This chapter is based on research sponsored by UCD Research. Portions of the work were done at the Center for Mathematical Sciences at the University of Madeira during Madeira Math Encounters XXXVI, sponsored by the European FP6-NEST-Adventure Programme (contract number 028875).

Appendix: author graph and publication graph

We calculated standard network measurements for the UCD co-authorship network. We considered two projection networks from the bipartite author–publication network. These are the author projection, where authors are linked if

Table 15.A1 Standard unweighted network measures for the two projections.

Network measure	Explanation	Author projection Author–(publication)–author	Publication projection Publication–(author)–publication
Vertex type	Basic entity in projection	Author	Publication
Edge type	Condition for linking	Co-authored publication	Same author
#vertices N	Size of population	642	7,911
#edges M	Total number of links	635	178,972
Edge density ρ	$\rho = M/(N*(N-1)/2)$	$0.0031 = 1/324.0$	$0.0057 = 1/174.8$
Max #triangles	In neighbourhood of vertex	42	23,436
Mean #triangles	Calculated over all vertices	1.83	1,810.2
Clustering coefficient	#triangles/#possible triangles, mean over all vertices	0.18	0.94
#components	#unconnected partitions	334	334
l.c. #vertices	Largest component (% of N)	234 (36%)	4,395 (56%)
l.c. #edges	Largest component (% of M)	540 (85%)	113,007 (63%)
l.c. edge density	Largest component	$0.0198 = 1/50.5$	$0.0117 = 1/85.4$
l.c. diameter	Longest geodesic path	12	13
l.c. mean pathlength	Of all geodesic pathlengths	4.69	5.05

they have co-authored a publication, and the publication projection, where publications are linked if they share a common author.

In Table 15.A1 we show properties of the two projections. Note that these classical SNA measurements (Wasserman and Faust 1994) are based on *unweighted projections*, meaning that, e.g. two authors are linked if they share publications, but the actual number of common publications is irrelevant. In Table 15.A2 we summarize properties of distributions derived from the network edges. The number of authors per publication and the number of publications per author are the bipartite degrees, while the others are taken from the projections. In the *weighted projection* each contains the number of *shared bipartite* neighbours, i.e. the edge weight between two authors carries the number of co-authored publications. The *degree* describes the size of the neighbourhood of a node, while the *strength* also takes those edge weights into account. The degree and strength distributions are strongly right-skewed, and look similar to the ubiquitous scale-free networks (Barabasi and Albert 2002), but the dataset is too small to proof scale-freeness.

About one-third (authors), and more than half (publications) of the nodes belong to the largest component. There are 334 components of authors (publications) which are not connected to each other – at least not inside the ISI–UCD dataset. In both projections we see a sparse graph. The clustering coefficient is high – the mean probability that collaborators of an author have published a common paper themselves is 18 per cent in UCD, which is lower than, for example, the 30–40 per cent found in extensive physics publication databases (Newman 2001) due to the restriction to UCD and ISI (omitting other links between authors).

Table 15.A2 Properties of distributions derived from the network edges.

Distribution		Mean	Std dev.	Skewness	Max
#UCD-authors per publication	Omitting non-UCD authors	1.19	0.51	3.11	5
Publication projection	Degrees (unweighted)	45.3	43.9	2.02	226
	Edge weights = #identical authors of two publications	1.05	0.23	5.41	4
	Strengths (weighted degree)	47.8	47.2	2.03	325
#publications per UCD-author	Represented in ISI database	14.7	19.6	3.61	218
Author projection	Degrees (unweighted)	2.0	3.16	4.56	18
	Edge weights #repeated co-authoring	3.04	4.67	5.88	63
	Strengths (weighted degree)	11.3	16.0	3.79	128

A 'mean UCD-author' has published around 15 papers mentioned in the ISI index, with the most prolific author having published 218 papers. The greatest number of co-authorships is 63, far above the average of three co-publications for collaborating pairs of authors at UCD. The most connected author has 18 UCD co-authors, in comparison to the mean of just two co-authors. All distributions are strongly right-skewed; most values are small, but very large values also appear – so all given mean values are to be treated with caution.

References

Barabasi, A.L. and Albert, R. (2002) 'Statistical Mechanics of Complex Networks', *Reviews of Modern Physics*, 74 (1), p. 47.

Barabasi, A.L., Jeong, H., Neda, Z., Ravasz, E., Schubert, A. and Vicsek, T. (2002) 'Evolution of the Social Network of Scientific Collaborations', *Physica A: Statistical Mechanics and its Applications*, 311 (3–4), 15, pp. 590–614.

Barber, M.J. (2007) 'Modularity and Community Detection in Bipartite Networks', *Physical Review E (Statistical, Nonlinear, and Soft Matter Physics)*, 76 (6), pp. 066102.

Borgatti, S.P. and Molina, J.-L. (2003) 'Ethical and Strategic Issues in Organizational Social Network Analysis', *The Journal of Applied Behavioral Science*, 39 (3), pp. 337–49.

Borgatti, S.P and Molina, J.-L. (2005) 'Toward Ethical Guidelines for Network Research in Organizations', *Social Networks*, 27 (2), pp. 107–17.

Bozeman, B. and Lee, S. (2005) 'The Impact of Research Collaboration on Scientific Productivity', *Social Science Studies*, 35 (5), pp. 673–702.

Burt, R.S. (2004) 'Structural Holes and Good Ideas', *American Journal of Sociology*, 110 (2), pp. 349–99

Clauset, A., Newman, M.E.J. and Moore, C. (2004) 'Finding Community Structure in Very Large Networks' *Physical Review E (Statistical, Nonlinear, and Soft Matter Physics)*, 70 (6), pp. 066111.

Danon, L., Diaz-Guilera, A., Duch, J. and Arenas, A. (2005) 'Comparing Community Structure Identification', *Journal of Statistical Mechanics*, p. P09008.

Edquist, C. (1997) *Systems of Innovation: Technologies, Institutions and Organizations*, New York and London: Pinter Publishers.

Etzkowitz, H. and Leydesdorff, L. (2000) 'The Dynamics of Innovation: From National Systems and "Mode 2" to a Triple Helix of University–Industry–Government Relations', *Research Policy*, 29 (2), pp. 109–23.

Fleming, Lee, Mingo, Santiago and Chen, David (2007) 'Collaborative Borkerage, Generative Creativity, and Creative Success', *Administrative Science Quarterly* 52 (3).

Fortunato, S. and Castellano, C. (2008) 'Community Structure in Graphs', in: *Encyclopedia of Complexity and System Science*, Heidelberg, London and New York: Springer.

Gibbons, M., Limoges, C., Nowotny, H., Schwartzman, S., Scott, P. and Trow, M. (1994) *The New Production of Knowledge: The Dynamics of Science and Research in Contemporary Societies*, London Sage.

Gloor, P. (2006) *Swarm Creativity, Competitive Advantage through Collaborative Innovation Networks*, Oxford: Oxford University Press.

Guimera, R., Uzzi, B., Spiro, J. and Amaral, L. (2005) Team Assembly Mechanisms Determine Collaboration Network Structure and Team Performance', *Science*, 308, pp. 697–702

Heinze, T., Shapira, P., Rogers, J.D. and Senker, J. (2009) 'Organizational and Institutional Influences on Creativity in Scientific Research', *Research Policy*, 38, (4), pp. 610–23.

Jones, B.F., Wuchty, S. and Uzzi, B. (2008) 'Multi-University Research Teams: Shifting Impact, Geography, and Stratification in Science', *Science*, 322 (5905), pp 1259–62.

Katz, J.S. and Hicks, D, (1997) *How Much is a Collaboration Worth? A Calibrated Bibliometric Model*. Proceedings on the Sixth Conference of the International Society for Scientometric and Informetric, Jerusalem, Israel, 16–19 June, pp. 163–75.

Katz, J.S. and Martin, B.R. (1997) 'What is Research Collaboration?' *Research Policy*, 26 (1), pp. 1–18.

Laudel, G. (2002) 'What do we Measure by Co-authorships?', *Research Evaluation*, 11, pp. 3–15.

Laudel, G. and Gläser, J. (2006) 'Tensions between Evaluations and Communication Practices', *Journal of Higher Education Policy and Management*, 28, (3), pp. 289–95.

Newman, M.E.J. (2001) 'Who is the Best Connected Scientist? A Study of Scientific Coauthorship Networks', *Physical Review*, E64 016131.

Newman, M.E.J. (2004a) 'Detecting Community Structure in Networks', *European Physical Journal B*, 38, pp. 321–30.

Newman, M.E.J. (2004b) 'Fast Algorithm for Detecting Community Structure in Networks', *Physica l Review E (Statistical, Nonlinear, and Soft Matter Physics)*, 69 (6), pp. 66133.

Newman, M.E.J. and Girvan, M. (2004) 'Finding and Evaluating Community Structure in Networks', *Physical Review E (Statistical, Nonlinear, and Soft Matter Physics)*, 69 (2), pp. 26113.

Olmeda-Gómez, C., Perianes-Rodríguez, A. and Ovalle-Perandones, M.-A. (2008) 'Comparative Analysis of University–Government-Enterprise Co-authorship Networks in Three Scientific Domains in the Region of Madrid', *Information Research*, 13 (3).

Raan, Anthony F.J. van (2005) 'Measurement of Central Aspects of Scientific Research: Performance, Interdisciplinarity', *Structure Measurement: Interdisciplinary Research and Perspective* 3:1, pp. 1–19.

Slaughter, S. and Leslie, L. (1997) *Academic Capitalism: Politics, Policies, and the Entrepreneurial University*, Baltimore: Johns Hokpins.

Uzzi, B. and Spiro, J. (2005) 'Collaboration and Creativity: The Small World Problem', *American Journal of Sociology*, 111, pp. 447–504.

Wasserman, S. and Faust, K. (1994) *Social Network Analysis: Methods and Applications*, Cambridge: Cambridge University Press.

Wuchty, S., Jones, B. and Uzzi, B. (2007) 'The Increasing Dominance of Teams in Production of Knowledge', *Science Express*, 316 (5827), pp. 1036 –9.

Yousefi-Nooraie, R., Akbari-Kamrani, M., Hanneman, R. and Etemadi, A. (2008) 'Association between Co-authorship Network and Scientific Productivity and Impact Indicators in Academic Medical Research Centers: A Case Study in Iran', *Health Research Policy and Systems*, 6, p. 9.

Ziman, J. (2000). *Real Science: What it is, and What it Means*, Cambridge: Cambridge University Press.

Part III

The systemic aspects of innovation (modeling)

16 Learning in innovation networks

Some simulation experiments

*Nigel Gilbert, Petra Ahrweiler and
Andreas Pyka*

Introduction

The SKIN model (Simulating Knowledge dynamics in Innovation Networks; for
a detailed introduction see Ahrweiler *et al*. 2004) is a multi-agent simulation of
firms that try to optimize their innovation performance in order to respond to the
requirements of a constantly changing environment. Simulated scenarios can
inform decision makers about the chances and risks of investing in different
learning activities while taking into account the firm's markets, its clients, competitors and partners, its external and internal resources, and its strategic
policies.

In this chapter, we suggest that the SKIN model can be linked to the body of
literature on 'organizational learning' (OL) (for an early overview, see Dodgson
1993 and for later surveys, Amable 2003, Bahlmann 1990, Lam 2003). Following Garvin's statement (Garvin 1993) that only learning that can be measured
will be useful to managers, the SKIN simulation shows the outcome of different
learning activities. The model embodies some theoretical ideas from the OL
literature and implements many OL concepts (e.g. from Argyris and Schön 1996,
Levinthal and March 1993, March and Olsen 1975, Senge 1990). Thus, the
SKIN model is not only interesting for managers and other practitioners responsible for empirical learning processes within firms but also for scientists testing
theories from the body of research on organizational learning.

The model

SKIN is a multi-agent model containing heterogeneous agents that act and interact in a complex and changing environment. The agents represent innovative
firms who try to sell their innovations to other agents and end users but who also
have to buy raw materials or more sophisticated inputs from other agents (or
material suppliers) in order to produce their outputs. This basic model of a
market is extended with a representation of the knowledge dynamics in and
between the firms. Each firm tries to improve its innovation performance and its
sales by improving its knowledge base through adaptation to user needs, incremental or radical learning, and co-operation and networking with other agents.

The elements and processes of the model will now be described in more detail, with an emphasis on the learning activities.

The core concept of the framework is the *knowledge,* which will manifest itself in the innovative production or delivery of manufactured and service products. The approach to knowledge representation used in the model is similar to Toulmin's evolutionary model of knowledge production (Toulmin 1967). This identified concepts, beliefs and interpretations as the 'genes' of scientific/technological development evolving over time in processes of selection, variation and retention. Ackermann interpreted the works of Kuhn and Popper according to this perspective allowing for different selection systems (Ackermann 1970). More recent studies (Dawkins 1989) discuss the idea of cultural replicators. A replicator is a unit which is copied – with random error – and which can in a way decide on the probability of its own replication. Although Dawkins does not compare his 'meme' as a cultural replicator with the gene as biological replicator, the implication seems obvious (Blackmore 1999, Dennett 1995). However, 'memes' are usually located in mental states. This cognitive aspect would seem to confine the concept to individuals or require that collectivities have a mental state. Instead, the SKIN model uses a similar concept, a 'kene', to represent the aggregate knowledge of an organization (Gilbert 1997).

The agents

The individual knowledge base of a SKIN agent, its kene, contains a number of 'units of knowledge'. Each unit is represented as a triple consisting of a firm's *capability C* in a scientific, technological or business domain (e.g. biochemistry), represented by an integer, its *ability A* to perform a certain application in this field (e.g. a synthesis procedure or filtering technique in the field of biochemistry), represented by a real number, and the *expertise level E* the firm has achieved with respect to this ability (represented by an integer). The firm's kene is its collection of C/A/E-triples.

When it is set up, each firm has also a stock of initial *capital*. It needs this capital to produce for the market and to improve its knowledge base, and it can increase its capital by selling products. The amount of capital owned by a firm is a measure of its size and also influences the amount of knowledge that it can support, represented by the number of triples in its kene. Most firms are initially given an amount of starting capital taken from a uniform distribution between zero and an initial maximum capital allocation, but in order to model differences in firm size, a few randomly chosen firms can be given extra capital.

The market

Firms apply their knowledge to create innovative products that have a chance of being successful in the market. The special focus of a firm, its potential innovation, is called an *innovation hypothesis*. In the model, the innovation hypothesis (IH) is derived from a subset of the firm's kene triples.

The underlying idea for an innovation, modeled by the innovation hypothesis, is the source an agent uses for its attempts to make profits on the market. Developing the innovation hypothesis into a product is a mapping procedure where the capabilities and abilities of the innovation hypothesis are used to compute an index number that represents the product.

A firm's product, P, is generated from its innovation hypothesis as

$$P = (C_1 * A_1) + (C_3 * A_3) + (C_4 * A_4) + \ldots \text{ modulus } N$$

where N is a constant.

The product has a certain quality, which is also computed from the innovation hypothesis in a similar way, by multiplying the abilities and the expertise levels for each triple in the innovation hypothesis and normalizing the result. In order to realize the product, the agent needs some materials. These can either come from outside the sector ('raw materials') or from other firms, which generated them as their products. What exactly an agent needs is also determined by the underlying innovation hypothesis: the kind of material required for an input is obtained by selecting subsets from the innovation hypotheses and applying the standard mapping function (see equation above).

These inputs are chosen so that each is different and differs from the firm's own product. In order to be able to engage in production, all the inputs need to be obtainable on the market, i.e. provided by other agents or available as raw materials. If the inputs are not available, the agent is not able to produce and has to give up this attempt to innovate. If there is more than one supplier for a certain input, the agent will choose the one at the cheapest price and, if there are several similar offers, the one with the highest quality.

If the agent can go into production, it has to find a price for its own product that takes account of the input prices it is paying and a possible profit margin. While the simulation starts with product prices set at random, as the simulation proceeds, a price adjustment mechanism increases the selling price if there is much demand, and reduces it (but no further than the total cost of production) if there are no customers. A range of products are considered to be 'end-user' products and are sold to customers outside the sector: there is always a demand for such end-user products provided that they are offered at or below a fixed end-user price. An agent will then buy the requested inputs from its suppliers using its capital to do so, produces its output and puts it on the market for others to purchase. Using the price adjustment mechanism, agents are able to adapt their prices to demand and in doing so learn by feedback.

In making a product, an agent applies the knowledge in its innovation hypothesis and this increases its expertise in this area. This is the way that learning by doing/using is modeled. The expertise levels of the triples in the innovation hypothesis are increased by 1 and the expertise levels of the other triples are decremented by 1. Unused triples in the kene eventually drop to an expertise level of 0 and are deleted from the kene; the corresponding abilities are 'forgotten' or 'dismissed' (Hedberg 1981).

Learning and co-operation: improving innovation performance

In trying to be successful on the market, the firms are dependent on their innovation hypothesis and thus on their kene. If a product does not meet any demand, the firm has to adapt its knowledge in order to produce something else for which there are customers (Duncan 1974). In the model, a firm has several ways of improving its performance, either alone or in co-operation, and in either an incremental or a more radical fashion. All strategies have in common that they are costly: the firm has to pay a 'tax' as the cost of applying an improvement strategy.

Incremental research

If a firm's previous innovation has been successful, i.e. it has found buyers, the firm will continue selling the same product in the next round. However, if there were no sales, it considers that it is time for change (evaluating feedback). If the firm still has enough capital, it will carry out 'incremental' research (R&D in the firm's labs).

Performing incremental research means that a firm tries to improve its product by altering one of the abilities chosen from the triples in its innovation hypothesis, while sticking to its focal capabilities. The ability in each triple is considered to be a point in the respective capability's action space. To move in the action space means to go up or down by an increment, thus allowing for two possible 'research directions'.

Initially, the research direction of a firm is set at random. Later it learns to adjust to success or failure: if a move in the action space has been successful the firm will continue with the same research direction within the same triple; if it has been a failure, the firm will randomly select a different triple from the innovation hypothesis and try again with a random research direction.

Radical research

A firm under serious pressure that is in danger of becoming bankrupt will turn to more radical measures, by exploring a completely different area of market opportunities. In the model, an agent under financial pressure turns to a new innovation hypothesis after first 'inventing' a new capability for its kene. This is done by randomly replacing a capability in the kene with a new one and then generating a new innovation hypothesis.

Partnerships

An agent in the model may consider partnerships (alliances, joint ventures etc.) in order to exploit external knowledge sources. The decision whether and with whom to co-operate is based on mutual observations of the firms, which estimate the chances and requirements coming from competitors, possible and past partners, and clients.

The information a firm can gather about other agents is provided by a marketing feature: to advertise its product, a firm publishes the capabilities used in its innovation hypothesis. (Capabilities not included in its innovation hypothesis and thus in its product, are not visible externally and cannot be used to select the firm as a partner.) The firm's advertisement is then the basis for decisions by other firms to form or reject co-operative arrangements.

In experimenting with the model, we can choose between two different partner search strategies, both of which compare the firm's own capabilities as used in its innovation hypothesis and the possible partner's capabilities as seen in its advertisement. Applying the conservative strategy, a firm will be attracted by a possible partner that has similar capabilities; using a progressive strategy the attraction is based on the difference between the capability sets.

Previously good experience with former contacts generally augurs well for renewing a partnership. This is mirrored in the model: to find a partner, the firm will look at previous partners first, then at its suppliers, customers and finally at all others. If there is a firm sufficiently attractive according to the chosen search strategy (i.e. with attractiveness above the 'attractiveness threshold'), it will stop its search and offer a partnership. If the possible partner wishes to return the partnership offer, the partnership is set up.

The model assumes that partners learn only about the knowledge being actively used by the other agent. Thus, to learn from a partner, a firm will add the triples of the partner's innovation hypothesis to its own. For capabilities that are new to it, the expertise levels of the triples taken from the partner are reduced by 1 in order to mirror the difficulty of integrating external knowledge (Cohen and Levinthal 1990). For partner's capabilities that are already known to it, if the partner has a higher expertise level, the firm will drop its own triple in favor of the partner's one; if the expertise level of a similar triple is lower, the firm will stick to its own version. Once the knowledge transfer has been completed, each firm continues to produce its own product, possibly with greater expertise as a result of acquiring skills from its partner.

Networks

If the firm's last innovation was successful, i.e. the amount of its profit in the previous round was above a threshold, and the firm has some partners at hand, it can initiate the formation of a network. This can increase its profits because the network will try to create innovations as an autonomous agent in addition to those created by its members and will distribute any rewards to its members who, in the meantime, can continue with their own attempts, thus providing a double chance for profits. However, the formation of networks is costly, which has two consequences: only firms with enough capital can form or join a network and no firm can be member of two networks at the same time.

Networks are 'normal' agents, i.e. they get the same amount of initial capital as other firms and can engage in all the activities available to other firms. The kene of a network is the union of the triples from the innovation hypotheses of all its participants. If a network is successful it will distribute any earnings above the amount of the initial capital to its members; if it fails and becomes bankrupt, it will be dissolved.

Start-ups

If a sector is successful, new firms will be attracted into it. This is modeled by adding a new firm to the population when any existing firm makes a substantial profit. The new firm is a clone of the successful firm, but with its kene triples restricted to those in the successful firm's advertisement, and an expertise level of 1. This models a new firm copying the characteristics of those seen to be successful in the market. As with all firms, the kene may also be restricted because the initial capital of a start-up is limited and may not be sufficient to support the copying of the whole of the successful firm's innovation hypothesis.

Organizational learning

In 1938, the American philosopher John Dewey introduced the concept of experiential learning as a permanent activity cycle (Dewey 1938), starting a discussion among educationalists about feedback learning and learning by doing. This discussion, however, referred mostly to individuals. It was in 1973 that Donald Michael introduced the idea of organizational learning (OL) (Michael 1973). Since then, the field of OL has grown steadily. In the 1970s, Argyris and Schön's influential monograph (Argyris and Schön 1996) proposed that a learning organization is one that is permanently changing its interpretation of the environment. In doing so, the organization learns new things and forgets old ones. Drawing on their background as action theorists, Argyris and Schön show how these interpretations are gained and how they are connected to different organizational behaviors.

We can apply the SKIN model to many of the ideas of the Argyris and Schön framework and use it to examine the assumption that, in the words of de Geus (de Geus 1997), the greatest competitive advantage for any firm is its ability to learn. As we have seen in the previous section, in the SKIN model, firms can:

1   use their capabilities (learning by doing/using);
2   learn to estimate their success via feedback from markets and clients (learning by feedback);
3   improve their own knowledge incrementally when the feedback is not satisfactory in order to adapt to rising technological and/or economic standards (adaptation learning, incremental learning);
4   radically change their capabilities in order to meet completely different requirements of markets and clients (innovative learning, radical learning).

Firms may also be also active on the meta level (called in the OL literature, the *double-loop learning* level) of the model. They can:

1　forget their capabilities (clean up their knowledge space);
2　decide on their individual learning strategies themselves (e.g. incremental or radical learning), constructing and changing the strategies according to their past experience and current context; the context consists of external factors such as the actions of clients, competitors and partners, and the availability of technical options, and internal factors such as their capital stock and the competencies available to them;
3　engage in networking and partnerships to absorb and exploit external knowledge sources, to imitate and emulate, and to use synergy effects (participative learning).

Furthermore, the SKIN simulation models some insights from empirical learning research, for example, addressing the difficulty of including external knowledge into the firm (e.g. Cantner and Pyka 1998), the ongoing diffusion of innovation-relevant knowledge in the market, and the influence of firms entering the market.

Models and simulation tools have long been used for strategic planning (cf. House 1983, Milling 1996, Monahan 2000, Scheer 1994 etc.). In addition, a lot of work has been done lately in the area of intelligent management and decision support systems (for a recent review see Gupta *et al.* 2006). In contrast to these, the SKIN approach allows strategic computational agents to act experimentally in complex knowledge-based scenarios. It is possible to track single firms in the simulation in order to observe their strategic learning behaviors in different situations and their outcomes, or to consider the effect of different forms of learning on a sector as a whole, the approach taken in this chapter.

Some experiments

The SKIN simulation may be used to test the impact of various learning activities on the survival and effectiveness of firms in a highly competitive and dynamic market. In the section, we present some simulations examining the effects of organizational learning on the performance of the model. In trying to vary and/or to stabilize their knowledge stocks by organizational learning, the simulated firms attempt to adapt to environmental requirements while the simulated market rewards those that are most successful.

We begin with some baseline experiments, in which no firm is able to learn. All firms start with a randomly generated kene and have no possibility of changing it through learning. Moreover, the prices of products are fixed and cannot be changed to suit the market. Not surprisingly, in this scenario, there is very little trading activity, because few firms are able to find suppliers for their needed inputs, and those that can are likely to find their requirements too expensive to be affordable. The market as a whole fails to 'take off', almost all firms fail to trade, and the consequence is that firms do not make a profit. Nevertheless all firms have to pay a tax just to remain in the market, which is taken from their starting capital. When it has used up all its capital, a firm 'dies' and is removed from the simulation. Eventually, the simulation ends with no remaining firms.

The pattern for scenarios when firms are only capable of learning by using and learning by feedback is similar. These types of learning operate at the market level, but do not affect the innovation performance of the firms. While adjustment of prices can help when a firm is making a product that is in demand, but which is too expensive for the customers, it is of no avail if the product is the wrong one – there is no demand for it – or the product cannot be made because the requisite inputs are not available. This is usually the situation in these basic scenarios, where the selection of products available is determined by the random chance of the initial kene configurations, and where there is no opportunity to engage in innovation to produce new, more desirable products. As Figure 16.1 shows, the population of firms decreases steadily until no more remain.

The picture changes once we allow adaptive learning by means of simulated research. Now the firms are able to modify their knowledge bases, generate new products and adapt the inputs that they require to make these products. For many firms, the learning that they are able to do will be unsuccessful and their fate will be the same as in the baseline scenario: little or no profit, a regular tax, although one larger than before to pay for the research they are doing, and eventual bankruptcy and removal from the simulation. Some firms, however, will strike lucky and find a product for which there are reasonably priced inputs available and a customer able to purchase. Profits accrue to these firms, which become steadily richer while their products remain in demand. However, the environment for these firms consists of all the other firms and their products. This is a highly dynamic environment and it is likely that before too long, either a firm from which it is buying an input or a firm to which it is selling its product will change its kene (and thus its inputs and product) or go out of business altogether. This will initiate a cascade of changes that may leave it without a viable product. However, it can live off its accrued capital while it searches for an alternative.

Thus, in a scenario in which adaptation of firms' kenes is possible, we obtain an initial sharp reduction in the number of firms as those that are not viable drop out, followed by recovery as firms accumulate capital from successful trading, and because of their success, attract newcomers into the market. A detailed look at the first 200 time periods (Figure 16.2) reveals that initially, overall, very few firms are capable of making products because the inputs they need are not available on the market. During the early steps, the firms 'redesign' their products to accommodate the inputs on the market and can then offer the products for sale. Then they redesign their products to match the requirements of their customers and many firms succeed in making sales. Those that either fail to adjust their innovation hypotheses so that they only require inputs that can be bought, or which make products that no other firm wants to buy, run out of capital and 'die'. The number of firms stabilizes at between 500 and 600 (measured in repeated runs under the same conditions). This can be seen in Figure 16.2, which plots the number of firms by time, in an experiment where all firms are capable of learning through incremental research, in addition to learning by doing and by feedback.

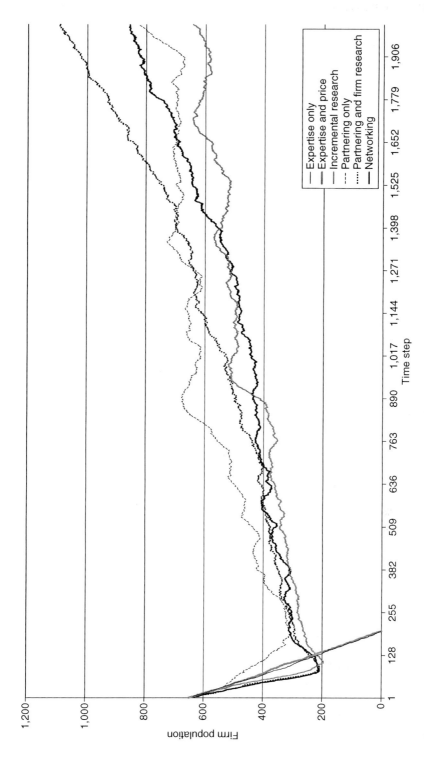

Figure 16.1 The population of firms over the first 2,000 time steps, for various combinations of organizational learning.

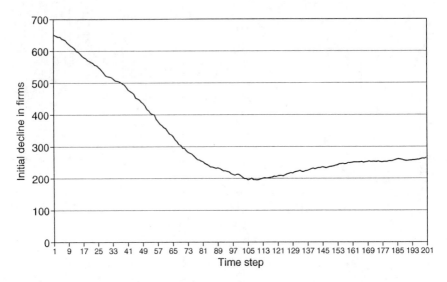

Figure 16.2 The first 200 time steps, showing the initial shake out of firms when the simulated sector can perform incremental research and price adjustment.

While incremental research does allow a firm to improve its product through varying its abilities, it is stuck with its initial, randomly assigned, basket of capabilities. Radical research (innovative learning) allows the firm to branch out, absorbing or creating new capabilities that can lead to completely new products. We allow the firms to engage in such radical research strategies only when they are close to running out of funds, and thus close to 'death'. If their remaining capital falls below a threshold value, they acquire a new capability in exchange for an existing one, then generate a new innovation hypothesis and, if they can find the required inputs and a customer, a new product. Such radical research is very risky and in the model, only succeeds about 5 percent of the time (i.e. the innovation hypothesis generated from a radical research effort only makes a surplus in about one in 20 cases). Nevertheless, the possibility of doing radical research has an effect on the overall success of the firms. The median number of steps before firms lose all their capital rises from about 40 to about 60 because of the opportunity for a 'fresh start' that undertaking a radical research program offers to those that manage to discover a successful innovation hypothesis (Figure 16.3).

A way of acquiring new knowledge in addition to incremental or radical research is to obtain it from other firms though some form of collaboration. Such participatory learning is modeled by the formation of partnerships between similar firms. Partners exchange capabilities, thus introducing new ideas into the firm from outside, rather than through internal research. Figure 16.2 also shows the changes in the population of firms when all the firms are able to engage in partnerships, but do no incremental or radical research. It will be seen that the

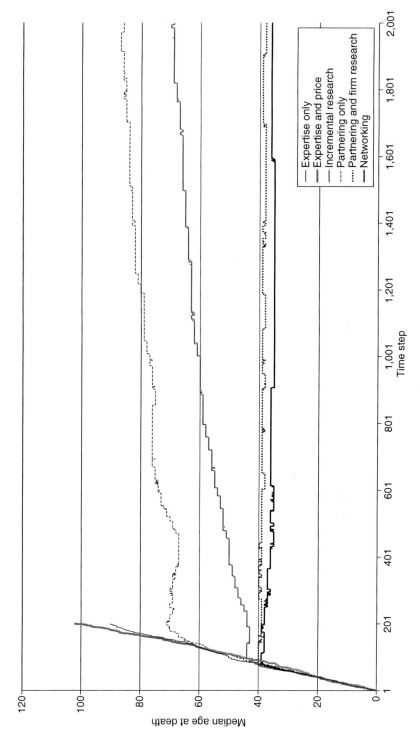

Figure 16.3 Median age of firms at the time step when they lose all their capital, by type of organizational learning.

innovations that are introduced through partnership are slightly more successful (as judged by the growth in the population of firms) than those that are obtained by internal research, perhaps because partnering allows the introduction of new capabilities from partners, while incremental research only improves the abilities for capabilities that the firm already possesses. The firms employ the 'conservative' strategy in these experiments, which means that they partner with other firms that have similar innovation hypotheses, thus limiting the amount of novelty. Moreover, firms preferentially partner with others that they have previously collaborated with, again limiting the possibility of novelty. Nevertheless, sufficient innovation is introduced that the sector as a whole is able to create a sustainable market and grow slightly faster than if each form operated with only its own innovation resources.

Allowing firms to engage in partnership formation together with incremental and radical research provides the best resources for innovation and yields the fastest growth in the number of firms. With this combination of types of learning, not only are firms able to acquire capabilities from partners, they are also able to improve the abilities using their own internal research activities.

Partnerships are dissolved at the end of each time step and only affect the knowledge base of the partners. However, experienced partners are also able to create networks that can make products on their own account, contributing any profit to the network members. Figure 16.2 also shows the population growth when all firms are allowed to become network members if they wish. The curve shows slower growth than partnering plus internal firm research because the network activities have the effect of generating additional competition for the firms. Figure 16.4 shows the typical near power law distribution of size of the networks, also found in real innovation networks.

Conclusions

In this chapter, a framework for modeling learning competence in firms is presented to improve the understanding of managing innovation. Focusing on their learning competencies, firms with different knowledge stocks attempt to improve their economic performance by choosing or suppressing radical or incremental innovation activities and by engaging in partnerships and networks with other firms. In trying to vary and/or to stabilize their knowledge stocks by organizational learning, they attempt to adapt to environmental requirements while the market strongly selects on the results. The simulation experiments show the impact of different learning activities, underlining the importance of innovation and learning. It demonstrates the importance of finding new capabilities from outside the firm, either through partnering or radical research. It also shows that the simpler forms of organizational learning, such as learning by doing and learning by feedback, are of limited value by themselves in the highly dynamic environment of modern knowledge based market sectors, although they are of significance when combined with other forms of learning.

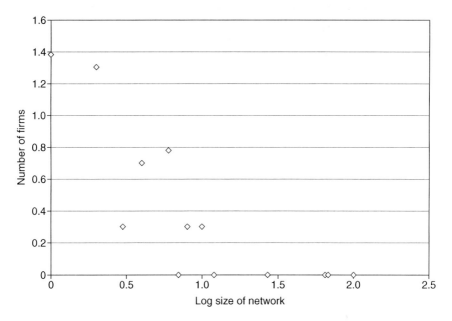

Figure 16.4 Log-log plot of network size (number of firms per network) after 2,000 time
steps.

The model also demonstrates the possibilities opened up by simulation for
carrying out experiments on a model that would be impossible to perform in the
real world. Clearly, it would not be practicable to constrain actual firms in their
learning capacities in order to study the effect on their success and determine the
causal consequences, still less to do this for a whole market sector. But as this
chapter has shown, it is possible and revealing to do so on a model of a sector.
This chapter has only made a start in examining the complex relationships
between firm and sector success and organizational learning; there are many
more possibilities that await investigation.

Appendix 16.1

The code for the NetLogo model on which this chapter is based is available on
request from the first author.

The simulation runs described in this chapter use the following parameter
settings:

- Initial capital of firms: taken from a uniform random distribution between 0
  and 20,000;
- Initial population of firms: 650;
- Number of large firms, with extra capital at the start: 0;
- Range of product index numbers in the sector: 0.0 to 100.0;

- Maximum difference between product and input index numbers for them to be considered substitutable: 1.0;
- All products with a product number below 5.0 are considered to be 'raw-materials' and all those with numbers above 95 are 'end-user' products;
- Price of raw materials: 1;
- Maximum price of end-user products: 1,000;
- Profit required to attract new start-ups: 400;
- Partnering search strategy: conservative;
- Attractiveness threshold to allow two firms to partner: 0.3;
- Capital cut-off below which firms do radical rather than incremental research: 1,000;
- Taxes: per time step: 200; per incremental research attempt: 100; per radical research attempt: 100; per partner: 100.

References

Ackermann, R. (1979) *The Philosophy of Science*, New York: Pegasus.

Ahrweiler, P., Pyka, A. and Gilbert, N. (2004) 'Simulating Knowledge Dynamics in Innovation Networks', in: Leombruni, R. and Richiardi, M. (eds) *Industry and Labor Dynamics: The Agent-based Computational Economics Approach*, Singapore: World Scientific, pp. 284–96.

Amable, B. (2003) *The Diversity of Modern Capitalism*, Oxford: Oxford University Press.

Argyris, C. and Schön, D.A. (1996) *Organizational Learning: A Theory of Action Perspective*, Reading, MA: Addison-Wesley.

Bahlmann, T. (1990) 'The Learning Organization in a Turbulent Environment', *Human Systems Management*, 9, pp. 249–56.

Blackmore, S. (1999) *The Meme Machine*, Oxford: Oxford University Press.

Cantner, U. and Pyka, A. (1998) 'Absorbing Technological Spillovers: Simulations in an Evolutionary Framework', *Industrial and Corporate Change*, 7, pp. 369–97.

Cohen, W.M. and Levinthal, D. (1989) 'Innovation and Learning: The Two Faces of R&D', *Economic Journal*, 99, pp. 569–96.

Cohen, W.M. and Levinthal, D. (1990) 'Absorptive Capacity: A New Perspective on Learning and Innovation', *Administrative Science Quarterly*, 35, pp. 128–52.

Dawkins, R. (1989) *The Selfish Gene*, Oxford: Oxford University Press.

Dennett, D.C. (1995) *Darwin's Dangerous Idea: Evolution and the Meanings of Life*, London: Penguin.

Dewey, J. (1938) *Experience and Education*, New York: Collier.

Dodgson, M. (1993) 'Organizational Learning: A Review of Some Literatures', *Organization Studies*, 14, pp. 375–94.

Duncan, R.B. (1974) 'Modifications in Decision Structure in Adapting to the Environment: Some Implications for Organizational Learning', *Decision Sciences*, 5, pp. 705–25.

Garvin, D.A. (1993) 'Building a Learning Organization', *Harvard Business Review*, pp. 81–91.

Geus, A. de (1997) *The Living Company*, London: Brealy.

Gilbert, G.N. (1997) 'A Simulation of the Structure of Academic Science', *Sociological Research Online*, 2. Online, available at: www.socresonline.org.uk/socresonline/2/2/3.html.

Gupta, J., Forgionne, G. and Mora, M. (2006) *Intelligent Decision-making Support Systems: Foundations, Applications, and Challenges*, Berlin: Springer.

Hedberg, B. (1981) 'How Organizations Learn and Unlearn', in: Nystrom, P.C. and Starbuck, W.H. (eds) *Handbook of Organizational Design*, Oxford: Oxford University Press.

House, W.C. (ed.) (1983) *Decision Support Systems: A Data-based, Model-oriented, User-developed Discipline*, New York: Petrocelli.

Lam, A. (2003) 'Organizational Learning in Multinationals: R&D Networks of Japanese and US MNEs in the UK', *Journal of Management Studies*, 40, pp. 673–703.

Levinthal, D.A. and March, J.G. (1993) 'The Myopia of Learning', *Strategic Management Journal*, 14, pp. 95–112.

March, J.G. and Olsen, J.P. (1975) 'The Uncertainty of the Past: Organizational Learning under Ambiguity', *European Journal of Political Research*, 3, pp. 147–71.

Michael, D.M. (1973) *On Learning to Plan and Planning to Learn*, Hoboken, NJ: Jossey-Bass.

Milling, P. (1996) 'Modeling Innovation Processes for Decision Support and Management Simulation', *System Dynamics Review*, 12, pp. 211–34.

Monahan, G.E. (2000) *Management Decision Making: Spreadsheet Modeling, Analysis and Application*, Cambridge, UK: Cambridge University Press.

Scheer, A.-W. (1994) *Business Process Engineering: Reference Models for Industrial Enterprises*, Berlin: Springer.

Senge, P.M. (1990) 'The Leader's New Work: Building Learning Organizations', *Sloan Management Review*, pp. 7–23.

Toulmin, S. (1967) *The Philosophy of Science: An Introduction* London: Hutchinson.

heory-based dynamical model of exaptive innovation processes

Marco Villani and Luca Ansaloni

Introduction

A major problem in research on innovations is the understanding of creativity, that is, the origins of innovations. A widespread interpretation is that the innovation processes are local recombinations of current capabilities, a claim originating on the Schumpeter researches (see also March 1991). On the other side, it is possible to consider creativity as a sudden blind insight. The latter perspective yields little theoretical and conception gain, but – in our opinion – also a theory explaining innovation as a pure recombination of already existing methodologies fails to treat the many cases of significant creativity it is possible to observe. In particular, while this perspective can explain the genesis of the 'better, faster and cheaper' innovations, it is not able to describe the so called radical innovation, whose origin and scope is not directly linked with the current events.

The key point seems to us that those approaches lack an appropriate ontology. In particular, the sudden blind insight perspective takes into consideration only human beings (and only as black boxes), whereas the second point of view emphasizes the role of artifacts, leaving the agents (human beings, or groups of human beings) a secondary role of pure recombinators: it becomes in such a way possible to ignore the role of agents and follow the pure technological trajectories of the artifacts.

On the contrary, we claim that both agents and artifacts matter. In fact, human agents – which are endowed with sophisticated cognitive and communication capabilities – can create and use artifacts in ways which are not obvious at all. Moreover, they are embedded in a web of relationships among themselves, with their organizations and with their environment which affect their ways of thinking and of using artifacts. Making (new) artifacts or generating (new) forms of organization in a society, is a multifaceted process that is based on recurrent sets of actions, in which a number of agents with different roles are involved. In this process, new concepts and objects are discovered and invented by simultaneously surveying the opportunities offered by the social and material environment, and adapting the existing categorizations of the agents. The (new) artifacts allow the development of the (new) recurrent sets of actions, that in time create (new) paths and opportunities.

In this context, innovation processes could be better comprehended if inserted within the so called Complex System Science, a discipline aiming to describe the behaviour of systems composed by many non linearly interacting parts. Parts, entities and relationships can change in time, often depending in a very significant way upon the initial conditions, each mutation being (or being not) stimulated by the mutation of other parts of the systems in an endless transformation. As outcome of these processes, communities or structures can emerge or disappear, influencing in such a way the behaviour of the lower level entities.

Within this vision, innovation is conceptualized as 'cascades of changes in the structure of agent-artifact space' (Lane and Maxfield 2005). This conceptualization highlights two key aspects of innovation processes. First, the single 'innovation' – as for example the introduction of a new artifact type – is not the unique object of analysis for innovation theory; rather it is the process whereby 'one (new) thing leads to another' that should provide the central focus for such a theory. Second, the relevant changes cannot be localized to artifact space alone (as is done, for example, with the theory of technological trajectories, or in the use of patents as the principal unit of empirical investigation). Rather there is a continuous relation among agents and artifacts, where the generation of new artifact types is mediated by the transformations happening in agents, and new artifact types may in turn mediate the transformations in agents. In innovation contexts, there is no relaxation to equilibrium. Rather, there is a continual production of novelties, in artifacts, functionalities and organizational forms, which makes it impossible to separate the 'effects' of any given one.

In this chapter, we propose a model where radical innovations are created by a process of 'exaptation' (Gould and Verba 1982, Ceruti 1995, Gould 2002), which according to our hypotheses represents a key aspect of innovation processes. By means of the explicit representation of artifacts and categories, the model eases the understanding of the exaptation phenomenon, seen in this context as a shift in terms of 'leading attributions', and allows the identification of the elements favouring its emergence.

Exaptation in biological sciences

In our opinion, the key point about the origin of radical innovation (and therefore not the better-faster-cheaper innovations) is the exaptation process.

The concept of exaptation has been introduced in recent years in order to explain the changes induced by the introduction of novelties, but the word and the idea originate in the biology field. In particular, the concept was already present in the Darwin big opera 'The origin of species' as one of the main mechanisms, complementary to adaptation, able to guide the evolution processes. The word 'exaptation' appears for the first time in Gould and Verba (1982) and probably the more clear definition is in Ceruti 1995: 'The processes whereby an organ, a part, a characteristic (behavioural, morphologic, biochemical) of an organism, which was originally developed for a certain task, is employed for carrying out tasks that are completely different from the original one.' One classical example of this process is represented by a line of feathered dinosaurs, arboreal or runners who developed the capability to take advantage of feathers for flying, when originally they were intended for thermoregulation purposes (Gould 2002).

Therefore, exaptation can provide a key to interpret the serendipity that characterizes the generation of novelties. In fact, the exaptation process highlights the fact that the functionalities for which a particular structure has been selected (or designed, in technological processes) are only a subset of the possibilities carried by the structure itself. Different from *adaptations*, which present functions for which they are designed, the exaptations generate effects that are not subject to pressures from the current selections, but potentially relevant later on.

Exaptation is not a simple side-effect. A simple side-effect is an unintended (and often undesirable) consequence of a particular change. Although the apparent affinity between the two notions, an exaptation is usually identifiable with respect to change of context (the thermoregulating appendages become wings), whereas we cannot say the same for a simple side-effect. Eventually, exaptation can be the outcome of recombination processes but is not a pure recombination process; again, it requires a change of context, not directly resultant from a simple mixture of existing elements.

The role of artifacts in social and technological exaptation

Technological innovations do not happen in isolation: they occur in big interconnected systems composed by agents and artifacts. In many cases the number of consequences generated in these systems by a new technology, a product, or a process can be incredibly large, realizing in such a way an exaptive potential whose size is practically unbounded. Thus, exaptation could represent the link connecting single technological progress to the emergence of recurrent patterns of interaction on the scale of the whole system. In this vision, exaptations are those characteristics of a certain technology that are co-opted by another origin or utility because of their current role.

Please note that in this context artifacts play a not secondary role, being a fundamental part of the emerging patterns of interaction. In particular they are able to convey information, becoming in such a way a key aspect of the whole process. This useful property could be explained by using the notion of decomposability, introduced by Simon (1996). An artifact can be defined as a hierarchical structure composed by subparts that are approximately independent in the short term, but connected by a global behaviour in the long term. Each subpart can carry a limited set of functionalities (this set being influenced by the presence/absence of other subparts), blocking of enhancing the pattern of recurrent set of actions the artifact is involved in. This 'switching' activity can change in time, since the subparts are selected and proliferate as a consequence of being only one among many aspects of the whole artifact. While some such subparts can have an important role with respect to the goal of the whole artifact, others remain latent waiting future activation.

Based on the above observations, we can say that an exaptation originates when one of the following situation holds:

1   A subpart is already providing a positive contribution to the functionality for which the technology was selected (only later and after a change of context does the subpart become the main component – for example, the phonograph invented by Edison in 1877: initially as an office dictaphone, its context dramatically changed leading to the exaptation of the phonograph gramophone, nowadays considered one of the most popular inventions made during the nineteenth century);
2   The subpart has no role in the overall performance of the system – the process of vitrification, originally developed to obtain the safe waste disposition of radio-active materials and now mainly exaptated for biologically dangerous processes (elimination of biochemical weapons);
3   The subpart initially provides a negative contribution to the overall performance of the system – the innovation in plastic production during the Second World War, based on sub products deriving from oil refinement (Dew *et al.* 2004).

Please note that usually artifacts own lots of subparts which have no role in its functionality (item 2); in this way, they convey much more information than the quantity explicitly manipulated by the artifact designer. These subparts initially are not associated with a particular use and at the same time do not generate substantial damages, thus avoiding elimination by selection (a phenomenon called in biology 'neutral mutation' (Kimura 1983). Nevertheless later on these subparts can gain significant roles, constituting in such a way an important resource for the artefact itself (the so called 'exaptive pool of possibilities', that is, the potential allowed for future selection episodes). A very elegant example of this phenomenon can be found in architecture, where spandrels, initially empty spaces between the vaults and arches in churches, have been later used as a support for paintings and mosaics. While their initial role was that of structural elements, spandrels thus became a key aesthetic feature (Gould and Lewontin 1979).

A macro view: the exaptive bootstrapping

Before introducing the model we desire to introduce the context in which exaptation plays a key role, by introducing the link between the macro level where the innovation processes occur and the micro level where the innovation originates.

We believe that the most important process is what we call exaptive bootstrapping. Exaptive bootstrapping processes focus on the exaptation of new functionalities, via new attributions of functionality, from existing patterns of interaction among agents and artifacts. It is the key element in innovative cascades – rather than adaptation of structures in agent-artifact space driven by pre-existing functional requirements and consequent selection mechanisms. 'Bootstrapping' refers to the process whereby the integration of new artifacts into new patterns of activity provides the points of view from which participants or observers can cognize new attributions of functionality and thus the basis for the exaptation of new functionality and its instantiation in more new artifacts.

Both in biological and in sociocultural innovation new functionality can emerge, as new structures give rise to new patterns of interaction among entities. But artifacts or organizational forms are *designed* to deliver a new functionality, and therefore the agents must generate new *attributions of functionality* in order to use these new objects. The capacity to generate new attributions of functionality is absent in biological evolutionary processes, but it is essential for the positive feedback dynamic in human sociocultural innovation.

The theory of exaptive bootstrapping provides a qualitative description of the positive feedback dynamic for artifact innovation. It posits the following stages for that dynamic:

1   New artifact types are designed to achieve some particular attribution of functionality;
2   Organizational transformations – new competence networks or scaffolding structures (Lane and Maxfield 2005) – are constructed to proliferate the use of tokens of the new type (that is, among other activities, to produce, exchange, install, or maintain them);
3   Novel patterns of human interaction emerge around those artifacts in use;
4   New attributions of functionality are generated – by participants or observers – to describe what the participants in these interactions are obtaining or might obtain from them;
5   New artifacts are conceived and designed to better instantiate the new attributed functionality.

Since the fifth stage concludes where the first begins, we have a bootstrapping dynamic that can produce cascades of changes in agent-artifact space. These cascades inextricably link innovations in artifacts, in organizational structure, and in attributions about artifact and organizational functionality.

Exaptation happens between the third and the fourth stage in this process, whereby new attributions of functionality arise from observing patterns of interaction among agents and already existing artifacts. The idea here is that artifacts gain their meaning through use, and not all the possible meanings that can arise when agents begin to incorporate new artifacts in patterns of use could have been anticipated by the designers and producers of those artifacts: the combinatory possibilities are simply too vast when a variety of different agents intent on carrying out a variety of different tasks have available a variety of different artifacts to use together with the new ones – not to mention that the designers and producers do not share the experiential base and the attributional space of all the agents that will use the artifact they produce, in ways that depend on their experience and attributions, not those of the artifact's designers and producers.

Meaning in use is one thing – the recognition that that meaning might represent a functional novelty is another. For this to happen, some participants in (or observers of) these patterns of interaction must come to understand that something more is being delivered – or could be delivered, with suitable modifications – to some class of agents (perhaps, but not necessarily, including themselves) than the participants were thinking to obtain through the interactions in which they were engaging – and which these agents might come to value. This 'uncovering' of the possibility of new functionality and value in the context of interactions around an artifact is very similar to the Lane and Maxfield (2005) analysis of the emergence of semantic uncertainty in the context of discourse.

Innovation cascades involve many cycles of the exaptive bootstrapping process. In addition, these cascades also include processes that are purely adaptive: given an attribution of functionality and an artifact that realizes it, it is possible to apply a known technology in order to improve the artifact or its method of production and make it better (according to the values associated with the given attribution of functionality), faster or cheaper. Such processes do not require the generation of new attributions of functionality.

In the model presented here, we focus on phenomena occurring at the micro-level (how individuals collect information about the external world, categorize it, and combine existing categories in order to create new ones) and meso-level (the exchange of information among individuals). However, we do not explicitly include the details concerning the macro-level events (the shared system of beliefs and the common physical and technological resources); these aspects are described in a deeper way in e.g. Lane *et al.* 2005, Villani *et al.* 2008.

The model

As previously explained, in order to understand the exaptation processes we need at least two main entities, agents and artifacts. Despite the fact that artifacts are passive objects, they carry more information than their designers have in their minds, and the ambiguity of language prevent a complete and perfect exchange of information about the artifacts among the agents. The final effect is that different agents can attribute different functionalities to the same artifacts, allowing the exaptation phenomena.

The peculiar aspect of EMIS (Exaptation Model in Innovation Studies) is the explicit description of artifacts, a feature till now not present in literature (for a model describing the feedbacks between artifact innovation and attributions of functionality, see Ferrari *et al.* 2009). In this chapter we focus our attention on the ontologies, and therefore provide only an essential description of the model and of the results, directing the interested readers to Villani *et al.* 2007 and Villani *et al.* 2009 for deeper details.

The basic version of EMIS involves two agents, a user and a producer; only the producer is able to build and modify the artifacts (one artifact for each category owned by the producer), and only the user can evaluate the artifacts.

Obviously, in real systems the agents (human beings or organizations) can play in time (or at the same time for different artifacts) both the roles.

The artifacts are 'goods', built by the producer and utilized by the user. The artifacts art_p are D-dimensional vectors, whose elements (indicated by the words 'characteristic', or 'feature') take values $\{0, 1\}$, where 1 indicates the presence of a given characteristic (feature) and 0 its absence. In order to avoid the production of the 'perfect artefact', where all the desired characteristics are present at the maximum level, we decided to bind to the threshold σ; the number of characteristics present simultaneously in the same artefact.

The tool the agents use in order to interact with the artifacts are the categories (Selby and El Guindi 1976, Markman and Gentner 2001, Macrae and Bodenhausen 2002), in our model D-dimensional vectors whose elements $C^{(i)}_x$ (indicated by the words 'characteristic', or 'feature') are discrete random variables taking values $\{1, 0, -1\}$. These values correspond to the '*functional attributions*' the user give to the corresponding artefact feature, 'the functionality carried out by the corresponding feature of the artifact I'm evaluating is useful/indifferent/damaging' (or 'I wish/don't care/don't wish to give to the artifact I'm building this particular functionality' if we are considering the producer). We assume that the number of relevant characteristics (that is, the characteristics corresponding to symbols '1' and '−1', (in the rest of the chapter indicated by the symbol $|1|$)) is only a fraction η of the total number of features, in order to prevent the presence of not plausible agents attributing relevance to each single cognitive detail (see Villani *et al.* 2007 for the details).

Dynamics

The producer processes already existing artifacts in order to build other artifacts; at the same time, it tries to add some desirable characteristics to the artifact it is producing. In order to do so, the producer selects one of its categories, and attaches the tag '1' to the artifact in correspondence to a '1' memorized for the chosen category, and a '0' tag in correspondence to a '−1' tag, preserving the values of the other features.

In order to evaluate an artifact the user computes its functionality by using all her/his categories and communicates the best result to the producer. In EMIS we define the 'functionality of an artifact with respect to a particular category', as an index measuring the level of user's satisfaction with the artifact. In order to evaluate an artifact, the agent must 'interpret' it by means of its categories. For each category the user calculates a functionality index of the artifact: where user's desires and producer's realization coincide ($A^i_p=1$ and $C^{(i)}_u=1$) the index increases, whereas user's desires and producer's realization do not coincide ($A^i_p=1$ and $C^{(i)}_u=-1$) the index decreases. All the other cases correspond to a neutral situation. The category having the highest index is the more appropriate one to interpret the artefact.

The user can complete the feedback with some additional information, indicating a subset of the features of the current artifact that give positive or negative contribution to the functionality (*actual information*), and/or about what the agent likes or dislikes (a subset of the features of the selected category that potentially have the highest contribution power when it is filtered by means of the selected category – *desired information*). When two categories are highly evaluating the artifact, the user can transmit to the producer some feature coming from both the best categories, giving rise to two different modalities of communication:

1 Symmetrical: both communicated categories provide the same amount of information 'actual' and 'desired';
2 Asymmetrical: the category that better interprets the artifact provides 'actual' information, while the second one provides 'desired' information.

The producer uses the received features to modify the features $C_p^{(i)}$ of their category employed to build the artifact. The new value of each characteristic is the ceil (or the floor for negative values) of the average between the received value and the already present one. All the remaining features are unchanged, except the admittance of some random noise.

How can EMIS detect the exaptation phenomena? The model simulates exchanges of products (artifacts) between producer and uses, the user evaluating the artifacts by means of their categories. In this context, an exaptation is a category change in interpreting the artifact, a user that after hundreds of steps changes the category that systematically was returning the best functionality. In a sense the best category is, for the user, the 'leading' category for this particular artifact; sometimes, but quite unlikely, after the last innovation(s) another category better evaluates the artifact and in consequence becomes the new referring category for the selected artifact. This can be interpreted as a variation of the utilization context of the artifact under exam; in this case, we observe an exaptation event.

Model dynamics

The knowledge space of the categories involved is $D=1,000$ features. The user utilizes five categories, while the producer owns only one category, corresponding to the artifact that it is building.

In order to create the initial categories we fix:

1 the threshold η, limiting the number of $|1|$'s in $C^{(i)}_i$, is set to 100;
2 the initial fraction of 1s (−1s) in $C^{(i)}_i$, is set to 0.05 (−0.05).

The value of η is relatively low with respect to the D value and indicates that the space of all the possible characteristics of a category is very large with respect to the actually realized ones.

We set $\sigma=200>\eta$, that is, we are assuming at the same time that the artifacts can carry out more characteristics than the planned ones and that the typical user focuses its attention only within a subset of the whole potentiality of the artifacts. Therefore, it is actually possible that an artifact carries out a number of functions larger than the number of functionalities for which it has been selected (exaptive pool of possibilities).

Figure 17.1 shows the evaluation the artifact received from the agent's categories during a simulation. Please note that at step 207 the second category suddenly increases its value and becomes the referring category.

Experimental results

We tested three main factors able to influencing the emergence of exaptation phenomena: communication among different agents; noise in communication and production processes; and evolution of the users' categories

Communication

Typically, the user transmits to the producer a (small) subset of the features extracted from the two categories that are returning the best functionality values. We analysed the behaviour of our model by varying the number of transmitted features, or bandwidth (B). In particular, in this set of experiments the total number of transmitted characteristics is set to be $B = \{20, 40, 60, 80, 100, 200\}$. The general observation is that the number of exaptations slightly increases as the

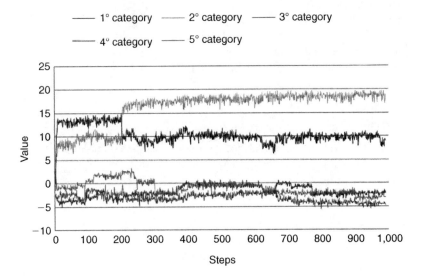

Figure 17.1 Artifacts evaluation.

Notes
At each step the user returns to the producer the best evaluation the artifact on the basis of its own categories. The figure shows the evaluation the artifact received from the 5 agent's categories. Please note that at step 207 the second category suddenly increases its value and becomes the referring category.

bandwidth augment, whereas the number of exaptations found by using the asymmetric modality is consistently larger than the number of exaptations found by using the symmetric modality. The highest part of exaptations happens within the first 50 steps, but there are few exaptations also after several hundreds of steps.

Noise

A second study concerns the analysis of two types of noise that can be possibly specified in the model, the *communication noise* (the value of the features communicated by the user agents is changed with probability α) and the *production noise* (the value of the characteristics of the artifact built by the producer is changed with probability β.

The communication noise does not affect the main behaviour of the model, although its presence slows down the processes and makes it more difficult to reach larger functionality values. The production noise worsens such a tendency, but at the same time increases the frequency of exaptation occurrences, both in the long period and in presence of low bandwidth communications (Figure 17.3). Some innovations, obtained because of this type of error, are able to foster a change of context for the whole artifact.

Learning

In this section we allow the user to modify her/his own categories (so far considered as fixed), through two different modalities.

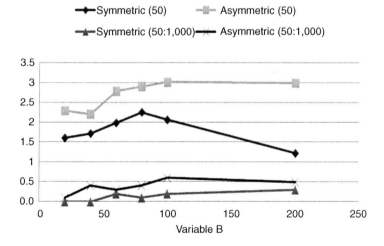

Figure 17.2 Effects of the communication modalities.

Notes
The figure shows the average number of exaptations for each simulation over 10 runs, for different quantities of features transmitted to the producer. The time interval of reference (from 50 to 1,000 steps) is also shown.

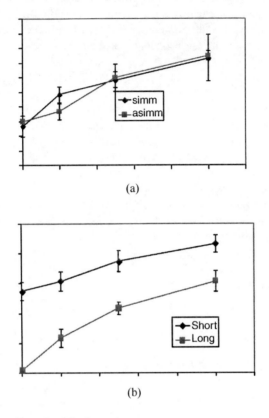

(a)

(b)

Figure 17.3 The effect of production noise.

Notes

a The average of exaptation occurrences versus the production error on ten runs for symmetrical and asymmetrical communication modalities.

b The average of exaptation occurrences in short and long runs.

The simplest way to modify the categories of a user is that of randomly creating a new category and substituting one of the already existing ones (in order to monitor only the not banal exaptations we have to choose among the categories not giving the highest functionality). This way reproduces the knowledge acquired by the user by interacting in new environments not explicitly present on the model.

A more sophisticated method is that of modifying a randomly selected category by using pieces of information coming from the artifact (in this situation also the category giving the highest functionality can be altered).

The rate of categories modification is measured by the *updating factor f*, the probability that at each time step a category of the user is selected for a substitution (updating). In case of updating, each category feature is changed (moving its preference toward a value matching the artifact feature) with probability P_{ch} (in our experiments fixed at 0.02).

The random modality shows a small increase in the number of exaptation events, for both symmetric and asymmetric communication. Higher updating rates f do not enhance this phenomenon. The second modality shows more interesting features. We find a bigger exaptation frequency, which increases with the growth of the adjournment rate f. Symmetric and asymmetric modalities show similar behaviours.

A real example of 'learned' exaptation could be that of the SMS (the short message service), initially introduced to send brief official messages from the telephone company and subsequently becoming a new means of communication. In this case the users succeeded in understanding the communicative potentiality of this system overcoming limits of space by means of the creation of a particular language, more 'assembled' and intuitive. As a result the phone companies initiated a market strategy based upon this new functionality.

Conclusion

Common claims in literature about innovation are that novelties are recombinations of current capabilities and that their search is local, or that creativity is constituted by sudden blind insight. Both views are unsatisfactory: the first hypothesis is not able to describe radical innovation; whereas the second provides little conceptual gain.

In this chapter, we propose that radical innovations are able to engage the so-called 'exaptive bootstrapping' process, and in particular that they are created by a process of 'exaptation'. We introduce an agent-based model designed to investigate the dynamics of some aspects of exaptation in a world populated by

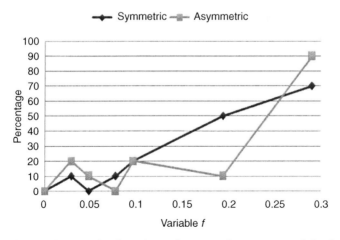

Figure 17.4 Learning from the environment: the percentage of simulations (over ten runs) with at least one exaptation, by varying the adjournment rate f.

agents, whose activity is organized around production and utilization of artifacts. The model (EMIS) explicitly includes agents and artifacts; as consequence, exaptations are easily modelled as shifts in terms of the 'leading attributions' (attributions corresponding to highest reward) that the agents assign to the artifacts.

The model doesn't claim to represent particular real cases; nevertheless, it relates with theory and ontologies, providing an interesting way to observe general behaviours and allowing the study of the not always evident consequences of the theoretical hypotheses.

The idea that the emergence of exaptation events is favoured by an asymmetrical communication (where evaluations and desires are differently expressed for different categories), by a high level of production noise and by the plasticity of the users' categories, the bandwidth (the number of cognitive features communicated among the agents) playing a secondary role seems to indicate that policies encouraging deep relationships among agents (human beings or groups of human beings) based on quality of information exchange (and not only on its quantity) and encouraging plasticity and flexibility in their minds could enhance radical new uses of artefacts, giving rise to new patterns of – positive, we hope – recurrent interactions.

This last theme introduces the analysis of the consequences at macro level of an exaptation occurrence, that is, the exaptive bootstrapping processes, objects of further researches.

Acknowledgement

The authors would like to acknowledge and extend their gratitude to David Lane, Roberto Serra, Davide Ferrari and Stefano Bonacini, who have made the completion of this work possible.

References

Ceruti, M. (1995) *Evoluzione senza Fondamenti*, Collana 'Il Nocciolo', Laterza: Bari.

Dew, N., Sarasvathy, S.D. and Venkataraman, S. (2004) 'The Economic Implications of Exaptation', *Journal of Evolutionary Economics*, 14 (1), pp. 69–84.

Ferrari, D., Read, D. and Leeuw, S. van der (2009) 'The Emergence of New Categories in Artifact Innovation', in: D. Lane, D. Pumain and S. van der Leeuw (eds) *Complexity Perspectives on Innovation and Social Change*, Berlin: Springer.

Gould, S.J. and Lewontin, R.C. (1979) *The Spandrels of San Marco and the Panglossian Paradigm*, Proceedings of the Royal Society of London, B. 205, pp. 581–98.

Gould, S.J. and Verba, E. (1982) 'Exaptation, A Missing Term in the Science of Form', *Paleobiology*, 8 (1).

Gould, J.S. (2002) *The Structure of Evolutionary Theory*, Cambridge, MA: Belknap Press.

Kimura, M. (1983) *The Neutral Theory of Molecular Evolution*, Cambridge, UK: Cambridge University Press.

Lane, D.A. and Maxfield, R.R. (2005) 'Ontological Uncertainty and Innovation', *Journal of Evolutionary Economics*.

Lane, D., Serra, R., Villani, M. and Ansaloni, L. (2005) 'A Theory-based Dynamical Model of Innovation Processes', *Complexus*, 2 (3–4), pp. 177–94.

March, J.G. (1991) 'Exploration and Exploitation in Organizational Learning Organisation', *Science*, 2 (1).

Macrae, N. and Bodenhausen, G. (2000) 'Social Cognition: Thinking Categorically about Others', *Annual Review of Psychology*, 51, pp. 93–120.

Markman, A.B. and Gentner, D. (2001) 'Thinking', *Annual Review of Psychology*, 52, pp. 223–47.

Selby, H.A. and El Guindi, F. (1976) 'Dialectics in Zapotec Thinking', in: K. Basso and H.A. Selby (eds) *Meaning in Anthropology*, Albuquerque, NM: University of New Mexico Press, pp. 181–96.

Simon, H.A. (1996) 'The Architecture of Complexity', in: *Sciences of the Artificial*, third edition, Cambridge, MA: MIT Press.

Villani, M., Bonacini, S., Ferrari, D., Serra, R. and Lane, D. (2007) 'An Agent-based Model of Exaptive Processes', *European Management Review*, 4, pp. 141–51.

Villani, M., Serra R., Ansaloni, A. and Lane, D. (2008) 'Global and Local Processes in a Model of Innovation', in: H. Umeo, S. Morishita and K. Nishinari (eds) *Cellular Automata*, 8th International Conference on Cellular Automata for Research and Industry Springer Lecture Notes in Computer Science 5191, pp. 401–8.

Villani, M., Bonacini, S., Ferrari, D. and Serra, R. (2009) 'Exaptive Processes: An Agent-based Model', in: D. Lane, D. Pumain and S. van der Leeuw (eds) *Complexity Perspectives on Innovation and Social Change*, Berlin: Springer.

18 Product architecture and firm organization

The role of interfaces[1]

Tommaso Ciarli, Riccardo Leoncini,
Sandro Montresor and Marco Valente

1 Introduction

The idea that the design of the products firms manufacture crucially affects their innovation capabilities and competitiveness is by now an established result (Henderson and Clark 1990, Ulrich 1995, Baldwin and Clark 2000). A number of research papers, starting from the seminal work by Sanchez and Mahoney (1996), have shown that this effect works mainly through the relationship between product architecture and firm organization, as their matching – or 'mirroring' – is responsible for the firm's capacity to compete with its rivals in developing new products and facing industry dynamics (e.g. Brusoni and Prencipe 2001).

Although such a relationship has received both theoretical and empirical support, the inner mechanisms through which it works are still far from fully understood. On the one hand, the standard argument according to which the architectural knowledge of the firm is embodied in the firm's organizational routines, and in its information filters and communication channels, such that changes in the organization are constrained by the information and communication structures (Henderson and Clark 1990), leaves the firm organization itself still quite 'black-boxed'. Indeed, important organizational choices, especially in terms of vertical integration and disintegration (i.e. firm's boundaries), are made also and above all to pursue a better exploration of the space of product designs, rather than to simply exploit them. On the other hand, the reference to specific case studies or, at most, to limited sector specific datasets, that the relevant literature employs to support the mirroring hypothesis, prevents the identification of general results.

In this chapter we provide three main contributions towards shedding some light on the relation between product architecture and firm organization. First, we suggest an integrated approach to the relation between product architecture, innovation and firm organization, which enables to retain the complex impact – that is, not simply 'dual' – that product architecture has on firms' competition: a complexity of which the empirical evidence shows several traces. Second, we claim that 'interfacing' components play a strategic role in managing such a complex relation, namely by effectively reducing the complexity. Third, we support our intuition on the role of interfaces as mediators between architecture and organization with the results of an agent-based simulation model that exploits the powerful NK metaphor, tailored to account for the deterministic choices of the firm and for the role of product interfaces. The model results provide an explanation for some regularities suggested in the literature on the way in which product architecture impacts on firms' competitiveness.

The remainder of the chapter is organized as follows. Section 2 briefly reviews the theoretical and empirical works which investigate the relationship between product and organizational architectures. Section 3 points to the complexity of product architecture and refers to a *pseudo*-NK model as a suitable tool to capture the implications for the firms' technological competition. Section 4 briefly summarizes the model of technological competition in Ciarli *et al.* (2008), extends it to analyze the key role of interfaces and modules, and discusses the simulation results. Section 5 concludes.

2 Product architecture and firm organization: theoretical and empirical accounts

Although quite intuitive, the notion of product architecture is far from unambiguously defined. According to a first interpretation, architecture would generally refer to the structure through which product components and subsystems integrate into a coherent mechanism. This is, for example, the idea Henderson and Clark (1990) refer to by pointing to the difficulties of recognizing and possibly introducing 'architectural innovations'. In particular, in their study of the photolithography industry, they show that, what they call 'architectural innovations', amounting to changes in the relationships between (unaltered) product components, dampen the incumbent firms' competitiveness as they bring to the front a new architectural knowledge, different from dominant one, incorporated in their own organization (Suarez and Utterback 1995). This and other related studies (e.g. Iansiti and Clark 1994, Iansiti 1995) point to the impact the organizational structure of the firm has on the product architecture it develops, supporting what in organization theory is known as 'Conway's Law' (Conway 1968), and more in general as the 'mirroring hypothesis'.

According to a different, and more specific, interpretation, the product architecture would refer to the 'specification of the interfaces among interacting physical components' (Ulrich 1995: 420), and to its impact on their mapping on the product functional elements. The impact of this specification on innovation and firm organization is different from the previous one. On the one hand, modular product architectures (whose components are related by decoupled interfaces[2]) have been found to require different organizations to be innovated and manufactured than integral architectures. The hypothesis that 'an integral organization is necessary for developing an integral product, while a modular organization is only capable of developing a modular product' (Colfer 2007: 2) has received lot of attention.

On the other hand, the impact that an *existing* product architecture has on the organization and on the organizational *changes* of the firm which manufactures and innovates the product itself has been investigated too. In this last respect, organization theory has mainly dealt with the hypothesis that organizational structures – in particular in terms of communication channels – perform relatively better when they 'match' the product design – in particular in terms of component interfaces and protocols. Sosa *et al.* (2004), for example, find evidence of the tendency of team communications to overlap with the design interfaces of a large jet engine project, and similar results are obtained with respect to other sectors, such as for example in the case of software products (see MacCormack *et al.* 2008 for a review).

Although quite useful, the 'mirroring hypothesis', which is the core of the latter approach to product architecture, appears to us a too simplistic interpretation. In particular, because it does not account for the complexity of the relationships between technological and organizational change, and for the role of product architecture in at least two respects. First of all, it should be recognized that the presence of strong interdependencies between the components of a complex product – i.e. an integral architecture – makes the choice of an innovative optimal design for it extremely difficult and sometimes even works as an 'obstacle of technological change' – such as in the case of manufacturing automation and fuel-cell technology (Buenstorf 2005: 223). On the other hand, modular architectures, while allowing for modifications of individual components, prevent innovations related to the 'global performance characteristics' of the product (Ulrich 1995: 432).

The second point, we will address in Section 4, is the fact that product architecture also affects the boundaries of firm organization and its degree of vertical disintegration. Also in the light of the different incentives the management has in favoring modular rather than integral architecture to deal with uncertainty.

In order to retain these and other elements, and to account for the complexity of product architecture, the mirroring hypothesis of organization theory needs to be integrated with other research strands, to which we now turn.

3 Product architecture, technological competition and industry organization: a complexity approach

Very often, the design and the manufacturing of a product architecture amounts to that of a *complex system*, as 'one made up of a large number of parts that interact in a non simple way' (Simon 1962: 468). In front of it, bounded rational economic agents thus need to resort to behavioral patterns of problem solving and search routines which have been proved to be suitably described by the so called *NK* methodology, developed by Kauffman (1993) with respect to the interdependencies of the genomes of organisms. Thinking of a product architecture as a string of N product components, each of which is interdependent with K others, and attaching to each component a 'fitness value' which measures its contribution to the overall (economic) fitness of the product, the firm's capacity to pursue the latter can be shown to depend on the product architecture itself, and on the size of K in particular.[3]

First of all, this kind of model provides an intuitive counterpart to the product architecture, which can be seen in the specification of the number of epistatic linkages, K, among the total product components, N. Second, these models allow retention of different strategies of fitness search (e.g. local versus global) in a rugged technological landscape, with interesting implications for different organizational structures (e.g. vertically integrated versus vertically disintegrated).

Although the explanatory power of this model structure has already been employed in dealing with a number of complex problems in economics of innovation (e.g. Auerswald *et al.* 2000, Kauffman *et al.* 2000, Fleming and Sorenson 2001), its standard version suffers from a number of restrictions – for example, the lack of learning effects in the search processes – that can be relaxed in the model we will propose in this chapter. What is more, following Frenken (2006), the *NK* model will be generalized to disentangle the differences which, in spite of some analogy, exist between *decomposability* and *modularity* in complex systems. Indeed, as several authors in the field have argued (e.g. Baldwin and Clark 2000, Langlois and Robertson 1992), complex products can be modular without being necessarily decomposable, at least in the sense Simon alludes by referring to the presence of independent sub-systems: personal computers and stereo sets are two clear examples of modular products whose sub-systems are far from independent. Even in non-decomposable systems, in fact, modularity can be obtained providing one or more product components work as 'standard interfaces', mediating the interaction with respect to all the others, each of which affects only one product function (Frenken 2006: 298). This kind of argument makes both the presence of technological standards and the 'battle' to establish them crucial for the process of technological competition, and not only with respect to the usual motivation of network externalities and co-ordination (Farrell and Saloner 1987, Katz and Shapiro 1985).

The last point is actually the starting one of the simulation-based part of the next section of the chapter. Following and extending our previous work (Ciarli *et al.* 2008), an industry will be investigated in which firms are involved in the technological competition for the improvement of a complex product. A new element will be introduced by referring to a more specific definition of product architecture, which distinguishes between 'components', 'modules' and 'interfaces'.

4 Competition in complex technological spaces: the role of interfaces

In order to better understand the organization and innovation impact of product architecture, we claim that the idea of 'interface' has to be defined by distinguishing between the product components and the bundles of components that can be grouped into modules according to well specified relationships. Accordingly, in a sort of eclectic view, a product architecture can be defined by means of a set of interfaces between product modules, or, equivalently, by determining the range of communication that can pass within and between different modules, in turn made up of sets of components. The set of interfaces may be physical, embodied in specialized elements, or only composed by a communication proto-col. The role of interfaces is to allow the firm to improve one module without evaluating each state of the others: that is, to deal with a pre-determined list of possible states of input and output relations between modules. As long as the module is able to interpret correctly all of the input messages, and to generate one of the output messages, there is no need to pay attention to what happens in the other modules. In practice, an interface insulates all the elements within con-nected modules, such that any modification within a module does not affect the functioning of the other modules, insofar as the output remains compatible with the interface.

Defined in this way, the set of interfaces may appear as an instrument for reducing the complexity of the research space. This is actually not the case, if we consider the usual method of assessing the level of complexity by the difficulty to reach the global peak of a problem space (Frenken *et al.* 1999). However, if the goal is not to reach the global theoretical peak, but to facilitate the improve-ment of the performance with respect to the current configuration of a product, then the use of interfaces hugely simplifies the process. In the following, we will show why this is the case, and will derive a few predictions that such perspective produces with respect to firms engaged in technological competition.

Consider a complex product, made of several elements each affecting each other's contribution to the product performance. A firm willing to improve the overall performance of such a product is forced to painstakingly test a large number of modifications of each element, and then see the final results. Given the complexity of the product structure, each modification of an element will affect the performance contribution of any other element, making the final outcome highly unpredictable.

Consider instead the same product in which a set of interfaces is introduced. All the elements are grouped into a number of modules, and the interactions of the elements in a module with those in other modules are no longer direct, but are mediated through an interface. Changing one element in a module will still affect the performance of the elements within the same module, but, as far as the changes respect the compatibility with the interface, they will not affect the con-tributions of other modules.

Notice that the new system is perfectly equivalent to the original one. That is, any possible configuration of the original system may be re-produced in the new system by an appropriate choice of elements' state and interface. Only, the system modularized via a set of interfaces permits to classify two types of an elements' modification: those that respect the compatibility with a given inter-face and those that do not.

The innovation pattern of a system organized in modules through interfaces has highly attractive features for firms facing a tough competitive environment. In fact, limiting the modifications to only those modules that respect the compat-ibility with the interface ensures that the between-module interactions are null, and therefore the firm needs to be concerned only with the within-module set of interactions. Consequently, the (apparent) complexity level of the modularized product is considerably reduced, and it is easier to generate a stream of perform-ance enhancing innovations. However, the limitation imposed by the need to maintain the compatibility with the interfaces constrains the number of modifica-tions permitted. It is therefore likely that a given set of interfaces will reach a point of 'saturation', where no more useful innovations may be introduced by respecting the compatibility. At this stage an innovation can take place only by modifying the very set of interfaces, which will open a new range of potential innovations within the (newly defined) modules.

From the organizational viewpoint, there is a marked difference between the innovators concerned only with module modifications and those concerned with interface innovations, or interface design. The former are allowed, by def-inition, to ignore the effects of a change of a module onto the other modules. Consequently, the firms concerned with module innovations may be 'special-ized' innovators, with a deep knowledge of the technology of modules and are allowed to know little of, or even fully ignore, the details concerning other modules: they can rely on the interface to guarantee the compatibility of their innovations to the other modules. Conversely, when a new set of interfaces is called for, the innovator is required to master many diverse skills, since the interfaces, by definition, determine the interactions between each and every element of the product. Moreover, the interface value is not apparent at the time of its introduction, but may strongly increase when module innovations push to the limit the constraints imposed by the interfaces. In other terms, while module innovators can immediately evaluate the goodness of a given innovation, interface innovators need to predict future innovation patterns, and design today a set of constraints that, they hope, will pose minimal limits to future developments.

The approach illustrated above provides a new perspective in the analysis of the relations between technological complexity, organization of production and competition. From such perspective, we are able to put forward a number of pre-dictions on the patterns of innovation in competitive markets, which we present as a set of conjectures. First, an architecture that makes use of interfaces is likely to induce a continuous stream of relatively small innovations, with respect to equivalent, integral systems, which can conversely advance by rare, and large, jumps of performance. Second, in products endowed with an interface architec-ture we expect to observe frequent innovations concerning the modules' ele-ments, but rare modifications of the interface elements. However, and third, interface innovations are likely to be much more productive in terms of perform-ance increment. Fourth, module innovators are more likely to be specialized innovators, with technological knowledge limited to a few areas, whereas inter-face innovators likely possess a wide range of technological skills.

In the rest of the chapter we present some preliminary results provided by the above perspective and supporting some of the above conjectures. The following paragraph briefly presents the model we use to explore the role of interfaces in technological competition. Next we present a summary of the main results obtained with the model.

4.1 A model for competition and complexity

In a previous work of ours (Ciarli *et al.* 2008), we investigated the interaction between industrial competition and innovation in a generic technological space (i.e. with no reference to one specific complexity structure). The particularity of the model stems from the endogenous determination of firms' production scope. That is, the model considers firms that face the 'make or buy' dilemma on each of the product components. The product overall performance depends both on the intrinsic quality of the components and on the strength and direction of their interaction. The more complex the technological space, the stronger the interaction, and consequently the more difficult is the adaptation of otherwise good components into a product with overall high performance.

A core result of the study is that competition forces firms to solve the trade-off between the economic gains due to specialization and the technological gains deriving from an integrated organization. Specialized firms enjoy economies of scale, being consequently able to offer a component at lower prices than competitors. However, only firms producing different components are able to take into account, in their technological search, the interaction among components. Consequently, the technological level of integral firms (covering a large number of components) is, on average, higher than specialized firms, which ignore the interactions among components. At the aggregate level of the industry, we showed that the tension between the two strategies of outsourcing and internalizing modules' production does not produce a stable pattern for industry organization, but rather generates a cyclical one, where periods with highly fragmented industrial structures (made of many different specialized producers) are followed by highly integrated ones (made of few firms producing most components in-house).

In that work we were interested in assessing whether increasing technological opportunities, represented by a moving technological frontier, affected the industrial structure, and which effects it produced. The modeling tool adopted is the *pseudo*-NK (Valente 2008), which allows, as in NK models, to represent complexity as interactions among the space dimensions, but enjoys many features that, contrary to NK, make it very useful for applications to economic models. First, *pseudo*-NK is defined over real-valued, instead of binary, variables, thus facilitating the representation of graded levels of modules' quality corresponding to graded values of independent variables. Second, *pseudo*-NK is deterministic, rather than stochastic, allowing modelers to determine the level of maximum fitness and the optimal point. Third, and crucial for our purposes, *pseudo*-NK determines which interactions are present among variables and their relative strength, and thus to distinguish the complexity stemming from the different interaction structures (which modules are connected).

For the purpose of the present work, we use all the mechanisms of that previous model,[4] but we modify the complexity structure of the technological space to simulate a product made of interfaces and modules, as indicated in the next paragraph.

4.2 Interfaces and modules: a two-stage technological race

By exploiting the flexibility of the *pseudo*-NK model, it is possible to generate a technological space made of interfaces and modules. The simplest structure of this type can be represented by one module being connected to all other modules (the interface), while the remaining modules are connected only to the interface. From the technical point this space shows the highest level of complexity, but, keeping constant the first dimension, the resulting sub-space is fully modular.

Adopting this technological space, we ran the competition model as in Ciarli *et al.* (2008), with respect to an industry composed of 50 firms, each producing a six-component product, only one of which (the interface) is related to all the other ones. In this way, we obtained highly interesting results that allow us to provide a preliminary illustration of the conjectures derived from the literature on the role of interfaces on firms' innovation and organization.

Figure 18.1a reports the average fitness increments per innovations for five independent runs and Figure 18.1b the number of successful innovations at the end of the 10,000 time simulation steps. The different grey series indicate the values for each module, while the black series refers to the interface module.

The figures show that the 'interfacing' module is modified more rarely than the modules having no connections (Figure 18.1b). However, the average products' fitness increment resulting from these fewer innovations is higher than the innovations generated by the other modules (Figure 18.1a). This result confirms that modules with a high degree of connections are much harder to modify, although if such an innovation succeeds, it is highly rewarding. Moreover, this result is perfectly in line with the perspective on interfaces that we described at the beginning of the section, confirming the robustness of its logic by means of simulation results.

We also found confirmation of our perspective on firms' organization. Figure 18.2a reports the number of firms actively producing each module (instead of buying it), out of the 50 firms in the industry, and Figure 18.2b the total amount of traded modules. Again, the results confirm the implications described above, the interface dimension results to be actively produced only by a small number of firms. However, the interface is traded in larger quantity, than other dimensions, showing that the emergence of a sort of 'standard' is generally imposed by a bunch of producers able to supply other firms with a given (working) format for the interface module.

5 Conclusions

In this chapter we have argued that the standard framework that organization theory has developed to investigate the implications of the product architecture for the firm performance and organization, though useful in some respects, might be too narrow in understanding the complex co-evolution between technology and organization. Indeed, there is much more than a simple mirroring phenomenon. This is particularly the case with respect to the impact that the degree of

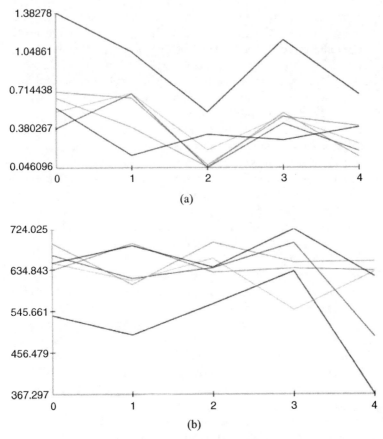

Figure 18.1 Innovation pattern per module: performance increase (a) and number of innovations (b) (the darker series is the interface, the *x* axis refers to 5 independent runs).

modularity of the product has on firm decision to tailor its organizational boundaries in searching for a better technological fit. Most crucially, 'interfacing' modules may play a strategic role in the dynamics of product innovation, and in the firms' ability to manage it.

As far as modularity is concerned, the chapter proposes to deal with it by referring to a sort of 'hardware' rather than 'software' idea of product interfaces among components. Rather than generic communication protocols between the product components, in the chapter the role of interfaces (both disembodied and embodied) is defined in such a way as to 'freeze' some modules from the others and to deal only with a pre-determined list of possible states of input and output relations.

The product architecture impinges on the technological competition between firms whose production scope is endogenously determined by their search of a superior product performance. More precisely, the modularity degree of the

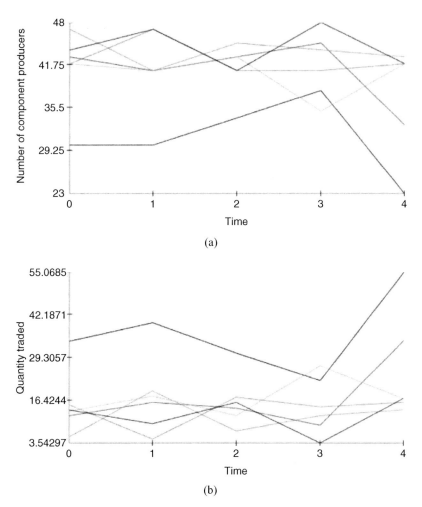

Figure 18.2 Number of firms producing and trading the interface (dark series) and the remaining modules (grey series), (the *x* axis refers to 5 independent runs).

product, which can be qualified by the use of interfaces, impacts on the intrinsic contribution of each component to the product, via the interaction with the related (non-modular) components. In turn, this architecture affects a firm's decision to change its organization in order to better deal with the trade-off between static and dynamic gains.

The simulation results show that a firm's performance, and its resulting organization, crucially depends on the way product architecture is designed and managed. In fact, production can be based on a set of modules 'coordinated' by an interface, in order to decrease the degree of interdependence between modules, thus allowing to explore their technological performance with higher degrees of freedom with respect to a situation characterized by higher level of interdependence. Firm organization and its performance will therefore vary with the degree of interdependence between the product components and with the number and the type of interfaces mediating the relationships between the components themselves.

Our main results show that if production can be organized by means of a set of 'interfacing' modules, the innovative activity carried out on the interface will record a more important increase in performance, while keeping more stable innovative dynamics in terms of number of innovations. The stability of the product architecture (due to the interface) turns out to favor more frequent and minor (incremental) innovations in the modules that will be carried out by a large number of integrated firms. Furthermore, the production of the interfaces is concentrated on a small number of specialized firms that have a stronger power in defining the product architecture standard. The innovative activity of those firms turns out to have a considerable impact on the production system.

Notes

1 R. Leoncini and S. Montresor gratefully acknowledge financial support by the MIUR-PRIN project 'Emerging Economic Regional Powers and Local Systems of Production: New Threats or New Opportunities?', and by the Province of Trento, as the sponsor of the OPENLOC research project under the call for proposals Major Projects 2006. The usual disclaimers apply.
2 Unlike an integral architecture, a modular one includes: (*a*) a one-to-one mapping from the functional elements and the physical components of the product, such as, for example, for a trailer in which the connecting function is accounted by the hitch pin only, and the same holds true for the cargo protection with respect to the box, and so on and so forth (Ulrich 1995); (*b*) a decoupling interface between components, such that a change in one component does not require a change in any other for the product to keep working.
3 The most important results, obtained by Kauffman by simulating the *NK* model, are pretty well-known. In brief, a local step-wise kind of search, that is component by component, leads to a global fitness maximum for the product with certainty if and only if the fitness contributions of the individual components are independent between them (i.e. with $K=0$). Conversely, as the system elements become progressively more interdependent, as K increases, a number of local maxima emerge (and increase) which set the individual search at work in a 'rugged landscape', with the possibility to end up in a local solution (Kauffman 1993, chapter 2).
4 In our previous work we considered (*a*) the structure of the technological space as characterized by maximum complexity (i.e. the performances of all components were influenced by the states of all the other components) and (*b*) we imposed a mild intensity to all the interactions between modules. The interested reader will find all the technical details in Ciarli *et al.* (2008) and (2010).

References

Auerswald, P., S. Kaufman, J. Lobo and K. Shell (2000) 'The Production Recipes Approach to Modeling Technological Innovation: An Application to Learning by Doing', *Journal of Economic Dynamics and Control*, 24 (3), pp. 389–450.

Baldwin, C. and K. Clark (2000) *Design Rules: The Power of Modularity*, Cambridge, MA: MIT Press.

Brusoni, S. and A. Prencipe (2001) 'Unpacking the Black Box of Modularity: Technologies, Products and Organizations', *Industrial and Corporate Change*, 10 (1), pp. 179–205.

Buenstorf, G. (2005) 'Sequential Production, Modularity and Technological Change', *Structural Change and Economic Dynamics*, 16, pp. 221–41.

Ciarli, T., R. Leoncini, S. Montresor and M. Valente (2008, August) 'Technological Change and the Vertical Organisation of Industries', *Journal of Evolutionary Economics*, 18 (3–4), pp. 367–87.

Ciarli, T., R. Leoncini, S. Montresor and M. Valente (2010) 'Linking Technological Change to Organisational Dynamics: Some Insights from a Pseudo-NK Model', in: R. Leoncini and N.D. Liso (eds) *Internationalization, Technological Change and the Theory of the Firm*, London: Routledge, forthcoming.

Colfer, L. (2007) 'The Mirroring Hypothesis: Theory and Evidence on the Correspondence between the Structure of Products and Organizations', Harvard Business School Working Paper.

Conway, M. (1968) 'How do Committees Invent', *Datamation*, 14, pp. 28–31.

Farrell, J. and G. Saloner (1987) 'Competition, Compatibility and Standards', in: H.L. Gable (ed.) *Product Standardization and Competitive Strategy*, Amsterdam: Elsevier.

Fleming, L. and O. Sorenson (2001) 'Technology as a Complex Adaptive System: Evidence from Patent Data', *Research Policy*, 30 (7), pp. 1019–39.

Frenken, K. (2006) 'A Fitness Landscape Approach to Technological Complexity, Modularity, and Vertical Disintegration', *Structural Change and Economic Dynamics*, 17 (3), pp. 288–305.

Frenken, K., L. Marengo and M. Valente (1999) 'Interdependencies, Nearly-decomposability and Adaptation', in: T. Brenner (ed.) *Computational Techniques for Modelling Learning in Economics*, Boston, Dordrecht and London: Kluwer.

Henderson, R.M. and K.B. Clark (1990) 'Architectural Innovation: the Reconfiguration of Existing Product Technologies and the Failure of Established Firms', *Administrative Science Quarterly*, 35, pp. 9–30.

Iansiti, M. (1995) 'Technology Integration: Managing Technological Evolution in a Complex Environment', *Research Policy*, 24, pp. 521–42.

Iansiti, M. and K. Clark (1994) 'Integration and Dynamic Capability: Evidence from Product Development in Automobiles and Mainframe Computers', *Industrial and Corporate Change*, 3, pp. 557–605.

Katz, M. and C. Shapiro (1985) 'Network Externalities, Competition, and Compatibility', *American Economic Review*, 75, pp. 424–40.

Kauffman, S.A. (1993) *The Origins of Order*, Oxford: Oxford University Press.

Kauffman, S.A., J. Lobo and W.G. Macready (2000) 'Optimal Search on a Technology Landscape', *Journal of Economic Behavior and Organization*, 43, pp. 141–66.

Langlois, R. and P. Robertson (1992) 'Networks and Innovation in a Modular System: Lessons from the Microcomputer and Stereo Component Industries', *Research Policy*, 21 (4), pp. 297–313.

MacCormack, A., J. Rusnak and C. Baldwin (2008) 'Exploring the Duality between Product and Organizational Architectures: A Test of the Mirroring Hypothesis', Working Paper 08–039, Harvard Business School.

Sanchez, R. and J.T. Mahoney (1996) 'Modularity, Flexibility and Knowledge Management in Product and Organisation Design', *Strategic Management Journal*, 17, pp. 63–76.

Simon, H. (1962) 'The Architecture of Complexity', *Proceedings of the American Philosophical Society*, 106, pp. 467–82.

Sosa, M., S. Eppinger and C. Rowles (2004) 'The Misalignment of Product Architecture and Organizational Structure in Complex Product Development', *Management Science*, 50, pp. 1674–89.

Suarez, F. and J. Utterback (1995) 'Dominant Designs and the Survival of Firms', *Strategic Management Journal*, 16, pp. 415–30.

Ulrich, K. (1995) 'The Role of Product Architecture in the Manufacturing Firm', *Research Policy*, 24, pp. 419–441.

Valente, M. (2008) 'Pseudo-NK: an Enhanced Model of Complexity', LEM Papers Series 2008/26, Laboratory of Economics and Management (LEM), Sant'Anna School of Advanced Studies, Pisa, Italy.

19 Social evaluations, innovation and networks[1]

Francesca Giardini and Federico Cecconi

1 Introduction

How individuals and institutions are affected by social relationships is the *vexata quaestio* of social sciences that has been puzzling scholars from different disciplines for many years. One option to face this challenge is to start by focusing on a more specific issue, like for instance social evaluations, whose connection to networks is hardly questionable. In fact, social evaluations, i.e. beliefs about other individuals, represent a constitutive part of our relational life and serve many different goals. Exchanging information about our peers is a pervasive activity through which two kinds of item are disseminated: either factual information, reporting actions and events involving other agents, or social evaluations, i.e. judgments about one's peers' competences and behaviors with regard to a given dimension (Miceli and Castelfranchi 1989).

In human groups, the exchange of evaluations serves as a means to create strong and weak ties (Granovetter 1973) and may be pivotal to either the creation or the enforcement of other kinds of relationships (friendship, acquaintances, business, etc.). On one hand, gathering information allows to make more accurate and complete evaluations of others, and people transmitting valuable information can achieve trust and respect. On the other hand, knowing facts about potential partners is pivotal to the establishment of new links that can bridge far and distant nodes, thus permitting to enlarge the group.

Social evaluations become critical in closed environments in which the web of relationships directly affects agents' behaviors, actions and results. This applies not only to human groups, but also to other kinds of networks, ranging from organizations to institutions and firms. In particular, industrial districts[2] in which firms collaborate to deliver a final product are characterized by a close interplay between the economic dimension and the social relationships. In the district, the form of production requires a high degree of cooperation between firms and the lack of formal agreements could lead actors to behave in an opportunistic manner, but the merging between social community and firms helps preventing this result. A number of studies have been confronted with the issue of innovation in industrial districts, and there is general agreement on the fact that physical proximity facilitates the process of innovation (Becattini 1990).

A complementary strand of research stresses the relational aspects of clusters, by analyzing the agents' behaviors as part of a network structure and explaining innovation in terms of network's topology and measures (Cowan and Jonard 2004).

We want to add to this picture in three ways:

- by adopting a cognitive perspective that differentiates between image and reputation in terms of individuals' belief configurations;
- by assuming that gossip may promote innovation intended as the exchange of useful information;
- by exploring the relationship between those two evaluations and the resulting networks, especially with regard to the strength of ties.

In our view social complex phenomena primarily emerge from agents' beliefs about the world and from their goals and plans, and they can be better understood and explained thanks to cognitive modeling. Cognition bridges the micro and the macro level, making macro-social phenomena, such as reputation, norms, and networks emerge, unintentionally, from cognitive representations and goals of single agents.

Here, we will devote our attention to the connection between social evaluations and networks with the purpose of unfolding this relationship. In Section 2 an essential background on both reputation and gossip in the social sciences, and agent-based simulation of industrial districts is provided. Section 3 describes the cognitive theory of reputation and gossip. In Section 4 the Repnet model, the related work and the main results are presented and discussed. The last section is devoted to discussing some of the limitations of the present work and future directions of research.

2 Background

2.1 Reputation and gossip in the social sciences

Gossiping, i.e. transmitting information and evaluations regarding a (usually absent) third party, is crucial for human societies in which it serves many different functions. Gluckman (1963) has been one of the first to stress the positive virtues of gossip, among which are the ability to maintain the unity, morals and values of social groups. Gossip contributes to stratification and social control, since it works as a tool for sanctioning deviant behaviors and for promoting, even through learning, those behaviors that are functional with respect to the group's goals and objectives, mainly norms and institutions (Wilson et al. 2000).

Gossip is a valuable source of information about the community, its members, its norms, values and habits, but it is also useful to map the social environment and to make inoffensive comparisons (Fine and Rosnow 1978). According to Ben-Ze'ev (1994), gossip is a pleasurable way to gather information that is otherwise hard to obtain, but it also serves to satisfy the so-called *tribal need*, namely, the need to belong to the group and to be accepted by it. Sommerfeld *et al.* (2007) consider gossip as a way to transfer social information within groups, alternative to direct observation. The possibility of being informed about things that are not directly observable, especially about other people's behaviors, is probably the most important function of gossip, as it is demonstrated by the huge amount of studies regarding reputation and pro-social behavior (for an introduction, see Fehr and Gachter 2000). Dunbar, in his study on the comparison between primate bonding mechanisms and human language, affirmed that:

> Language, of course, has a further benefit over conventional primate bonding mechanisms: the fact that we can use it to exchange information about ourselves and other members of our social network. Primates can only know what they see for themselves ... We can monitor what is going else-where within the network, allowing continuous updating of our knowledge of the matrix of relationships.
>
> (2001: 291)

Theories of indirect reciprocity explain large scale human cooperation in terms of conditional helping by individuals who want to uphold a reputation and then to be included in future cooperation (Panchanathan and Boyd 2004). Economic experiments from traditional societies all over the world have provided strong support to the idea that reputational concerns can support cooperation and altruistic behaviors (Henrich *et al.* 2004; Gintis *et al.* 2001). Reputational information can also help to solve the *tragedy of the commons*, a social dilemma referring to the fact that a public good will be overused if everybody is allowed to do so. Allowing people to build up a reputation, prevented the public resource from being overused.

In our view, a more detailed account of reputation and gossip is required, due to the complexity of the phenomenon at stake, both at the individual and at the social level. Exchanging social information is a two-fold activity. It is a purposive act leading to specific and intended consequences, but it is also an *emergent* social phenomenon that has not been previously predicted and purposefully realized. In other words, its consequences can be unintended and functionally maintained, i.e. what emerged is independent of the agents' awareness and decisions, but it constrains their actions and determines their efficacy (Castelfranchi 1998). Once created and transmitted, social evaluations influence other agents' minds, changing their behaviors and goals. This can happen either intentionally, because the gossipee has the goal of influencing the receiver or even the target agent, or unintentionally, so that gossip effects are unintended but can equally give rise to functional phenomena. In this latter case, gossip has a two-fold nature: it emerges from collective behaviors of information spreading but it needs to be represented into individuals' minds in order to work effectively.

2.2 Agent-based simulations of industrial clusters

Porter defined clusters as 'geographic concentrations of interconnected companies' (1998: 197), in which proximity to partners and competitors provides those opportunities for innovation that constitute a major drive of the district's success. The idea that clustered firms promote innovation is well expressed in the notion of 'milieux innovateur' (Maillat et al. 1995) that consider the firm's geographical and cultural proximity as a major force driving firms toward innovative products and technologies. Being embedded into the same socio-institutional structure stimulates cooperation among partners and favors dissemination and accumulation of new knowledge (Bathelt 2008). Studies on innovation show that the likelihood of innovation and the related degree of novelty increases as interactions and learning increase (Landry et al. 2002, Hoen 2001).

Pittaway et al. (2004) present a systematic review of research linking the networking behavior of firms with their innovative capacity. They analyze several studies to establish a clear relationship between innovation and different aspects of networking (structure, relations, types of linkages), with the aim of elaborating suggestions to policy makers to sustain clusters' innovation and increase their competitiveness.

Albino et al. (2006) stressed the importance of learning processes as exchange of information and knowledge within the district and distinguish three kinds of learning: learning by specializing, by interacting, and by localizing. To investigate how different types of learning may affect innovation processes they used agent based simulation techniques developing an artificial cluster in which firms interact and exchange goods and knowledge.

Computational simulations applied to the study of industrial districts are widely used and the results obtained span organizational, financial and social aspects of the clusters. The complexity of the phenomenon is mainly due to the high number of dimensions interacting in a non-linear fashion, so that traditional methodologies, like laboratory experiments and direct observation, are not sufficient. Artificial experiments permit to overcome this obstacle, since they are especially suited to deal with complexity. In Gilbert's terms (2004), there are two distinct but complementary ways of designing models: *descriptive* and *abstract*. The former class of models incorporates many details of the real situation with the aim to build up detailed and realistic models that could be used to analyze an existing districts' processes and results (Brenner 2001, Zhang 2003). On the other hand, in the abstract approach, the paucity of realism is counterbalanced with the introduction of cognitive and social factors that lead to findings not directly descriptive of or applicable to any real industrial district, but extremely useful to explore the consequences of various assumptions and initial conditions (Biggiero, Sevi 2009, Giardini et al. 2008a, 2008b, Squazzoni, Boero 2002). Here we are not interested in going into details about these different traditions, but it is useful to say that our model is an abstract one aimed to highlight the role of informal social ties in industrial districts, in which they promote innovation and favor competitiveness.

Before introducing the model we will explain how individual behaviors and collective effects can be reconciled into a cognitive perspective.

3 A cognitive theory of reputation and gossip

Cognition works as a bond between the micro and the macro level, the so-called *micro–macro link* (Conte and Castelfranchi 1995): macro-social phenomena may emerge, unintentionally, from micro-elements and their interactions. In this view, social evaluations and networks derive from apparently autonomous social behaviors whose bases stand in the individual minds and in the relationships people are engaged in. In order to understand how social evaluations may affect social networks, we need to investigate how agents collect information about their peers, how they transform information in beliefs, and which goals drive agents' actions, both when they are looking for information and when they pass on that news. Answering these questions is pivotal to the understanding of the relationship between evaluations and networks.

When dealing with social evaluations, a preliminary distinction is needed. Following Conte and Paolucci (2002), we distinguish between *image* and *reputation*: the former refers to an evaluation regarding another agent's competence, behaviors and attitudes in which the source is clearly identified, whereas the latter designates an evaluation in which the source is missing. This difference is not inconsequential. In fact, spreading an inaccurate image, also involuntarily, exposes the evaluator to the risk of being reciprocated with false or inaccurate information or, even worse, of being ostracized. On the contrary, reputation is anonymous in itself, it circulates in the social network but its origin is unknown. Therefore, reputation is spread easier than image, and it is also more difficult to modify. Image, even when it is broadly transmitted, remains an evaluation coming from a specific and identified source that is responsible for it, whereas reputation becomes an intangible mark floating within the social network. In cognitive terms, an image is a belief about a target coming from an identified source ('According to me, John is a nice guy'), while reputation necessitates a belief about the target, but also the belief that other agents believe that there is a specific evaluation about a given target ('People say that John is a nice guy').

Gossip has a triadic structure in which we can distinguish:

- A *gossiper* is an agent who has the goal to spread information. Informing another agent can be the only purpose of gossip, or it can be instrumental to other goals (influencing the receiver, punishing the gossipee, promoting gossipee's image, enhancing groups' feelings, ostracize someone, etc.), more or less hidden.
- A *target* or third party: an agent whose behaviors, attitudes, choices and emotions are the topic of the communication. The target belongs to the same group of gossiper and receiver and she is judged according to the group's rules and habits. Topic of the gossip talk can be an evaluation about an agent, not necessarily a report on her behaviors.

- A *receiver* (or more than one): one or more agents chosen from the gossiper to be told about the target. Receivers belong to the same social network, sharing the same knowledge and values of gossiper and gossipee. Choosing the receiver is pivotal to achieve gossiper's goals: the receiver can be the actual target of communication or she can serve as a vehicle to reach the intended target.

Relationships among the three roles above are neither symmetrical nor equal. First, there is an asymmetry of power: the gossipee may find herself in a position where she is helpless and vulnerable. DeSousa (1994) challenged this view considering gossip as a subversive form of power used by the weak, in this case the gossiper, to protect herself against more conventional powers. Looking at the relationship between the gossiper and the receiver, we find another asymmetry: the former influences the latter, providing the receiver with new knowledge that may change her goals, beliefs and intentions.

Once we have described what happens at the lowest level, we can put forward some hypotheses about social evaluations and networks at the macro level. The cognitive model of reputation and gossip can be related to social networks under several respects, but here the focus will be on the difference between networks based on reputation and networks based on image. In fact, the distinction between image and reputation appears to be relevant also in terms of network's structure and enlargement potentiality, as we will see in the next section.

4 The model

4.1 Related work

Using *multi-agent based social simulation* (MABSS) we can isolate and test for single mechanisms and effects that in the real world cannot be unraveled, while, on the other hand, the simulation findings help to better understand the real dynamics of the given phenomenon (Gilbert, Doran 1994). This methodology allows to create models of artificial societies, i.e. sets of autonomous agents planning, acting and evaluating their actions in an artificial environment with given features. Before describing the Repnet model we want to briefly introduce some related results about the effects of image and reputation in an artificial market and in a simulated cluster of firms.

The difference between image-based and reputation-based networks has been proposed by Conte *et al.* (2008). They developed a computational system, RepAge (Sabater *et al.* 2006), implemented on an agent architecture and tested in an artificial market. In this setting, agents were allowed to exchange image only or image plus reputation. Their results showed that social networks based upon image perform more poorly than networks based upon reputation at least when partner selection is a common goal of the network members. This is mainly due to two distinct effects. On the one side, when only image, i.e. evaluation coming from an identified agent, is available, the presence of cheaters triggers a mutual defeat strategy, leading the system to collapse. This happens also with informational error: once agents find out that they received a false information, the informer cannot be trusted any more. On the other hand, results coming from simulation experiments with RepAge show that reputation-based networks are more flexible. Since evaluations were not immediately tested and they could not be attributed to any specific agent, the chain of retaliations was prevented and the network was more error tolerant.

Giardini *et al.* (2008a, 2008b) modeled firms as agents organized into three different layers. In their model, called SOCRATE, firms performed two kinds of exchanges: informational exchange and material exchange. Agents had to choose the best available supplier in order to deliver high quality products. When their known suppliers are not available, they relied on informers, i.e. other agents transmitting evaluations, in order to avoid the costs of direct interaction and to acquire useful information. Two different settings were tested: an image only setting, in which agents transmitted their own evaluations and retaliation against bad informers was allowed, and a reputation setting. In this latter condition, evaluations were not tested immediately and evaluator's identity was undisclosed, so that agents exchanged reputation without the fear of retaliation. The results showed that the quality of production was higher in the cluster with reputational information, compared to the cluster with image, for the same percentages of cheating. In other words, social information gave rise to different network configurations: image-based clusters performed better when cheaters were few but quality of production dramatically decreased for higher levels of cheating. Conversely, clusters in which agents exchanged reputation were more flexible and resistant when the number of cheaters was high, and the cluster's quality of production was only partially affected. In addition, agents in these networks explored the environment both spreading and using untested evaluations, so that innovation was promoted through the inclusion of new partners and informers.

4.2 The Repnet model and results

Encouraged from the results obtained with the SOCRATE model, we moved one step further and designed a very simple environment in order to investigate whether image and reputation produce different effects on a network of interacting agents. In this work, we were not interested in the effects on the economic performance and quality of production, therefore we simplified the structural aspects of the cluster, focusing our attention on the network's features.

Moving from the assumptions that gossip encourages the creation of ties among individuals and that these ties can be positively related to higher innovation spreading, we expected to see distinct networks emerging. More specifically, each time agents with complementary competences created a link they achieved innovative results and the resulting networks differed according to the kind of evaluation transmitted. The model was developed in NetLogo[3] (Wilensky 1999), and contained 60 agents interacting for a fixed time period of 500 ticks.

Agents were organized in three groups of 20 (A, B, C) and agents belonging to each group were endowed with a specific competence (a, b, c).

Competences were complementary

$$(Aa \subset Bb; Bb \subset Cc; Cc \subset Aa)$$

therefore agents needed to find partners with the matching competence in order to conclude a transaction. In order to find a partner, agents had to ask their peers about the ID of a complementary agent. Informers could either cheat their peers, suggesting them a non-complementary agent, or be honest and provide the name of a matching agent. When an agent failed to find a matching partner, the link between those two agents was not created and innovative knowledge was not transmitted.

We tested two experimental conditions, the one with image and the other with reputation. In the former condition agents retaliated against bad informers, excluding them from their network of contacts, whereas in the reputation condition retaliation was not allowed. Other things being equal, we expected to see two distinct network configurations, especially with regard to:

• image network: higher number of strong ties among honest agents;
• reputation network: higher number of weak ties among agents.

Results obtained from the NetLogo simulations have been loaded into ORA,[4] a software for network analysis developed at the Center for Computational Analysis of Social and Organizational Systems (CASOS), Carnegie Mellon University.

In Figure 19.1 the network of transactions among agents are showed. As expected, there are three separate networks of agents (A, B, C) interacting with their matching peers.

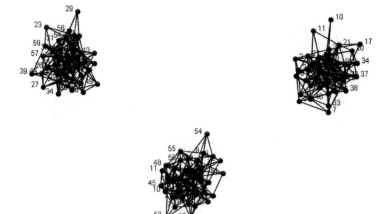

powered by ORA, CASOS Center @ CMU

Figure 19.1 Transaction network (at the end of the simulation there are three groups of mathing peers).

In Figure 19.2 we can see the Image Network in which only strong links (weight > 20) are displayed. Notwithstanding retaliation against bad informers, the network remained connected and the overall number of links, with weight ranging from 20 to 44, was 99. The network structure included both strong and weak ties, showing that, even when agents received bad information, they isolated cheaters and trusted honest partners, thus maintaining a connected network.

On the contrary, the Reputation Network (Figure 19.3) was poorly connected, compared to the former one, and its structure appeared to be really different. There was a single node serving as a hub and few other edges between nodes. The total number of ties with (weight > 20) was 42, thus there were fewer strong ties compared to the image condition but there were more weak ties (weight > 5): 644 links in the Reputation Network compared to the 574 in the Image Network. In this condition, bad informers were easily identified and the progressive lack of trust led to disgretation of many links.

5 Discussion and future work

Agents living in social systems are actually embedded in complex networks of relationships, having different sizes and configurations. The agents' position in a network can be more or less peripheral, and the agents' roles partially depend on the strength and the number of the nodes they are linked to.

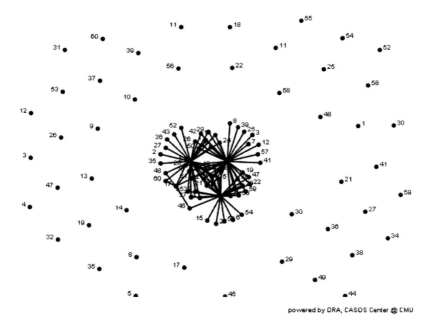

Figure 19.2 Image network with tie strength >20.

Figure 19.3 Reputation network with tie strength >20.

In this work we have explored through the techniques of MABSS whether and how social evaluations affected the networks' size and which types of tie among stylized firms emerged.

In our model, agents had to find a complementary partner in order to create an innovative product; on the other hand, they exchanged social evaluations in order to easily find matching peers. Although preliminary, the obtained results showed that the transmission of different kinds of information affected the network's structure and gave rise to two distinct configurations, lending further support to the idea that the way in which socially relevant information about one's peers are transmitted make a difference on the network structure, and then on the network's potentiality for innovation. Innovation requires passing information or knowledge from one agent to another and the possibility of knowing one's potential partners through others' evaluations is critical, especially for those environments like industrial districts, in which the interplay between economic dimensions and social relationships is very close and the informal connections among people working in the cluster play a relevant role.

This network of relationships can be enlarged or reduced thanks to evaluations about agents involved in the exchange of useful competences. Reputation and gossip are pivotal to the establishment of strong and weak ties, two kinds of equally important linkages. As Granovetter pointed out, both weak and strong ties are needed for social groups to work effectively: 'social systems lacking in weak ties will be fragmented and incoherent. New ideas will spread slowly, scientific endeavors will be handicapped, and subgroups separated by race, ethnicity, geography, or other characteristics will have difficulty reaching a modus vivendi' (1973: 202).

This work is far from providing a definitive answer on innovation, and future work will be concerned with testing the robustness of the network to different levels of cheating and improving the realism of the model, trying to enrich both the economic and the social aspects of the cluster. What this work actually highlights is the fact that unfolding the dynamics of innovation and the role of social evaluations is an ambitious travail that requires interdisciplinary research and a considerable effort but that can be really helpful to address the issue of how individuals and institutions are affected by social relationships.

Notes

1 This work has been partially supported by the European Community under the FP6 programme (eRep Project, contract number CIT5-028575), and by the Italian Ministry of University and Scientific Research under the FIRB programme (Socrate Project, contract number RBNE03Y338). We would like to thank Rosaria Conte for her helpful comments and all the people involved in the SOCRATE project for their comments on earlier draft of this chapter.
2 In this work the terms 'cluster' and 'district' are used as synonyms.
3 For further information about NetLogo, please visit the website. Online, available at: http://ccl.northwestern.edu/netlogo.
4 For further information about ORA, please visit the website. Online, available at: www.casos.cs.cmu.edu/projects/ora.

References

Albino, V., Carbonara, N. and Giannoccaro, I. (2006) 'Innovation in Industrial Districts: An Agent-Based Simulation Model', *International Journal of Production Economics*, 104, pp. 30–45.
Bathelt, H. (2008) 'Knowledge-based Clusters: Regional Multiplier Models and the Role of "Buzz" and "Pipelines"', in: C. Karlsson (ed.) *Handbook of Research on Cluster Theory*, Cheltenham, UK: Edward Elgar Publishing.
Becattini, G. (1990) 'The Marshallian Industrial District as a Socio-Economic Notion', in F. Pyke, G. Becattini and W. Sengenberger (eds) *Industrial Districts and Inter-firm Cooperation in Italy*, Geneva: IILS.
Ben-Ze'ev, A. (1994) 'The Vindication of Gossip', in: R.F. Goodman and A. Ben-Ze'ev (eds) *Good Gossip*, Westbrook Circle Lawrence: University Press of Kansas, pp. 11–23.
Biggiero, L. and Sevi, E. (2009) *Opportunism by Cheating and its Effects on Industry Profitability: The CIOPS Model, Computational and Mathematical Organization Theory*, 17 June. Online, available at: www.springerlink.com/content/k27562j5112041 62/?p=a42bb774845f45f6860b8bd57aee1fbf&pi=0.

Brenner, T. (2001) 'Simulating the Evolution of Localized Industrial Clusters – An Identification of the Basic Mechanism', *Journal of Artificial Societies and Social Simulation*, 4 (3). Online, available at: http://jass.soc.surrey.ac.uk/4/3/4.html.

Castelfranchi, C. (1998) 'Through the Minds of the Agents', *Journal of Artificial Societies and Social Simulation*, 1 (1). Online, available at: www.soc.surrey.ac.uk/JASSS/1/1/5.html.

Conte, R. and Castelfranchi, C. (1995) *Cognitive and Social Action*, London: UCL Press.

Conte, R. and Paolucci, M. (2002) *Reputation in Artificial Societies: Social Beliefs for Social Order*, Heidelberg: Springer.

Conte, R., Paolucci, M. and Sabater Mir, J. (2008) 'Reputation for Innovating Social Networks', *Advances in Complex Systems*, 11 (2), pp. 303–20.

Cowan, R. and Jonard, N. (2004) 'Network Structure and the Diffusion of Knowledge', *Journal of Economic Dynamics and Control*, 28 (8), pp. 1557–75.

Doran, J. and Gilbert, N. (1994) 'Simulating Societies: An Introduction', in: G.N. Gilbert and J. Doran (eds) *Simulating Societies: The Computer Simulation of Social Phenomena*, London: UCL Press.

Dunbar, R. (2001) 'Brains on Two Legs', in: F. De Waal (ed.) *Tree of Origin*, Cambridge, MA: Harvard University Press, pp. 173–92.

Fehr, E. and Gachter, S. (2000) 'Fairness and Retaliation: The Economics of Reciprocity', *Journal of Economic Perspectives*, 14 (3), pp. 159–81.

Fine, G. and Rosnow, R. (1978) 'Gossip, Gossipers, Gossiping', *Personality and Social Psychology Bulletin*, 4, pp. 161–8.

Giardini, F., Di Tosto, G. and Conte, R. (2008a) 'A Model for Simulating Reputation Dynamics in Industrial Districts', *Simulation Modelling Practice and Theory*, 16 (2), pp. 231–41.

Giardini, F., Di Tosto, G. and Conte, R. (2008b) 'Reputation and Economic Performance in Industrial Districts: Modelling Social Complexity through Multi-agent Systems', in: C. Cioffi Revilla and G. Deffuant (eds) *The Second World Congress on Social Simulation* (WCSS08), 14–17 July, Fairfax, VA: George Mason University.

Gilbert, N. (2004) *Agent-based Social Simulation: Dealing with Complexity*. Online, available at: www.soc.surrey.ac.uk/staff/ngilbert/ngpub/paper165_NG.pdf.

Gilbert, N., Pyka, A. and Ahrweiler, P. (2001) 'Innovation Networks – A Simulation Approach', *Journal of Artificial Societies and Social Simulation*, 4 (3). Online, available at: http://jass.soc.surrey.ac.uk/4/3/8.html.

Gintis, H., Smith, E.A. and Bowles, S. (2001) 'Costly Signaling and Cooperation', *Journal of Theoretical Biology*, 213 (1), pp. 103–19.

Gluckman, M. (1963) 'Gossip and Scandal', *Current Anthropology*, 4, pp. 307–16.

Granovetter, M. (1973) 'The Strength of Weak Ties', *American Journal of Sociology*, 78 (6), May, pp. 1360–80.

Granovetter, M. (1985) 'Economic Action and Social Structure: The Problem of Embeddedness', *American Journal of Sociology*, 91 (3), pp. 481–510.

Henrich, J., Boyd, R., Bowles, S., Camerer, C., Fehr, E., Gintis, H., McElreath, R., Alvard, M., Barr, A., Ensminger, J., Henrich, N.S., Hill, K., Gil-White, F., Gurven, M., Marlowe, F.W., Patton J.Q. and Tracer, D.,(2005) ' "Economic Man" in Cross-Cultural Perspective: Ethnography and Experiments from 15 Small-scale Societies', *Behavioral and Brain Sciences*, 28, pp. 795–855.

Hoen, A. (2001) *Clusters: Determinants and Effects*, The Hague: CPB Netherlands Bureau for Economic Policy Analysis.

Landry, R., Amara, N. and Lamari, M. (2002) 'Does Social Capital Determine Innovation? To what Extent?' *Technological Forecasting and Social Change*, 69 (7), pp. 681–701.

Maillat, D., Lecoq, B., Nemeti, F. and Pfister, M. (1995) 'Technology District and Innovation: The Case of the Swiss Jura', *Archives of Regional Studies*, 29 (3), pp. 251–63.

Miceli, M. and Castelfranchi, C. (1989) 'A Cognitive Approach to Values', *Journal for the Theory of Social Behaviour*, 19, pp. 169–93.

Panchanathan, K. and Boyd, R. (2004) 'Indirect Reciprocity can Stabilize Cooperation without the Second-order Free rider Problem', *Nature*, 432 (7016), pp. 499–502.

Pittaway, L., Robertson, M., Kamal, M. and Denyer, D. (2004) *Networking and Innovation: A Systematic Review of the Evidence*, Lancaster University Management School Working Paper 2004/016.

Porter, M. (1998) 'Clusters and the New Economics of Competition', *Harvard Business Review*, 76, pp. 77–90.

Sabater, J., Paolucci, M. and Conte, R. (2006) 'RepAge: Reputation and Image among Limited Autonomous Partners', *Journal of Artificial Societies and Social Simulation*, 9, (2). Online, available at: http://jasss.soc.surrey.ac.uk/9/2/3.html.

Sommerfeld, R.D., Krambeck, H., Semmann, D. and Milinski, M. (2007) 'Gossip as an Alternative for Direct Observation in Games of Indirect Reciprocity', *Proceedings of the National Academy of Sciences of the United States of America*, 104, pp. 17435–40.

Sousa, R. de (1994) 'In Praise of Gossip: Indiscretion as a Saintly Virtue', in: R.F. Goodman and A. Ben-Ze'ev (eds) *Good Gossip*, Lawrence, KS: University Press of Kansas, pp. 25–33.

Squazzoni, F. and Boero, R. (2002) 'Economic Performance, Inter-Firm Relations and Local Institutional Engineering in a Computational Prototype of Industrial Districts', *Journal of Artificial Societies and Social Simulation*, 5 (1), p. 1. Online, available at: http://jasss.soc.surrey.ac.uk/5/1/1.html.

Wilensky, U. (1999). NetLogo, Center for Connected Learning and Computer-Based Modeling, Evanston, IL: Northwestern University. Online, available at: http://ccl.northwestern.edu/netlogo.

Wilson, D.S. and Wilczynski, C., Wells, A. and Weiser, L. (2000) 'Gossip and other Aspects of Language as Group-level Adaptations', in: C. Heyes and L. Huber (eds) *The Evolution of Cognition*, Cambridge, MA: MIT Press.

Zhang, J. (2003) 'Growing Silicon Valley on a Landscape: An Agent-based Approach to High-tech Industrial Clusters', *Journal of Evolutionary Economics*, 13, pp. 529–49.

20 Complexity-friendly policy modelling

Flaminio Squazzoni and Riccardo Boero

Today, innovation has gained momentum to the detriment of history, stability, and tradition (Lane *et al.* 2009). The increasing social demand for innovation in the globalization era, both at a firm and a national system level, contrasts with the complex, multifaceted, systemic and unpredictable nature of innovation processes. Indeed, innovation is created by a network of dispersed interactions between many heterogeneous agents (i.e. firms, consumers, venture capitalists, research institutions and institutional agencies) and it largely depends on a particular and contextual combination of different aspects (i.e. social, economic, institutional, and cultural aspects). It is also profoundly nonlinear and is subject to path dependence.

It is therefore difficult to predict and impossible to plan. Since it often materializes into 'out-of-equilibrium' aggregative dynamics, traditional equilibrium solution-oriented models and standard statistical methods used for policy forecasting and decisions find it difficult to capture its essence. The unpredictability of innovation can therefore cause a sense of frustration both in scientists and policy makers, given the pressurizing social expectations from taxpayers and public opinion that science-informed policy makers should do something to improve either business innovative capability, or the local/regional/national system. This is because innovation is the most crucial component of the competitive advantage of firms, regions and countries.

Unfortunately, our present situation provides many examples of how traditional policy tools fail to deal with the complexity of socio-economic systems especially for innovation (e.g. Rossi and Russo 2009). There are two kinds of deficiency. First, standard equilibrium solution-informed and standard statistical models are poorly equipped to understand the micro/local details that make a real difference for social-economic outcomes. As a matter of fact, there have been many puzzling outcomes that simply do not appear on the policy makers' radar screen given that they lay outside the domain of traditional science or even the law of large numbers (Moss 2002, Miller and Page 2007).

Second, traditional forecast-oriented models which prescribe *ex ante* solutions and recipes dramatically underestimate the entire process of policies, including the reaction of agents to policy decisions, the aggregate effect of their interactions and their systemic consequences on large spatial-temporal scales. Standard policy making models consider agents as atomized entities possessing rational expectations which individually react to a set of incentives, do not consider interactions or the mutual influence between agents and seem to take place 'off-line' and outside the particular system involved (Finch and Orillard 2004: 5).

This chapter aims first to question the state-of-the-art of policy making, in a complexity perspective. There is evidence that policy making is currently not equipped to tackle the challenge of the complexity of the innovation process. So, together with its relevant impact on scientific endeavour, the complexity perspective needs a new approach to policy issues.

The second aim is to provide an overview on how agent-based models (ABMs) can change policy making in a more complexity-friendly perspective. While standard policy making is an attempt to reduce or eliminate complexity, ABMs allow us to understand and 'harness' complexity (e.g. Axelrod and Cohen 1999). In a certain sense, when complexity is presumed and not eliminated from the policy makers' radar, the quest for innovation policy and the quest for innovation in policy become essentially the same thing.

The structure of this chapter is as follows: In the first part, we illustrate the background and present the complexity perspective and its challenge to conventional policy making. The second part introduces applications of ABMs for policy purposes, by categorizing these approaches and illustrating some examples. We have identified two types of policy ABMs, i.e. 'prescriptive models' and 'participatory modelling', discussed their constituencies and provided two examples. Finally, we emphasize the innovative flavour that ABMs can bring to the policy arena.

The challenge of complexity for policy making

Durlauf (1998) summarized the main ideas of complexity which make understanding and prediction of socio-economic systems so challenging from a standard economic theory viewpoint. These are: interaction and positive feedback, increasing returns, path dependency, nonlinearity and evolving processes. All these make the toolkit of the conventional economist, i.e. rational expectations, representative agent and equilibrium solution, a blank firing gun.

The same inspiration is found in Axelrod and Cohen (1999), who insisted on the impact of complexity for understanding, explanation and prediction. They also outlined a new perspective on decision making to show that the complexity of socio-economic systems cannot be reduced or eliminated but rather should be 'harnessed'. More recently, Lane *et al.* (2009) focused on innovation in complex social systems, suggesting a new theoretical perspective which links agents, relations, and artifacts that explain the continuous emergence of innovative artifacts in the economy and society which also makes relevant implications at a policy making level.

The complexity perspective is in fact full of important implications for policy making. First, when the role of interaction between various and heterogeneous agents is seriously taken into account, it is expected that: (*a*) 'multiple types of internally consistent aggregate behaviour' can easily occur which are difficult to predict; (*b*) socio-economic systems can be stuck in 'undesirable steady states', such as social inequality and segregation, prevalence of sub-optimal institutions or anti-social conventions; (*c*) the consequences of any given policy depend critically on the nature of such interdependences; (*d*) the effect of different policies might have a non-linear nature, 'rendering history a poor guide to evaluating policy effectiveness'; (*e*) the identification of effective policy options depends less on abstract conventional theoretical models and more on empirical studies that look at the fine details of the social context (Durlauf 1998).

Second, agents are not usually viewed as fully rational utility maximizers who behave independently of each other, but rather as adaptive agents who are context dependent and follow heterogeneous threshold preferences. It is therefore difficult to predict how small changes in incentives can cause a given behaviour rather than the opposite one at a micro level. Hence aggregate outcomes are highly sensitive to unpredictable details that are impossible to understand within *ex ante* statistical data and macro analyses (Moss 2002).

Innovation makes these peculiarities of complex socio-economic systems even more important so understanding, prediction and intervention are really a complicated job lot. Taking into account complexity means that the prediction of extreme events or complex social outcomes by macro and external observers is infeasible and in-depth models are needed to understand the generative process responsible for any particular event or outcome, no matter whether this analytical effort makes a perfect prediction or not (Moss 2002).

What makes prediction so difficult in complex systems and therefore policy forecasts and *ex ante* decisions, is that action at the micro level does not add up in a simple, linear and system-wide manner. So, particular details can have dramatic consequences on future events. Moreover, since innovation dramatically increases the relevance of social reflexivity, uncertainty and the imperfect knowledge of agents, expectations on future events and social outcomes might have a crucial and heterogeneous influence on agents' behaviour.

The consequence is that, even if an agent's behaviour is in principle predictable at a given time, concrete interactions among multiple agents can cause unexpected outcomes, such as self-fulfilling prophecies and local optimum traps (e.g. Frydman and Goldberg 2007). Therefore, the challenge for policy in a complex world is not to predict the future state of a given system, but to understand the system's properties and dissect its generative mechanisms and processes, so that policy decisions can be better informed and embedded within the system's behaviour, thus becoming part of it.

Furthermore, if the purpose of a policy maker is not simply to learn but also to negotiate and manage solutions and problems are un-structured, i.e. there is little consensus about values and goals, policy analysis is expected to be participative to reduce information asymmetries, increase mutual learning and leave room for grounded practical and pragmatic innovation (e.g. Duijn *et al.* 2003). As mentioned by Byrne (1998: 139–56) in his review on the impact of complexity on urban planning and governance, the complexity perspective can provide policy makers with a framework to understand policy as 'reflexive social action' which, once based on the participation of stakeholders, makes unstructured problems more manageable.

In conclusion, complexity challenges traditional policy making especially on two points. First, it focuses more on the pivotal role of modelling and understanding rather than on prediction, forecasting and planning. In the latter the most important part is the capacity of policy makers to understand a system's properties. As Axelrod and Cohen (1999) argued, 'while complex systems may be hard to predict, they may also have a good deal of structure and permit improvement by thoughtful intervention'.

Second, by emphasizing that social systems are composed of heterogeneous human agents, which are 'reflexive' and influenced by the macro outcomes they cause, the complexity perspective allows us to reframe policy as a component of the behaviour of a particular system. Once viewed as something that is embedded within the system, rather than taking place before and off-line, policy starts to be practiced as a crucial component which interacts with other components in a constitutive process.

The ABMs as a new frontier for innovation policy

An ABM is a computational model where a given outcome is understood and modelled as the result of the interaction among heterogeneous, adaptive and localized agents (e.g. Epstein and Axtell 1996, Gilbert 2008). It is the standard technique around which new promising and overlapping research fields, such as computational social sciences, agent-based computational economics, and social simulation, now revolve (e.g. respectively Epstein 2006, Tesfatsion and Judd 2006, Edmonds *et al.* 2007).

Although most agent-based model makers privilege analytical and theoretical enquiry, there are many examples of ABMs for policy purposes. We can distinguish two types, i.e. 'prescriptive models' and 'participatory modelling'. The inspiration of the former comes from positivism, i.e. the idea that since science can achieve causal explanations and objective knowledge about the world, one of the most important social functions of the scientist is to make a system's behaviour understandable and to inform policy makers with appropriate prescriptions and recommendations. The latter are more inclined towards subjectivism and social constructivism, i.e. the idea that researchers, policy makers and stakeholders should participate in a mutual learning process, so that policy making is completely embedded in a system, becoming a constitutive part of it (Bousquet and Trébuil 2005: 4).

'Prescriptive models'

In this case, ABMs aim to find hidden mechanisms that can determine outcomes, by using theories and scientific evidence to provide implications and solutions before policy making takes place. The goal of the modeller is to formulate some *ex ante* scenario analysis that can help policy makers to understand a system's behaviour and identify the best policy option, given their objective-functions and constraints. The motto could be 'learn before deciding'.

There are many examples of ABMs of this type. They can be made to support *prospective scenario analyses*, such as in Rouchier and Thoyer (2003), where different EU decision making procedures are tested to promote an agreement between countries e.g. on the formation of a market of genetically modified products. They can also be *retrospective analyses*, such as the well-known and pioneering example of Lansing and Kremer (1993). In this case, a model of the Bali irrigation system was made that integrated social and natural aspects to show that past institutions, strongly defended by local farmers, were capable of generating a more sustainable self-organized development better than new mass agriculture technologies, inserted some years previously under Green Revolution principles. The ability of the model to reveal hidden socio-cultural mechanisms that regulated the co-evolutionary link of environment, agriculture and social structure eventually convinced policy makers to give up the Green Revolution principles and to contextualize the policy into the peculiar features of Bali.

A brilliant example in the field of innovation policy is presented in Rosewell *et al.* (2008). The authors developed an empirical data-based ABM to study the relevance of links between firms for innovation in certain strategic sectors of Manchester City region, affected by a period of deep crisis and restructuring. The purpose of the analysis was to make the system understandable for policy makers, draw informative implications and provide clear solutions for appropriate policy support for the regional growth. The analysis was based on a model that combines quantitative data from a large survey on 1,500 firms, qualitative evidence drawn from interviews conducted on businesses, trade associations and agencies, and an on-line panel survey on business networks.

Organizations (i.e. firms and public agencies) were the model's agents and represented each relevant Manchester City region industrial sector, i.e. engineering and textiles, creative/communication, financial/professional services and life sciences. They had three main parameters: innovation, absorptive/secrecy, and network density indexes. Agents were able to generate an innovation, had a threshold function that determines their propensity to adopt innovations of others and were located in empirically traced networks so that parameters were tuned to empirical data. Patterns from each sector were analysed and compared.

The evidence showed that innovation was more probable and spread more easily as a cascade among firms networked in supply chains, than across localized clusters or between groups of competitors. Links and connections that criss-cross complementary and heterogeneous firms matter more than anything else. The policy implication is that the conventional approach of policy makers to networking, i.e. promoting sector clusters, did not guarantee the innovativeness of Manchester City region. The analysis suggested: (*a*) incorporating as many as possible firms with no trading links into the city region, (*b*) providing strong support to create business networks among firms, though there was no chance for policy makers to affect what is going on there; (*c*) not focusing simply on standard knowledge input (e.g. incentives for R&D), but applying a creative scheme to stimulate new relationships between creative and non-creative businesses.

'Participatory modelling'

The second type is 'participatory modelling', where policy models are designed in collaboration with stakeholders as knowledge bearers in model building, calibration, and validation. In this case, models are also used as a means to promote discussion, negotiation and to define decisions in a participative bottom-up way (e.g. Ramanath and Gilbert 2004, Moss and Edmonds 2005, Nguyen-Duc and Drogoul 2007).

The idea is that since policy analysis is aimed at targeting existing social systems,

> Stakeholders and independent domain experts can provide descriptions of the goals and actions of the relevant actors as well as patterns and modes of interaction among them. They can also evaluate the plausibility of the models designed to incorporate those descriptions in the software code that constitute each agent ... Stakeholder participation entails not only validation by domain experts but also a more organic process of development of the models in which the stakeholders both explicate and refine their understanding of the target systems and use the models to investigate alternative policy or other strategic options.
>
> (Moss 2002: 7273)

Unlike modelling from pre-established theoretical constructs, here models are devised on the basis of observation and empirical validation where informative description of generative processes can be made. The point is that, given the irreducible sources of complexity, 'confrontation between field and modelling processes has to be permanent because of openness and uncertainty features of these systems' (Barreteau *et al.* 2003). In this case, the goal of the modeller is not to put scientific theories and models into practice, but to reduce the information asymmetry that is supposed to penalize any scientific observation. This is done by building models together with stakeholders involved in the system to approximate empirical reality in due detail (e.g. Downing *et al.* 2003). The criteria followed for modelling the target are 'descriptive adequacy' rather than 'analytical tractability', and 'concretization' rather than 'isolation' (Moss 2008). As a rich representation of a system, the resulting model can also be used to support highly contextualized decision processes (Janssen 2002). In this case, the motto could be 'stay on the ground and design with people'.

There are several examples of ABMs of the second type, in particular in the field of common-pool resource and land use management where social and ecological aspects are both taken in account (e.g. Costanza and Ruth 1998, Jannsen 2002, Becu *et al.* 2003, D'Aquino *et al.* 2003). Although not directly addressed to socio-economic innovation issues, this field shows the advantage of this perspective at its best.

A brilliant example of this is Etienne *et al.* (2003), where an ABM is used to provide a scenario analysis on different strategies of natural resource management in Causse Méjan, a limestone plateau in southern France dominated by a rare grassland ecosystem endangered by pine invasion. The model helps stakeholders to discuss alternative long-term management strategies for sheep farms and woodlands, by designing simulations with different space scales and long-term dynamics.

The building blocks of the model were developed step-by-step through the direct involvement of stakeholders and the recourse to detailed empirical data. The involvement of the stakeholders throughout the modelling process was crucial for; (*a*) deriving both information of their resources and the entities they are used to manage; (*b*) having their viewpoint and opinions; (*c*) formulating the indicators to monitor the dynamics of the system and (*d*) formulating and discussing scenarios and prospective management rules to tackle the pine-encroachment problem. The involvement started from the local and specific aspects of the situation and progressively moved towards global collective issues.

In this model, there are three types of agents, namely sheep farmers, forestry workers and conservationists, located on a realistic landscape. Each agent is represented by data-driven parameters. For example, sheep farmers are represented by empirical data on their production system, grazing pressure on ranges, environmental awareness, chain mower availability, attitude towards pine encroachment, year of retirement and most probable production system after retirement. Their grazing management strategy is the result of their production system (meat or milk, intensive or extensive), grazing pressure, distance to shed, and pine canopy cover on rangelands. Their behaviour, when faced with pine encroachment is affected by labour time available to cut pine trees, the amount of land tenure tax, the amount of timber harvested and range productivity.

Thanks to the simulation of eight different scenarios, elaborated according to stakeholders' viewpoints and objectives and to the intuitive space visualization of long-term outcomes, the model allows stakeholders to achieve a collective agreement on sustainable development paths based on cooperation. This is possible because the model allows an explicit reconstruction of the heterogeneous (conflicting) viewpoints through which agents perceive the natural resource, the socio-ecological system, their objectives and other agents.

The representation of these viewpoints during the simulation process makes it possible to inform stakeholders of the issues that divide them and their common dependence upon a solution to the pine encroachment problem. Simulations help them to understand and negotiate the acceptability of possible systemic trends caused by their behaviour and interaction. As the authors outline, this example shows that 'beyond the classical use of modelling as a decision support to control a system, agent-based models are also powerful supports to adaptive learning processes' (Etienne *et al.* 2003: 8.3).

In conclusion, both types of models mentioned above are of paramount importance so that policy makers do not follow common sense and intuitive models, on one hand, or very abstracted 'off-line' macro predictions, on the other.

Indeed, ABMs are the only technique available today to formalize models based on micro foundations, such as agents' beliefs and behaviour and social interactions, all aspects that we know are of a certain importance to understand macro outcomes, also for policy purposes (e.g. Epstein 2006). In both examples, the added value of modelling is not for forecasting but on the empirical accuracy that helps harness the complexity of reality. A stylization could be that the former ABMs are more suitable when policy makers need to *learn from science* about the complexity of systems where their decision is needed, while the latter ABMs are more suitable when policy makers need to *find and negotiate certain concrete* ad hoc *solutions*, so that policy becomes part of a complex process of management that is internal to the system itself.

Concluding remarks

Our aim was to provide an overview of the complexity of policy modelling, with special attention to the relevance of ABMs. The added value of ABMs is that they combine the advantages of formal modelling and the possibility to achieve a realistic picture of any particular systems, where heterogeneous agents and interactions are explicitly taken into account and space-time outcomes are simulated and visualized. By going in deep at the micro level, these models allow us to understand the detailed mechanisms responsible for the macro behaviour of systems and to support realistic scenario analysis. This is why ABMs are fundamental for a complexity-friendly perspective on policy modelling.

Having said this, complexity perspective also requires 'innovating' innovation policy. First, although a large amount of taxpayers' money has been channelled into innovation policy recently, there is widespread opinion that it requires serious reconsideration concerning approaches, methods, and tools to support and evaluate policy making (e.g. Lane *et al.* 2009, Rossi and Russo 2009).

If the complexity of socio-economic systems is to be taken seriously, a new perspective is needed that calls for: (*a*) a new attitude by scientists and policy makers to emphasize modelling, understanding and the explanation of systemic processes rather than precise predictions and forecasts of a system's future trajectories; (*b*) new means to understand and manage the interaction processes between agents involved in a system and policies, and include agents' interreactions as one of the most challenging source of unpredictability of outcomes; (*c*) consequently, a new conceptualization of policy as part of a system itself; (*d*) given the relevance of time and the mechanisms of innovation diffusion, the inclusion of different time scales to evaluate the policy impact, out of any shorttermism (e.g. Rossi and Russo 2009).

Pushed to an extreme, the complexity perspective allows us to reframe policy as a process that takes place within a system, is a component of behaviour of the system itself, interacts with other components and consequently needs to be flexible and adaptive to the behaviour of the other system's components. Translating this approach into concrete examples, as well as building and sharing new best practices to form a new toolkit for the twenty-first-century policy-oriented scientists, is in our view the complexity-friendly frontier of policy modelling.

References

d'Aquino, P., Le Page, C., Bousquet, F. and Bah, A. (2003) 'Using Self-Designed Role-playing Games and a Multi-agent System to Empower a Local Decision-making Process for Land Use Management: The SelfCormas Experiment in Senegal', *Journal of Artificial Societies and Social Simulation*, 6 (3). Online, available at: http://jasss.soc.surrey.ac.uk/6/3/5.html.

Axelrod, R. and Cohen, M.D. (1999) *Harnessing Complexity. Organizational Implications of a Scientific Frontier*, New York: Free Press.

Barreteau, O., Antona, M., D'Aquino, P., Aubert, S., Boissau, S., Bousquet, F., Daré, W., Etienne, M., Le Page, C., Mathevet, R., Trébuil, G. and Weber, J. (2003) 'Our Companion Modelling', *Journal of Artificial Societies and Social Simulation*, 6 (1). Online, available at: http://jasss.soc.surrey.ac.uk/6/2/1.html.

Becu, N., Bousquet, F., Barreteau, O., Perez, P. and Walker, A. (2003) 'A Methodology for Eliciting and Modelling Stakeholders' Representations with Agent-based Modelling', in: Hales, D., Edmonds, B., Norling, E. and Rouchier, J. (eds) *Multi-agent-based Simulation III*, Berlin, Heidelberg: Springer Verlag.

Bousquet, F. and Trébuil, G. (2005) 'Introduction to Companion Modeling and Multi-agent Systems for Integrated Natural Resource Management in Asia', in: Bousquet, F., Trébuil, G. and Hardy, B. (eds) *Companion Modeling and Multi-agent Systems for Integrated Natural Resource Management in Asia*, Los Banos, Philippines: Metro Manila, IRRI, pp. 1–17.

Byrne, D. (1998) *Complexity Theory and the Social Sciences. An Introduction*, London: Routledge.

Costanza, R. and Ruth, M. (1998) 'Using Dynamic Modelling to Scope Environmental Problems and Build Consensus', *Environmental Management*, 22, pp. 183–95.

Downing, T.E., Moss, S. and Pahl-Wostl, C. (2003) 'Understanding Climate Policy using Participatory Agent-based Social Simulation', in: Moss, S. and Davidsson, P. (ed.) *Multi Agent Based Social Simulation*, Berlin, Heidelberg: Springer, pp. 198–213.

Duijn, M., Immers, L.H., Waaldijk, F.A. and Stoelhorst, H.J. (2003) 'Gaming Approach Route 26: A Combination of Computer Simulation, Design Tools and Social Interaction', *Journal of Artificial Societies and Social Simulation*, 6 (3). Online, available at: http://jasss.soc.surrey.ac.uk/6/3/7.html.

Durlauf, S.N. (1998) 'What Should Policymakers Know About Economic Complexity?', *Washington Quarterly*, 21 (1), winter, pp. 157–65.

Edmonds, B., Hernandez, C.I. and Troitzsch, K.G. (2007) (eds) *Social Simulation: Technologies, Advances and New Discoveries*, New York: IGI Global.

Epstein, J.M. (2006) *Generative Social Science. Studies in Agent-based Computational Modeling*, Princeton: Princeton University Press.

Epstein, J.M. and Axtell, R. (1996) *Growing Artificial Societies: Social Science from the Bottom Up*, Cambridge, MA: MIT Press.

Etienne, M., Le Page, C. and Cohen, M. (2003) 'A Step-by-step Approach to Building Land Management Scenarios Based on Multiple Viewpoints on Multi-agent System Simulation', *Journal of Artificial Societies and Social Simulation*, 6 (2). Online, available at: http://jasss.soc.surrey.ac.uk/6/2/2.html.

Finch, J. and Orillard, M. (2004) 'Introduction: The Scope of Complexity and its Implications for Policy', in (eds) *Complexity and the Economy. Implications for Economic Policy*, Cheltenham, UK, and Northampton, MA: Edward Elgar.

Frydman, R. and Goldberg, M.D. (2007) *Imperfect Knowledge Economics: Exchange Rates and Risk*, Princeton: Princeton University Press.

Gilbert, N. (2008) *Agent-based Models*, London: Sage Publications.

Janssen, M. (ed.) (2002) *Complexity and Ecosystem Management: The Theory and Practice of Multi-agent Approaches*, Cheltenham, UK, and Northampton, MA: Edward Elgar.

Lane, D., Pumian, D. and van der Leeuw, S. (2009a) 'Conclusion', in: Lane, D., Pumain, D., van der Leeuw, S. and West, G. (eds), *Complexity Perspectives in Innovation and Social Change*, Berlin, Heidelberg: Springer Verlag.

Lane, D., Maxfield, R., Read, D. and van der Leeuw, S. (2009b) 'From Population to Organization Theory', in: Lane, D., Pumain, D., van der Leeuw, S. and West, G. (eds) *Complexity Perspectives in Innovation and Social Change*, Berlin, Heidelberg: Springer Verlag.

Lansing, J.S. and Kremer, J.N. (1993) 'Emergent Properties of Balinese Water Temple Networks: Coadaptation on a Rugged Fitness Landscape', *American Anthropologist*, 95 (1), pp. 97–114.

Miller, J.H. and Page, S.E. (2007) *Complex Adaptive System. An Introduction to Computational Models of Social Life*, Princeton: Princeton University Press.

Moss, S. (2002) 'Policy Analysis from First Principles', *PNAS*, 99 (3), 14 May, pp. 7267–74.

Moss, S. (2008) 'Alternative Approaches to the Empirical Validation of Agent-based Models', *Journal of Artificial Societies and Social Simulation*, 11 (1). Online, available at: http://jasss.soc.surrey.ac.uk/11/1/5.html.

Moss, S. and Edmonds, B. (2005) 'Sociology and Simulation: Statistical and Qualitative Cross-validation', *American Journal of Sociology*, 110 (4), pp. 1095–131.

Nguyen-Duc, M. and Drogoul, A. (2007) 'Using Computational Models to Design Participatory Social Simulations', *Journal of Artificial Societies and Social Simulation*, 10 (4). Online, available at: http://jasss.soc.surrey.ac.uk/10/4/5.html.

Ramanath, A.M. and Gilbert, N. (2004) 'The Design of Participatory Agent-based Social Simulations, *Journal of Artificial Societies and Social Simulation*, 7 (4). Online, available at: http://jasss.soc.surrey.ac.uk/7/4/1.html.

Rosewell, B., Wiltshire, G., Owens, P., Naylor, R. and Metcalfe, S. (2008) 'Innovation, Trade and Connectivity', *Manchester Independent Economic Review*. Online, available at: www.manchester-review.org.uk/projects/view/?id=719.

Rossi, F. and Russo, M. (2009) 'Innovation Policy: Levels and Levers', in: Lane, D., Pumain, D., van der Leeuw, S. and West, G. (eds) *Complexity Perspectives in Innovation and Social Change*, Berlin, Heidelberg: Springer Verlag.

Rouchier, J. and Thoyer, S. (2003) 'Modelling a European Decision Making Process with Heterogeneous Public Opinion and Lobbying: The Case of the Authorization Procedure for Placing Genetically Modified Organisms on Market', in: Hales, D., Edmonds, B., Norling, E. and Rouchier, J. (eds) *Multi-Agent-based Simulation III*, Berlin, Heidelberg: Springer Verlag.

Testfatsion, L. and Judd, K.L. (eds) (2006) *Handbook of Computational Economics: Agent-based Computational Economics: Volume II*, Amsterdam: North Holland.

21 The agent-based NEMO model (SKEIN)

Simulating European Framework Programmes

Ramon Scholz, Terhi Nokkala, Petra Ahrweiler,
Andreas Pyka and Nigel Gilbert

1 Understanding collaborative European R&D efforts

Since 1984, the European Union has established seven Framework Programmes for research and technological development. The principal objective of the Framework Programmes is to contribute to the 'creation of a genuine European Research Area (ERA) by fostering more integration and coordination in Europe's previously fragmented research sector' (Communication from the Commission no date). The Framework Programmes aim to create durable cooperation links and better integration and coordination of research efforts across the European Union by supporting cooperation between the organizations producing and exploiting knowledge, namely universities, industry, research centers and public authorities (see e.g. European Commission CORDIS web page).

The Framework Programmes are organized into a number of sub-programs, under which specific calls are published in order to attract applications for collaborative research projects. These collaborative projects have to meet the main criteria of scientific excellence and cross-border collaboration and must in general be so complex as to require joint efforts at the European level. The collaborative projects may be funded through different funding instruments, catering for different types of research with different types and sizes of consortia.

The FP6-funded research project NEMO – Network Models, Governance and R&D collaboration networks – investigates the interaction between the political governance, structures and functions of R&D collaboration networks that are induced by policy programs, in particular the networks that have emerged in the European Framework Programmes (FPs). The goal is to identify ways to evaluate efficient network structures for R&D collaboration networks and to develop policy designs for their creation. A better understanding of the structures and dynamics of networks allows for improvements to the efficacy of network-based R&D policy instruments, and thus promotes the knowledge-based economy in Europe. R&D collaboration networks are considered to be an advantageous organizational form for knowledge creation and diffusion processes, and allow for the exploration and exploitation of the dispersed knowledge in the European Union.

In NEMO's interdisciplinary approach, the formation of R&D collaboration networks is investigated with numerical models and computer simulations. For empirical calibration of these models, data on real-world R&D collaborations in the FPs, qualitative empirical insight into the rules and processes of consortium formation, and theoretical considerations constitute the common conceptual framework.

In this chapter, we introduce an agent-based model SKEIN (Simulating Knowledge dynamics in EU-funded Innovation Networks) aimed at simulating the emergence of collaboration networks and knowledge production in EU-funded R&D collaboration projects. The model benefits from the insights of innovation sociology and innovation economics (Antonelli 2005, Caloghirou *et al.* 2003, Goyal and Moraga-Gonzalez 2001, Malerba 2004, Powell *et al.* 2005). It builds partly on previous work (e.g. Ahrweiler *et al.* 2004, Gilbert *et al.* 2001, 2007, Pyka *et al.* 2007), which focused on innovation networks among firms in knowledge-intensive industries. Agent-based models (ABMs) have the advantage of allowing for a representation of R&D collaboration networks that is very close to real world collaborations (e.g. Axtell 2000, Deichsel and Pyka 2009). Agent-based models are able to address the complexity of innovation and knowledge production processes in a manner not captured by more traditional research approaches. Knowledge production takes place in largely self-organizing networks involving many different types of actors, embedded in various institutional frameworks. ABMs allow us to detect the emergence and development of various collaboration patterns while providing an opportunity to modify the settings in which these processes take place.

The aim of the SKEIN model is to study the impact of different sets of policy rules, or incentive-led R&D policy, on the emerging network structures. This chapter will therefore try to answer the following research questions:

1   Does incentive-led R&D policy influence the structure of the science network?
2   Does incentive-led R&D policy influence the outcomes of research?

The agents in our model are the various actors engaged in knowledge production, such as university and research institutes departments, and research divisions of firms. The agents are modeled as applying for research projects following the different calls for projects by the European Commission to finance their collaborative research efforts. In the following sections we first describe the empirical research that facilitated the construction of the model. We then describe the model and finally test it with an experiment relevant to European Union R&D policy.

2 Empirical contributions to construction of the SKEIN model

The model is informed by empirical analysis of, on the one hand, the external governance rules set by the European Commission, and on the other hand, the internal collaboration rules defining the consortium formation and knowledge production processes. The external governance rules are explicit rules commanding compliance. Internal collaboration rules comprise norms and patterns of behavior describing either what participants have done or what they would do under certain circumstances pertaining to the different stages of the collaboration process.

In order to operationalize the EU influence on the development of complex network structures we analyzed European Union documents related to the Framework Programmes. These documents contain the European Commission rules guiding the call and project selection procedures. The rules concerning Framework Programmes 1–7 were coded into the so-called EURuleD database. In the EURuleD database, detailed empirical data on the different rule sets steering each Framework Programme were obtained, sorted, and analyzed. The external rules from the database were implemented in the SKEIN model by means of a 'scenario file', in which default parameter values were preset for the model, and the intervals from which valid values can be taken were defined. Additionally, a graphic user interface between the EURuleD database and the SKEIN model was created to allow a calibration to real world settings and to enable the users of the model to run experiments with different types of policy scenarios. This they could do through the user interface by changing the parameter values for the governance rules, such as the size and duration of projects. This makes it possible to create scenarios for all different Framework Programmes, as well as for certain European instruments within them. In addition the simulation of the history of all FPs can be easily executed (Kruckenberg *et al.* 2008).

In contrast to the external governance rules, the internal collaboration rules guiding consortium formation and collaboration within project consortia have been integrated into the basic SKEIN model. The internal collaborations rules were devised from an analysis of three sets of quantitative and qualitative data from the fifth and sixth Framework Programmes. Quantitative data from a representative survey of FP5 participants at the researcher level provided a broad perspective of the patterns of collaboration and joint knowledge production. The data were derived from a 2007 on-line survey carried out by the Austrian Institute of Technology (formerly Austrian Research Centers GmbH). The dataset included 1,686 valid responses, representing 3 percent of all relevant FP5 participants, and covering 1,089 (12 percent of all relevant) FP5 projects. Two primarily qualitative datasets from FP6 included 22 in-depth qualitative interviews with researchers in seven projects in the EU's New and Emerging Science and Technology program (NEST), and five case-studies of Integrated Projects (IPs) in three thematic areas: information society technology, sustainable development, and aerospace. The qualitative data allowed us to concentrate on the in-depth analysis of individual collaboration histories, motivations for cooperation and non-cooperation with specific potential partners, and descriptions of interaction within the projects. These three different dataset were used to construct 35 rules pertaining to six stages of the collaboration process: consortium formation; proposal submission; funding decision; task division; intra-project collaboration structures, processes and context; and finally future collaboration and its framework conditions. These were then operationalized as parameters of the SKEIN model (Nokkala *et al.* 2008, Nokkala 2009).

3 The SKEIN model

In NEMO's multi-agent simulation, SKEIN, the agents are research divisions of firms, departments of universities and research groups of public research institutes. For simplicity, we refer to all agents as 'institutes'. The institutes are modeled as trying to find collaboration partners with complementary knowledge, and then applying for funding to the Framework Programmes of the European Union to finance their research. The Framework Programmes therefore have to be considered as a part of the institutes' operational environment. In the present version of the simulation, policy actors are exogenous and can exogenously modify, adapt and change the rules of the Framework Programmes. Therefore, currently the focus of the model is not on policy learning but rather the behavior of the institutes and the resulting network structures.

Knowledge generation and diffusion processes are extremely complex: technologies are usually combined and require the interplay of different technological approaches, as well as well-designed interfaces between them. Inter-, cross- and trans-disciplinarity shape modern sciences. For basic as well as applied research both an increasing speed as well as an increasing differentiation can be observed. This complexity of knowledge creation and diffusion processes is complemented by the design of the Framework Programmes: because of the conditions to receive funding, institutes engage in partnerships and networks; no single institute can expect to fulfill all the necessary requirements in isolation. With the exception of the FP7, there is no funding available for single agents in the Framework Programmes. These complexities of European R&D collaboration are taken into account in the construction of the SKEIN model: the model requires agents to form partnerships in order to be able to compete for funding.

Collaborative research projects allow for the combination of heterogeneous knowledge-bases following the requirements of the calls for proposals as well as the requirements of present day knowledge generation and diffusion processes. By engaging in collaborative research, institutes increase their flexibility, their chances of being successful in applying for FP-funding, and the probability of the successful introduction of a high impact novelty. To give an idea of the structure of the ABM developed, Figure 21.1 depicts one cycle of the simulation.

Every period (i.e. every iteration in the simulation) all institutes are engaged in the preparation of research proposals to submit to the EU Commission in order to receive funding in the Framework Programmes. Because a single institute cannot submit a proposal on its own, the institutes are also engaged in partner search. To simplify, each institute can start one proposal per period.[1] In order to set up a proposal, institutes search for as many best fitting partners for the project as necessary to constitute a consortium or proposal group following the guidelines of the Framework Programmes. Therefore they are undergoing several search and fitting-routines. In the first step, they are looking for suitable partners in their current research network from previous collaborations, and, in the second step, asking their partners for possible additional collaboration partners. In the third step, previously unknown suitable partners are invited. This

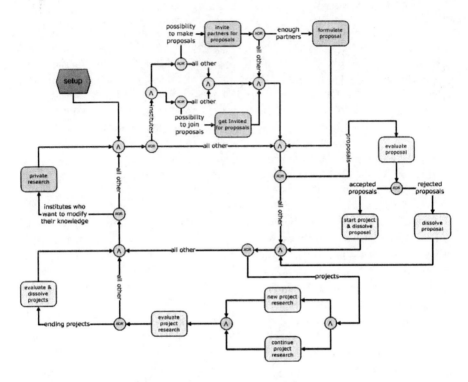

Figure 21.1 The structure of the model.

order of invitation is based on one of the rules formulated based on the empirical studies. The fit of the different institutes is determined by several aspects. The most important criteria are the formal rules of the Framework Programme, the compatibility of their cognitive distance[2] and their previous experience with each other. Every institute can participate in several proposals if it is invited to join a consortium by other institutes. If the necessary requirements for a proposal are fulfilled, the consortium submits the proposal to the European Commission and the evaluation process starts.

In the evaluation process, the European Commission selects a predefined number of proposals, those that fit best to the funding criteria. If a proposal does not match the requirements of the funding agency, it will be rejected. In this case, the proposal network dissolves and the actors try to create new groups and to submit new proposals in one of the following periods. If a proposal is competitive and matches the criteria set by the European Commission as described in the particular Framework Programme, the proposal is accepted. In this case, the institutes involved in the proposal receive funding and can start the project, producing for example journal publications or patents. At the end of the project a final report has to be delivered to the Commission. Intermediate and final results will be evaluated by the European Commission during and at the end of the project.

Institutes perform research in the various projects in which they are participating. The research is done in small subgroups (e.g. work packages). Every period during the project a new subgroup is established. The subgroups try to introduce novelties and to improve them until the project ends. Successful cooperation between the partners strengthens their network ties. The European Commission evaluates the performance of project consortia only at the end of projects.

Institutes not only engage in collaborative research projects but also try to improve their knowledge base through their own research. Therefore, in our model the institutes are permanently engaged in R&D leading to changing knowledge bases.

A more formal and in-depth description of the model, the agents, the knowledge space and metrics we are using, and the research process, can be found in Scholz *et al.* (2009).

4 Results

The standard simulation is used as a point of reference for comparison with the real world networks as well as with our policy driven experiments. Thus first we will summarize the results of the standard simulation we reported in Scholz *et al.* (2009) and then compare these with the results of the policy driven experiments.

4.1 The standard simulation

The settings we have used to create the standard scenario are shown in Scholz *et al.* (2009) and Table 21.3 in Appendix 21.1. These settings have been collected through the aforementioned empirical studies of recent FPs, and by sensitivity analyses for the model specific parameters.

The artificial networks created by the simulation have two important differences to the usual network analyses created from the real world data about the Framework Programmes and FP-funded projects. First, our network represents the entire science landscape, which consists both of the institutes participating in the FPs and of the ones who do not have a financed project. Second, we do not create a link between two organizations when they are in the same project. Links between institutes are only created out of repeated successful cooperation. Due to these two differences our network data differ mainly in three aspects: (*a*) due to the non-participating institutes we have a smaller share in the largest component of the vertices; (*b*) the institutes do not connect to all others in a consortium, which results in a smaller average mean degree; (*c*) due to fewer links within a project we also get a smaller clustering coefficient.

Data from two exemplary runs of the standard simulation are shown in Table 21.1 and some real world network statistics from FP7 for comparison are shown in Appendix 21.1 and in Scholz *et al.* (2009). The standard simulation is designed as a Monte Carlo experiment averaging the results of 20 runs.

The advantage of our method of creating the artificial networks is that we do not include formal links without knowledge flow between two actors. This

Table 21.1 Network results of the standard scenario[1]

		Stand 1 P80	Stand 2 P80
	Total network	919	916
Number of vertices	Largest component	474	491
	Share in largest component (%)	51.58	53.60
	Total network	1,080	1,064
Number of edges	Largest component	1,021	1,028
	Share in largest component (%)	94.54	96.62
Clustering coefficient		0.63	0.63
Diameter of largest component		4	4
Mean degree		2.35	2.31
Fraction of vertices with degree above the mean (%)		33.19	33.73
Standard deviation		3.629	3.421

Note
1 The data in table are two exemplary runs from the Monte Carlo simulation.

allows us to measure the changes and the influence of the policy driven experiments more realistically. In this way we can observe changes in the extent of the participation of actors in the FPs, as well as changes in the number of participants attracted to the FPs. Furthermore we have the opportunity to assess the value[3] of a link to identify the core centers of the cooperation within the network. The resulting network structure is shown in Figure 21.5 in Appendix 21.1.

The model also collects input and output data about the cooperation between the actors. However, due to the absence of corresponding real world data we are only able to compare these data with the equivalent data obtained from our policy experiments. Figure 21.2 shows the data for a Monte Carlo simulation of the standard scenario. The cross-hatched, solid line represents the number of sponsored projects, which form the input into the simulation. The number of formulated proposals as a measure of possible successful project cooperation opportunities is represented by the solid line without cross-hatching. The cross-hatched, dashed line measures the quantity of scientific output, represented by the number of successful cooperations, and the dashed line without marks signals the quality of the scientific output. The first 30 periods are needed for the initialization of the simulation.

The output data of our model indicates that through continuous cooperation with the same (most preferred) partners, the sheer number of projects increases, while the quality of the output shows a decreasing trend. If we look at the micro level, we still can observe some cooperation groups that are more successful in the long run. These groups have some core actors and several additional actors that change over time.

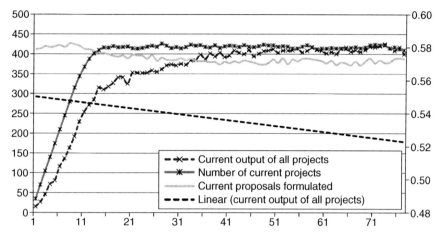

Figure 21.2 Aggregate results averaged over 20 runs from the standard scenario.

4.2 The policy-driven experiment

Because of the limited space in this chapter we will present only one policy driven experiment from the many possible. We have selected an experiment where large projects have been included into the standard simulation, along with geographical location as one of the selection criteria used by the European Commission to select projects. With large projects we implement the policy rule that project consortia of this kind have to be of a minimum size which is larger than 2.5 times a normal project. Furthermore, the consortia members have to originate in at least four different geographical areas. Table 21.2 shows the most important results from this simulation.

When we compare these results with the standard simulation we are able to identify important structural changes in the network. The share of the largest component has increased. This is a result of two factors. On the one hand institutes need to find more suitable partners for their project and on the other hand they are not allowed to use all preferred partners as before, because of the additional restrictions. Furthermore the mean degree increases. This indicates that actors have bigger ego nets[4] including the new, previously non-preferred partners. But this increase in connections carries with it the opportunity cost of a less good and more difficult to assemble network, as indicated by the bigger diameter of the largest component and a lower network clustering coefficient.

Figure 21.3 represents the structure of the network. This figure is not as clear and readable as the one of the standard simulation. In the second glance we can see that less actors have only a few ways to access the knowledge of the science network (connections to the largest component). The number of sub networks connected only by one actor has decreased. So the denser and more interlinked network has relatively fewer highly connected actors (larger dark grey discs) and

Table 21.2 Network results of the policy-driven experiment

		ALL 1 P80	*ALL 2 P80*
Number of vertices	Total network	920	920
	Largest component	671	636
	Share in largest component (%)	72.93	69.13
Number of edges	Total network	2,768	2,773
	Largest component	2,765	2,763
	Share in largest component (%)	99.89	99.64
Clustering coefficient		0.57	0.59
Diameter of largest component		6	7
Mean degree		6.02	6.03
Fraction of vertices with degree above the mean (%)		33.70	36.74
Standard deviation		7.983	8.292

relatively fewer cut points[5] (large pale grey discs) compared to the population of the largest component within the standard scenario. Thus the number of actors filling structural holes in the network increases. This shows as a higher value of integration of more actors into the international science landscape and relatively fewer actors now being dependent on a single actor to get access to the science network.

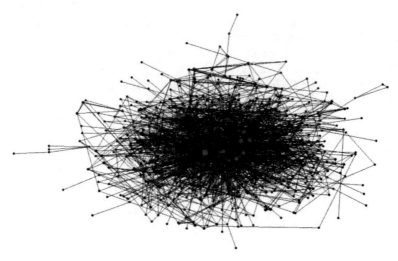

Figure 21.3 Visualisation of the policy-driven scenario (the size of a node indicates its degree (the number of links to that node) and the large pale grey discs stands for a cut point).

Figure 21.4 shows the aggregate trends from the simulation. By comparing the output data of the policy-driven experiment with the standard scenario, we can see a drastic decrease in the number of proposals. This results from the additional restrictions for formulating a successful proposal. By including requirements that encourage giant projects and constrain the geographical dispersion, we can see that the quality of the cooperation slightly increases or remains constant, while the quantity of the scientific output increases. This effect mainly results from the higher change rate of partners not part of the core groups. So the projects profit from the ideas and knowledge of the additional attracted partners. But the value of the output is not increasing as much as the number of the outputs. This is a result of the smaller 'absorptive capacities' of the newly joint actors. They are not as well aware of the new partner and their way of acting. Therefore the projects need more time to increase the value of their output. So in the quality of the scientific output still has a decreasing trend.

5 Conclusions

Over the last few decades, collaborative R&D arrangements have become an increasingly common way of producing knowledge (e.g. Gibbons *et al.* 1994, Hagedoorn and van Kranenburg 2003) and have been encouraged by national and international funding agencies and policy makers (Benner and Sandström 2000). Since 1984 the European Union has funded collaborative research through Framework Programmes; the currently running seventh Framework Programme provides altogether €53 billion for European R&D efforts. Understanding how collaborative arrangements are born and structured, how they can be encouraged, and what kinds of networks are most effective in the creation, transfer and distribution of knowledge, has become increasingly important in the

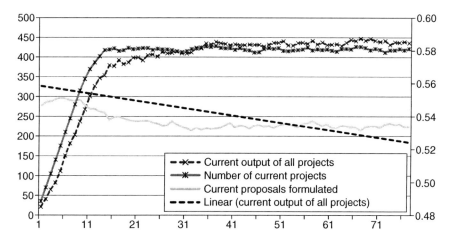

Figure 21.4 Aggregate results averaged over 20 runs from the policy-driven scenario.

context of a competitive knowledge society. Being able to encourage and appraise optimally functioning network-structures is essential for the efficient and effective use of the European resources and R&D infrastructure.

On bases of the results of our experiment we can answer our research questions. We asked first whether EU's incentive-led R&D policy influences the structure of the science network, and second, whether it influences the outcomes of research conducted in the network. As shown in Section 4.2 above, the structure of the science network is strongly influenced by the political rules, resulting in a denser and more interlinked network. Similarly, it seems that the outcomes of research are also influenced by the policy rules, albeit not as strongly as the network structure.

This chapter also shows that agent-based modeling is a powerful tool in simulating the scientific knowledge flows within R&D collaborations in the EU-funded Framework Programmes. It gives the user the opportunity to learn about the complex processes in research networks as well as offering a first intuition of the ongoing processes. An advantage of an agent-based model is that the macro-level behavior of a complex system emerging from micro- and meso-dynamics within the simulation can be examined. These dynamics include the interdependency of single actors in the system as well as interactions with the user. The SKEIN model also provides a user-friendly interface, which allows policy makers to test different kinds of policy scenarios on the simulation, without having to master the complexities of programming a simulation model themselves.

The results of the policy experiment described above suggest that, by introducing additional criteria, such as encouraging large projects or geographical dispersion, the European Framework Programmes are able to achieve some of their goals (increased, high-quality collaborations and denser networks with less structural holes), although such policy changes may jeopardize some other goals (e.g. fast, efficient and comprehensive knowledge transfer between the different actors and regions in the European Union). Running these kinds of scenarios on the agent-based model thus allows for an *ex ante* evaluation of changes to the Framework Programmes. It also provides a valuable tool for discussing focal policy considerations, such as how to encourage the emergence of key players in the R&D field, and how to minimize the number of actors unconnected to European R&D collaboration networks. This will help both national and international policymakers in improving the effectiveness and efficiency of R&D funding instruments, and thus help promote the emergence of European Knowledge Society.

Appendix 21.1

Table 21.3 Settings of the standard simulation

Parameter	Value
Starttime of Rule-Period	0
Stoptime of Rule-Period	100
Min-Duration of Projects	10
Kene-share Firms	0.6
Kene-share Research Institutes	0.6
Kene-share Universities	0.6
Max-nCandidates	10
Max-long-Distance (conservative)	60
Max-short-Distance (conservative)	20
Cooperation threshold	0.5
Min-long-Distance (progressive)	40
Min-short-Distance (progressive)	10
negative Proposal Increment	0.05
nConservative-Proposals	1.5
nRadical-Proposals	18
nKene-Elements from Institutes	1
nKene-Elements from Research Project	2
nFirms (initialized)	600
nResearch-Institutes (initialized)	300
nUniversities (initialized)	300
Fraction of prefered Partners	0.8
positive Proposal Increment	0.1
Project Revenue	120
Max-Duration of Projects	16
General Tax	100
Max-nProposals	2
Max-nProjects	5
Border-short-Distance (conservative)	20
Percentage-short-Distance (conservative)	0.4
Border-long-Distance (conservative)	70
Percentage-long-Distance (conservative)	0.2
Border-short-Distance (progressive)	20
Percentage-short-Distance (progressive)	0.2
Border-long-Distance (progressive)	70
Percentage-long-Distance (progressive)	0.4
nBig-Firms	30
nBig-Research-Institutes	300
nMembers of Big-Agents	8

Table 21.4 Network results of FP6 networks

		FP6-Aerospace	FP6-Innovation	FP6-NEST
Number of vertices	Total network	1,135	815	400
	Largest component	1,116	746	289
	Share in largest component (%)	98.33	91.53	72.25
Number of edges	Total network	22,682	5,971	1,470
	Largest component	22,637	5,801	1,184
	Share in largest component (%)	99.8	97.15	80.54
Clustering coefficient		0.88	0.89	0.93
Diameter of largest component		4	11	10
Mean vertices size		2.020	1.383	1.208
Fraction of vertices with degree above the mean (%)		35.1	32.9	22.3
Standard deviation		3.681	1.041	0.616

Figure 21.5 Largest component in the standard scenario.

Acknowledgments

We acknowledge support from the European FP6-NEST-Adventure Programme, contract no. 028875, Network Models, Governance and R&D Collaboration Networks (NEMO).

We are grateful to the members of the NEMO consortium, especially to Cedrick Ansorge, Andreas Brandes, Hannes Brauckmann, Barbara Heller-Schuh, Lena Kruckenberg, Manfred Paier, Karsten Pötschke, Marian Schmidt and Petra Wagner-Luptacik, who contributed to the empirical work which enabled the construction of the SKEIN model.

An earlier version of this chapter has been presented at the Sixth conference of the European Social Simulation Association, 14–16 September 2009.

Notes

1 The restrictions for starting a proposal are on the one hand the number of already running projects and on the other hand the number of currently joint proposals. These restrictions represent the time constraints of the institutes for the different periods.

2 Institutes can only cooperate if they have enough similarities and dissimilarities (cf. Nooteboom 1999). The required degree of similarity is defined by their search strategy.

3 The value of a link is a numerical result that represents the weight of previous cooperative activities of two institutes. It decreases over time and increases in case of successful joint work.

4 The neighborhood of an actor is the set of actors they are connected to together with the actors that are connected to them. An ego centered network is the subgraph induced by the set of neighbors. That is the network that consists of all the neighbors and the connections between them. The idea of an ego network can be extended to a group of actors and the neighborhood is simply the union of the neighborhoods of the group. This procedure returns the adjacency matrix of the ego network and provides an option to include or exclude ego(s) from the network.

Borgatti (2002)

5 A cut point in a network represents an actor that isolates parts of the network, if it is removed.

References

Ahrweiler, P., Gilbert, N. and Pyka, A. (2004) 'Simulating Knowledge Dynamics in Innovation Networks (SKIN)', in: R. Leombruni and M. Richiardi (eds) *The Agent-based Computational Approach*, Singapore: World Scientific Press.

Antonelli, C. (2005) 'Models of Knowledge and Systems of Governance', *Journal of Institutional Economics*, 1 (1), pp. 51–73.

Axtell, R. (2000) *Why Agents? On the Varied Motivations for Agent Computing in Social Sciences*, Center on Social and Economic Dynamics, Working Paper no. 17.

Benner, M. and Sandström, U. (2000) 'Institutionalizing the Triple Helix: Research Funding and Norms in the Academic System', *Research Policy*, 29, pp. 291–301.

Borgatti, S.P., Everett, M.G. and Freeman, L.C. (2002) *Ucinet 6 for Windows*, Lexington, KY: Analytic Technologies.

Caloghirou, Y., Ioannides, S. and Vonortas, N.S. (2003) 'Research Joint Ventures', *Journal of Economic Surveys*, 17 (4), pp. 541–70.

Communication from the Commission (no date) *Mobilising the Brainpower of Europe: Enabling Universities to Make their Full Contribution to the Lisbon Strategy.* COM (2005)152 final.

Deichsel, S. and Pyka, A. (2009). 'A Pragmatic Reading of Friedman's Methodological Essay and What It Tells Us for the Discussion of ABMs', *Journal of Artificial Societies and Social Simulation*, 12 (4), p. 6. Online, available at: http://jasss.soc.surrey.ac. uk/12/4/6.html.

European Commission CORDIS web page. Online, available at: http://cordis.europa.eu/ home_en.html

Gibbons, M., Limoges, C., Nowotny, H., Schwarzman, S., Scott, P. and Trow, M. (1994) *The New Production of Knowledge – The Dynamics of Science and Research in Contemporary Societies*, London: Sage.

Gilbert, N., Pyka, A. and Ahrweiler, P. (2001) 'Innovation Networks – A Simulation Approach', *Journal of Artificial Societies and Social Simulation*, 4 (3), 8.

Gilbert, N., Pyka, A. and Ahrweiler, P. (2007) 'Learning in Innovation Networks – Some Simulation Experiments', *Physica A: Statistical Mechanics and its Applications*, 374 (1), pp. 100–9.

Goyal, S. and Moraga-Gonzalez, J.L. (2001) 'R&D networks', *R&D Journal of Economics*, 32 (4), pp. 686–707.

Hagedoorn, J. and van Kranenburg, H. (2003) 'Growth Patterns in R&D Partnerships: An Exploratory Statistical Study', *International Journal of Industrial Organization*, 21, pp. 517–31.

Kruckenberg, L., Brandes, A. and Ahrweiler, P. (2008) *R&D Governance Rules of the EU Framework Programmes: The EURuleD Archive.*

Malerba, F. (2004) 'Sectoral Systems: How and Why Innovation Differs across Sectors', in: J. Fagerberg, D.C. Mowery and R.R. Nelson (eds) *The Oxford Handbook of Innovation*, Oxford: Oxford University Press, pp. 380–406.

Nokkala, T., Heller-Schuh, B., Paier, M. and Wagner-Luptacik, P. (2008) *Internal Integration and Collaboration in European R&D Projects*, Paper presented at the First ICC Conference on Network Modelling and Economic Systems. ISEG, Lisbon, Portugal, 9–11 October 2008.

Nokkala, T. (2009) *Internal Collaboration Rules in International R&D Collaboration Projects – Analysis of Seven NEST Projects*, NEMO Working Paper no. 15.

Nooteboom, B. (1999) *Inter-firm Alliances: Purpose, Design and Management: International Analysis and Design*, New York: Routledge Chapman & Hall.

Powell, W.W., White, D.R., Koput, K.W. and Owen-Smith, J. (2005) 'Network Dynamics and Field Evolution: The Growth of Interorganizational Collaboration in the Life Sciences', *American Journal of Sociology*, 110 (4), pp. 1132–205.

Pyka, A., Ahrweiler, P. and Gilbert, N. (2007) 'Simulating Knowledge-Generation and Distribution Processes in Innovation Collaborations and Networks', *Cybernetics and Systems*, 38, pp. 667–93.

Scholz, R., Pyka, A., Ahrweiler, P. and Gilbert, N. (2009) *Simulating Knowledge Dynamics in EU-founded Innovation Networks (SKEIN)*. Online, available at: www.wiwi.uni-bremen.de/traub/index-Dateien/institut.htm.

22 Innovation in complex social systems

Some conclusions

Petra Ahrweiler

Policymakers have identified innovation as one of the most important policy targets for dealing with challenges such as the current economic and financial crisis. Innovation policy frameworks, e.g. the Lisbon Agenda of turning Europe into a knowledge-based economy, are of ever increasing importance:

> Looking at the R&D and innovation components in greater detail, the EU has urged its member states to increase planned investments in education and R&D (consistent with national R&D targets) and consider ways to increase private sector R&D investments, for example, by providing fiscal incentives, grants and/or subsidies. Priorities of the Lisbon Agenda are again of increased relevance (large research infrastructures, knowledge transfer schemes, joint R&D programmes, mobility of researchers and international co-operation).
>
> (OECD 2009a: 28)

The role of innovation for modern economies is immediately obvious looking at income distributions and the share of knowledge-intensive industries in different world regions: the correlation is significant – high-tech regions match with high-income regions (cf. Krueger *et al.* 2004). 'R&D expenditures and intensity have been found to have a significant effect on per capita GDP growth' (OECD 2009b: 5). The extensive evidence for this correlation has long been monitored and documented by international and national institutions in much detail (e.g. OECD 2009a, 2009b, European Commission 2002, 2008). The positive impact of R&D investment and innovation is analyzed and confirmed on many levels. For example, in relation to firms the 2008 Innovation Scoreboard Report of the European Commission summarizes its data as follows:

> The role of R&D investment as an input factor for a company and its impact on performance parameters such as profits, net sales and market share are analyzed. Some descriptive statistics are presented to illustrate this issue for sectors with a high reliance on R&D. The analysis shows how the Scoreboard may be a useful tool to compare the relative performance and behaviour of enterprise groups. The relationship between R&D investment, sales

and market shares is illustrated by descriptive statistics for automobiles and parts, pharmaceuticals and the car manufacturing sector.

(European Commission 2008: 202)

However, these analyses mostly provide evidence in using correlations from econometric data. They do not tell much about causal chains and mechanisms, about the traceable line from investment to result. Empirical evidence proving a direct and immediate profitability of R&D investment is scarce. There are even studies suggesting that R&D intensity is negatively associated with innovation and economic growth (Jordan and O'Leary 2007).[1]

Especially in a time of diminished public resources and difficult capital markets, this is not acceptable. The strong need for justifying public and private investments produces a tendency in innovation policy, business management and public discourse to expect that the current investments in R&D, higher education institutions, science-industry networks etc. will immediately produce a flow of products and processes with high commercial returns. The requirement is to see value for money, and that is money for money. If there is a considerable investment as input, there must be a considerable, beneficial, and short-term output, which can be directly traced to this input.

This expectation still feeds from one of the first conceptual policy frameworks, the so-called linear model of innovation, which was fundamental to postwar innovation policy: it assumed that innovation – like through an input–output pipe – could be directly triggered by investing into basic scientific research, which would be immediately followed by applied research and technology development, and would end with production and diffusion bringing products and services to the market (cf. Bush 1945).

However, too often policymakers who had put large amounts of money into the R&D end of this pipe and sat at the output end waiting for the benefits, had been disappointed. Due to this situation of 'market failure', the linear model and theoretical frameworks favoring it were heavily criticized, e.g. in discussing the principal-agent theory of policymaking (van der Meulen 1998, Kassim 2003), or in promoting the garbage-can-model of policymaking (Mucciaroni 1992).

On the practitioner side and for the critical public mind, the disappointments and legitimatory problems arising from missing outputs were considerable and showed the limits of steering, control and policy functions. If not a principle apprehension against the importance of knowledge and innovation (Jordan and O'Leary 2007), the responsible innovation managers mention a frustration with the too messy and complicated features of the innovation process, which simply 'does not seem to compute'. If economic profits cannot be directly traced to knowledge, R&D and technology production – how can policymaking and innovation management be more than a kind of roulette in a game without rules? This frustration has already led to a kind of 'innovation bashing', which could seriously damage the recovery of economies (cf. Ahrweiler *et al.* 2010).

Manuel Castells (2000: 84ff.) discussed an interesting analogy to the current debate referring to the role of information and communication technologies

(ICTs) in his famous book *The Network Society*. He reports some puzzlement in the 1970s and 1980s, when the IT revolution – undisputedly changing our world – did not reflect in growth rates of the Western economies, the so-called *productivity paradox* (cf. Maddison 1984, Krugman 1994). On the contrary, there was an observed decreasing productivity level responding to the increase of information technology. Was IT economically insignificant as a technology – or even counter-productive – because we could not see any economic growth responding to the technological revolution? There were some growth theorists at that time saying so.

Knowledge has always been a challenge to economic growth theory (cf. Hanusch/Pyka in this volume). First it figured as a residual variable to work and capital (Solow 1956, 1957), then 'knowledge and innovation' advanced as a new productivity factor, e.g. in the new growth theory of Paul Romer which was used by the OECD in their famous 1996 paper on the knowledge-based economy (OECD 1996). In new growth theory (e.g. Romer 1990, Grossman and Helpman 1991), the continuously increasing factor human capital, i.e. the sum of all technological capabilities of human beings in the production process, secured the usage of capital with constant marginal productivity – leading to limitless and continuous growth. This framework would have offered the best fit to the mentioned expectations: we invest in technology, research and learning, and we will get direct and ever increasing economic returns. However, empirical economic research quickly falsified the general applicability of framing the relation between technological innovation and economic growth like this.

In empirical reality, growth processes are never continuous: they are specific to technologies and sectors showing multiple layers of small cycles, they stagnate, they slow down, they are characterized by time-delays, they break up, they go on – sometimes incrementally, sometimes in radical jumps. Neo-Schumpeterian approaches in economics concluded: if we are interested in this fine-granulation of growth processes, we have to look deep into the real dynamics of innovation, i.e. on the micro-level: this is because success and failure of empirical innovation processes determine the movements in productivity (Nelson and Winter 1982). Accordingly, economic growth can be observed on the macro-level, but an explanation for growth cannot be found there.

This quick look into the recent history of growth theory already directs us towards doubting the justification of the expectation of immediate economic returns for R&D investments. Castells continued his analogous discussion concerning the economic profitability issues of the ICT revolution with further elaboration on the reasons why there is no direct input–output relation:

1   We have to account for the *lag effects* – knowledge and new technology need quite a time to enter the market and to diffuse widely;
2   We have to account for serious productivity measurement problems, especially where the service sector is concerned. 'The focus on non-technological innovation has been most prominent in the services sector, which now accounts for more than 70 percent of GDP in OECD countries.

Indeed, empirical evidence shows that innovation in this sector takes different forms than in the manufacturing sector. Services firms innovate through informal R&D, the purchasing and application of existing technologies, as well as the introduction of new business models. There is a growing recognition that innovation encompasses a wide range of intangible activities, in addition to R&D. Efforts to improve measures of such innovative activity, or show that R&D needs to be supported by a complementary range of other investments, are still underway' (OECD 2009b: 7). The same holds for qualitative improvements of technologies, as in most cases not the amount of output produced increases but the technical features improve offering a higher quality;

3   We have to account for sector-specific productivity – aggregated productivity figures might not tell the true story.

These deficiencies of growth theory and measurement are commonly acknowledged (see e.g. European Commission 2002, Saviotti and Pyka 2004, cf. Hanusch/Pyka in this volume) and currently researched: 'The OECD is working with the international research and statistical community to produce a better measure of investment in innovation and its impact at the macro-economic level' (OECD 2009b: 5).

More important still is to look at the reasons why empirical growth rates depart from a linear relationship with R&D investment, and what the consequences are for innovation policy and innovation management. Refuting a simple causal connection between innovation and productivity does not imply that there is none at all. We just have to take up the challenge to investigate it as what it is: a complex empirical phenomenon. The task is to enter the turbulent layers of small innovation cycles and the innovation dynamics of innovation networks.

Given the contributions of this book, it would be indeed surprising to see immediate and easily measurable output for any improvements in the area of knowledge, research, and learning. The insight, which might have been provided by our book is that – though growth can be observed on the systems level – growth cannot be explained and controlled on the systems level. We have to investigate the non-linearities and path-dependencies of sector-specific productivity located in institutional contexts (see Saviotti in this volume). There is a strong influence of geography that matters (cf. Cooke *et al.* 2004; Ebersberger/ Becke in this volume). And ultimately, growth as a system level phenomenon is produced by a complex interaction pattern on the micro level of innovative actors in networks (see Allen *et al.* in this volume). This is why we have to investigate the role of collaborative arrangements in innovation.

For business innovation management this means difficult decisions: the true uncertainty (Knight 1921) of knowledge availability, access, and transfer, of technology absorption, of financial risk, of regulatory barriers, institutional impediments, of market access and profitability counteracts all predictability (Ahrweiler and Pyka 2008). The characteristics of firm innovation in complex social systems, for example, – be it for big multinationals (see Narula and

Michel, and Heidenreich *et al.* in this volume) or for small and medium businesses (see Asheim in this volume) – leave much remaining uncertainty on the shoulders of innovation managers. Though the European Commission principally confirms that: 'the link between R&D investment and company size and profitability is examined', it is still:

> A difficult question for a company ... to establish what is the optimum level of R&D to maximize return on investment. At sector level, it seems that there is a standard set by the major R&D players in the sector. Large companies increasing their R&D intensity beyond this level may run the risk that this additional effort will be inefficient. In contrast, a higher-than-average R&D intensity in smaller firms may mean that they rely more on R&D to grow and increase market share.
>
> (European Commission 2008: 202)

University management faces a whole range of new challenges which cannot be met without sound understanding about the central morphology universities are not only part of but are composed of themselves (see Etzkowitz and Ranga in this volume). Universities have a 'third mission', i.e. providing innovation and entrepreneurship, which requires new infrastructures and processes (see Lacy, Magin and von Kortzfleisch, Allen and O'Shea in this volume). Furthermore, universities need to provide cutting-edge research for enabling new knowledge and technologies. Here, academic research today means large interdisciplinary teams (see Obermeier *et al.* in this volume) working in international and interorganizational collaborative projects on complex research problems. Since collaborative research has become the dominant and most promising way to produce high-quality output (Bozeman and Lee 2005), these collaboration structures are again a target for university management.

Innovation policy needs to accept and handle the complex features of innovation (see Rossi *et al.* in this volume). This implies resisting the temptation of false expectations concerning short-term economic rewards for R&D investment:

> Governments ... need to focus on medium to long-term actions to strengthen innovation. A broad range of policy reforms will be needed in OECD economies and non-OECD economies to respond to the changing nature of the innovation process and strengthen innovation performance to foster sustainable growth and address key global challenges.
>
> (OECD 2009a: 12)

The task at hand is to develop a complexity-adapted way to support, on the one hand, innovation policy design and analysis (see Squazzoni and Boero in this volume), and, on the other, to understand and analyze the self-organizing coordination mechanisms (see Casti in this volume), which arise in and between participating innovative actors in R&D networks. Though the linear model is still ghost-riding through practitioner needs and public discussions (see above), innovation policymakers have long since favored the so-called 'neo-liberal model', which includes issues of open innovation (Chesbrough 2003), innovation networks etc.: 'the linear model of innovation, that assumes that research leads directly to innovation, has proved to be insufficient to explain innovation performance and to design appropriate innovation policy responses' (European Parliament 2006: 18). However, it seems as if this change has not yet been radical enough (see above).

It is understandable that the linear model still influences the expectation structures of policymakers, business managers and the public. Simple messages about causes and effects are always well received where there is a need for control – however, in this case it has been made very clear over the past decades that 'they do not compute' (*New York Times* 2008).

This book has now introduced an alternative to the linear type of thinking. We applied a broad thematic and methodological framework using the perspective that innovation happens in complex social systems. Combining empirical research from social science with a 'hard science approach' to innovation, the complexity of our target did not imply that we could not get a good handle on it. We used a combination of empirical research and modeling to understand innovation structures and processes, both on the micro and on the systems level. We applied and further developed methods from empirical socio-economic innovation research, computational social network analysis, and agent-based simulation to understand and manage the complex features of innovation in social systems. The potential for a roll-out of this methodology suggests that our research program could ultimately create methodological foundations for the sustainable transformation of innovative economies.

By providing a conceptual framework for the combined application of empirical research, computational network analysis, and agent-based modeling, our research can develop a novel methodology that allows for a fully integrated and more comprehensive understanding of innovation in complex social systems than has hitherto been achieved. By creating simulation platforms, which reflect the interaction of technologies and actors (Villani and Ansaloni in this volume), of products and organizations (Ciarli *et al.* in this volume), the evolution of innovation networks in knowledge-intensive industries (Gilbert *et al.* in this volume), and of the formation and development of clusters (Giardini and Cessoni in this volume), our research allows the modeling of interactions between existing policies and business practices, future policy scenarios, and alternative business strategies.

Agent-based modeling of innovation processes can advise actors of innovation networks in a way, which makes innovation indeed computable (cf. *Nature 460* 2009). This applies to governmental actors at all levels and private sector actors in various business sectors. Developing computational designs for innovation policy laboratories *in silico* can help to inform policymakers and business managers about optimal network structures for collaborative research and innovation (Scholz *et al.* in this volume) to set up and implement efficient strategies for establishing an innovative, competitive and efficient economy.

Notes

1 This study stated that the innovation output of Irish business may be negatively affected by R&D links to universities, and that future national prosperity may be undermined by continued investment in university research and university-industry collaboration. The resolution of the study was that attempts to build local linkages and clusters were a wasted effort.

References

Ahrweiler, P., Pyka, A. and Gilbert, N. (2010) 'A New Model for University–Industry Links in Knowledge-based Economies', *Journal of Product Innovation Management* (forthcoming).

Ahrweiler, P. and Pyka, A. (eds) (2008) 'Innovation Networks', *International Journal of Foresight and Innovation Policy*, 4 (3/4).

Bozeman, B. and Lee, S. (2005) 'The Impact of Research Collaboration on Scientific Productivity', *Social Science Studies*, 35 (5), pp. 673–702.

Bush, V. (1945) *Science, the Endless Frontier. A Report to the President on a Program for Postwar Scientific Research*, July 1945, Washington, DC.

Castells, M. (2000) *The Rise of the Network Society*, Malden, MA: Blackwell.

Chesbrough, H. (2003) *Open Innovation: The New Imperative for Creating and Profiting from Technology*, Boston, MA: Harvard Business School Press.

Commission of the European Communities (2002) 'Benchmarking National RTD Policies: First Results', *Commission Staff Working Paper*, Brussels SEC, 129.

Cooke, P., Heidenreich, M. and Braczyk, H.-J. (2004): *Regional Innovation Systems: The Role of Governance in a Globalized World*, London: Routledge.

European Commission/EuroStat (2008) *Science, Technology and Innovation in Europe*, Brussels.

European Commission/DG Research (2002) *Benchmarking National Research Policies: The Impact of RTD on Competitiveness and Employment (IRCSE)*, Brussels.

European Parliament (2006) 'Decision No 1639/2006/EC of the European Parliament and of the Council of 24 October 2006 Establishing a Competitiveness and Innovation Framework Programme' (2007 to 2013), *Official Journal of the European Union*, L310, pp. 15–40.

Grossman, G.M. and Helpman, E. (1991) 'Quality Ladder and Product Cycles', *Quarterly Journal of Economics*, 106, pp. 557–86.

Jordan, D. and O'Leary, E. (2007) 'Is Irish Innovation Policy Working? Evidence from High-Technology Businesses', *Journal of Statistical and Social Inquiry Society of Ireland*.

Kassim, H. (2003) 'The Principal–agent Approach and the Study of the European Union: Promise Unfulfilled?', *Journal of European Public Policy*, 10 (1), pp. 121–39.

Knight, F.H. (1921) *Risk, Uncertainty and Profit*, Boston, MA: Hart, Schaffner & Marx.

Krueger, J., Cantner, U., Ebersberger, B., Hanusch, H. and Pyka, A. (2004) 'Twin Peaks in National Income: Parametric and Nonparametric Estimates', *Revue Économique*, 55 (6), pp. 1127–44.

Krugman, P. (1994) *Peddling Prosperity: Economic Sense and Nonsense in the Age of Diminished Expectations*, New York: W.W. Norton.

Maddison, A. (1984) 'Comparative Analysis of the Productivity Situation in the Advanced Capitalist Countries', in: J.W. Kendrick (ed.) *International Comparison of Productivity and Causes of the Slowdown*, Cambridge, MA: Ballinger.

Meulen, B. van der (1998) 'Science Policies as Principal–agent Games: Institutionaliza-tion and Path Dependency in the Relation between Government and Science', *Research Policy*, 27 (4), pp. 397–414.

Mucciaroni, G. (1992) 'The Garbage Can Model and the Study of Policy Making: A Cri-tique', *Polity*, 24 (3), pp. 459–82.

Nelson, R.R. and Winter, S. (1982) *An Evolutionary Theory of Economic Change*, Cam-bridge, MA: Harvard University Press.

OECD (1996) *The Knowledge-Based Economy. General Distribution OCDE/GD*, (96) 102, Paris.

OECD (2009a) *Policy Responses to the Economic Crisis: Investing in Innovation for Long-term Growth*, June, Paris.

OECD (2009b) *Interim Report on the OECD Innovation Strategy*, June, Paris.

Romer, P. (1990) 'Endogenous Technical Change', *Journal of Political Economy*, 98 (5), pp. 71–102.

Saviotti, P.P. and Pyka, A. (2004) 'Economic Development by the Creation of New Sectors', *Journal of Evolutionary Economics*, 14 (1), pp. 1–36.

Solow, R.M. (1956) 'A Contribution to the Theory of Economic Growth', *Quarterly Journal of Economics*, 70, pp. 65–94.

Solow, R.M. (1957) 'Technical Change and the Aggregate Production Function', *Revue of Economics and Statistics*, 39, pp. 214–31.

'This Economy Does Not Compute' (2008) *New York Times*, 10 January: A29.

'The Economy Needs Agent-Based Modelling' (2009) *Nature 460*, 08 June, pp. 685f.

Index